The Search for
Anti-Inflammatory Drugs

*Case Histories
from Concept to Clinic*

The cover figure depicts metabolism of endogenous and exogenous substrates by 5-lipoxygenase (5-LO) in human neutrophils. In the bottom half of the figure, a stimulus provokes the release of endogenous arachidonic acid (AA) from membrane phospholipids (PL) and activates the 5-LO/5-LO activator protein (FLAP) system which then converts free AA to leukotriene B4 (LTB4). In the top half of the figure, the activated 5-LO/FLAP system is shown converting exogenous AA and 15-HpETE to DiHETE and 5,15-DiHete, respectively.

Cover illustration and figures by Leigh Rondano with assistance from Ann Hoffman, Carol Homon and Dr. Tom Parks, Boehringer Ingelheim Pharmaceuticals, Inc.

Cover graphic design by David Gardner, Boston, MA.

The Search for Anti-Inflammatory Drugs

Case Histories from Concept to Clinic

Vincent J. Merluzzi

Julian Adams

Editors

Birkhäuser
Boston • Basel • Berlin

Vincent J. Merluzzi
Boehringer Ingelheim
    Pharmaceuticals, Inc.
900 Ridgebury Road
Ridgefield, CT 06877, USA

Julian Adams
MyoGenics, Inc.
1 Kendall Square, Bldg 200
Cambridge, MA 02139, USA

Library of Congress Cataloging In-Publication Data

The Search for anti-inflammatory drugs: case histories from concept to clinic /
    Vincent J. Merluzzi, Julian Adams, editors
      p.   cm.
    Includes bibliographical references and index.

    1. Anti-inflammatory agents--Reserach--History.   I. Merluzzi, Vincent J.,
1949-   . II. Adams, Julian, 1954-   .
[DNLM: 1. Anti-Inflammatory Agents.   2. Technology, Pharmaceutical--
history.]
QV 247 S439   1995   95-1583
615'.7--dc20   CIP

Printed on acid-free paper
© 1995 Birkhäuser Boston   *Birkhäuser*

Softcover reprint of the hardcover 1st edition 1995

ISBN-13: 978-1-4615-9848-0   e-ISBN-13: 978-1-4615-9846-6
DOI: 10.1007/978-1-4615-9846-6
Typeset by Martin Stock, Cambridge, MA

9 8 7 6 5 4 3 2 1

Contents

Preface

Perspectives on Anti-Inflammatory Drugs

Inflammation is a very complicated process of interrelated events and cas cades that does not allow for an easily defined, focused attack for drug discovery. It is evident from years of research and development that certain classes of compounds (e.g., NSAIDs, steroids, and so on) have had a measure of success in alleviating pain and even dampening cellular/hormonal mechanisms involved in the process. Clear, mechanism-related therapies (e.g., for arthritis) and targeted drugs (e.g., for transplantation) have not been available in the past and, in reality, research in inflammation has relied on more phenomenological approaches for resolving symptoms or on blatant cytoreductive approaches in cases like organ transplantation.

In the last decade, approaches that have revealed novel cellular pathways in which intervention is possible for lymphocyte regulation (for example, cyclosporine and FK506) and small molecular weight mediators (e.g., leukotriene inhibitors) are now either standard therapy or will be in a short time. These latter approaches have been the result of research from the 1970s up to the present.

The chapters in this book give a clear background on how some of these current drugs were discovered and some of the problems and highlights that took place in the development process. The chapters on cyclosporine and FK506 provide a rich history of how these important chemical entities were discovered and how they have made it on the long and arduous path into the market place. They are classic examples of the first generation of mechanism-related, immunosuppressive, anti-inflammatory drugs.

Inhibition of leukotrienes, either by direct enzyme inhibition, translocation, or antagonism at the level of the mediators, has been a focus for the last ten to fifteen years as well. These inhibitors are currently reaching the level of large clinical efficacy trials in asthma and other indications. It has been a long journey for these compounds but nevertheless a fruitful one to understanding these mediators, the mechanisms involved in their synthesis, release, and ultimately in their biological effects. It is not surprising therefore to see a wealth of data on antagonism of these pathways. These chapters, along with newer approaches (e.g., azaspirines), provide interesting examples on the discovery of anti-inflammatory drugs.

The purpose of this book is to bring forward in both narrative and technical format the true nature of the discovery and development process for anti-inflammatory drugs. Ultimate success or failure in the clinic is not the principle focus or issue here, but rather an understanding of the ways concepts come into being and all of the thinking, nuance, and prejudice as well as the scientific and political processes that come into play in drug discovery. We hope that you will enjoy, as we have, the stories presented in this book on the specific discovery and development of new chemical entities.

The introductory chapter by Dr. Ivan Otterness sets the stage for the following chapters. It presents the appropriate background and history to the important but complicated field of research and development concerning arthritis. Arthritis is a disease with complex processes that ultimately transgress all areas of inflammatory research. It is anticipated that over the next decade we may be reading a similar treatise on events that have led to new concepts in biotechnological products for inflammation, and perhaps on small organic molecules whose mechanism of action is related to pathways in cellular activation at the molecular level of gene expression. One aspect that will probably remain constant, though, is the decision-making that allows projects to go forward or retreat. We hope that this aspect of this book will distinguish it from others that are usually collections of purely technical articles.

Vincent Jay Merluzzi
Julian Adams

Spring 1995

Contributors

Alison M. Badger, SmithKline Beecham Pharmaceuticals, 709 Swedeland Road, King of Prussia, PA 19406, USA

T. Beveridge, Department of Clinical Research, Sandoz Pharma Ltd., Lichtstrasse 35, CH-4002, Basel, Switzerland

Jean F. Borel, Department of Preclinical Research, Sandoz Pharma Ltd., Lichtstrasse 35, CH-4002, Basel, Switzerland

Dee W. Brooks, Immunoscience Research Area, Abbott Laboratories, D-47K, AP-10, One Hundred Abbott Park Road, Abbott Park, IL 60064, USA

Frederick J. Brown, Department of Medicinal Chemistry, Zeneca Pharmaceuticals Group, a business unit of Zeneca, Inc., 1800 Concord Pike, Wilmington, DE 19897, USA

George W. Carter, Immunoscience Research Area, Abbott Laboratories, D-462, AP-9, One Hundred Abbott Park Road, Abbott Park, IL 60064, USA

G.C. Crawley, Zeneca Pharmaceuticals, Mereside, Alderley Park, Macclesfield, Cheshire SK10 4TG, England

Peter R. Farina, Department of Inflammatory Diseases, Boehringer Ingelheim Pharmaceuticals, Inc., 900 Ridgebury Road, P.O. Box 368, Ridgefield, CT 06877, USA

S.J. Foster, Zeneca Pharmaceuticals, Mereside, Alderley Park, Macclesfield, Cheshire SK10 4TG, England

Carol Ann Homon, Department of Inflammatory Diseases, Boehringer Ingelheim Pharmaceuticals, Inc., 900 Ridgebury Road, P.O. Box 368, Ridgefield, CT 06877, USA

Shizue Izumi, Department of Pharmacology, Pharmacological Research Laboratories, Fujisawa Pharmaceutical Co., Ltd., 2-1-6, Kashima, Yodogawa-ku, Osaka, Japan

Z.L. Kis, Department of Preclinical Research, Sandoz Pharma Ltd., Lichtstrasse 35, CH-4002, Basel, Switzerland

Edward S. Lazer, Department of Medicinal Chemistry, Boehringer Ingelheim Pharmaceuticals, Inc., 900 Ridgebury Road, P.O. Box 368, Ridgefield, CT 06877, USA

R.M. McMillan, Vascular Inflammatory and Musculoskeletal Research Department, Zeneca Pharmaceuticals, Mereside, Alderley Park, Macclesfield, Cheshire SK10 4TG, England

Michihisa Nishiyama, R&D Planning and Coordination, Fujisawa USA, Inc., Three Parkway North, Deerfield, IL 60015-2548, USA

Masakuni Okuhara, Exploratory Research Laboratories, Fujisawa Pharmaceutical Co., Ltd., 5-2-3, Tokodai, Tsukuba, Ibaraki, Japan

Ivan G. Otterness, Department of Immunology and Infectious Disease, Pfizer, Inc., Central Research, 558 Eastern Point Road, Groton, CT 06340, USA

Thomas P. Parks, Department of Inflammatory Diseases, Boehringer Ingelheim Pharmaceuticals, Inc., 900 Ridgebury Road, P.O. Box 368, Ridgefield, CT 06877, USA

Petpiboon Prasit, Department of Medicinal Chemistry, Merck Frosst Centre for Therapeutic Research, P.O. Box 1005, Pointe-Claire Dorval, Quebec H9R 4P8, Canada

Tsung Ying Shen, Department of Chemistry, University of Virginia, McCormick Road, Charlottesville, VA 22901, USA

Philip J. Vickers, Merck Frosst Centre for Therapeutic Research, P.O. Box 1005, Pointe-Claire Dorval, Quebec H9R 4P8, Canada

E.R.H. Walker, Vascular Inflammatory and Musculoskeletal Research Department, Zeneca Pharmaceuticals, Mereside, Alderley Park, Macclesfield, Cheshire SK10 4TG, England

1

The Discovery of Drugs to Treat Arthritis: A Historical View

Ivan G. Otterness

The discovery of drugs to treat arthritis mirrors in many ways the discoveries and developments that have taken place in all areas of pharmacology. Yet in other ways, the discovery of drugs for arthritis remains unique. Sodium salicylate was the first useful synthetic drug for arthritis, and the discovery of new drugs has continued unabated to the present, making this one of the most commercially important therapeutic areas. The first analgesic/anti-inflammatory drugs set the tone—competition was between unique chemical entities protected by patents. Thus, competition became focused on structural differentiation which could lead to improvements in safety or efficacy. Yet in spite of the many commercial successes and the discovery resources poured into the area over the last century, arthritis therapy has remained a bastion of empirically discovered drugs. With the exception of the cyclooxygenase inhibitors, even today the mechanism of action of the principal drugs still is not fully understood. It is not clear if this counter-intuitive outcome arises from the history of the subject, or whether it is based on the difficulty of the therapeutic target. In the absence of a cure for arthritis, all improvements in therapy, however small, have been welcomed enthusiastically. It can be hoped that advances in cellular physiology, in receptor biology, ion channels, enzymology, and molecular biology will bring a rationalization of drug discovery and, with it, bring us a great deal closer to the ideal therapeutic for arthritis.

Early Therapy

Drug discovery did not start with the modern chemical/pharmaceutical company; it began many years ago in the most primitive communities. Shamans,

The Search for Anti-Inflammatory Drugs
Vincent J. Merluzzi and Julian Adams, Editors
© Birkhäuser Boston 1995

witch doctors, medicine men, healers, folk doctors—the names conjure up visions of the holders of the healing wisdom of primitive peoples. Some of their knowledge was of value but much of the healing lore was undoubtedly a placebo, with its value emanating from the ceremony accompanying its administration. How did primitive humans discover useful plant drugs? Some suggestive evidence comes from studies in zoopharmacognosy (use of medicinal plants by animals) and ethnopharmacology (essentially an anthropology of healing plants). In humans as well as in animals, the state of being unwell appears to stimulate the empirical search for something that will improve the sense of well-being, whether it is a change in diet or ingestion of new substances not previously a part of the diet.

Galen's Impact

The medical philosophy of Galen (129–200 c.e.) dominated not only medicine but therapeutics for more than a millennium after the Roman era. It was an abstract system of humors and their qualities (heat, cold, dry, moist) with sickness attributed to an excess of humors giving an imbalance in these qualities. Specific diseases were not recognized. To treat sickness, the therapy had to be designed to restore the balance of these qualities and bring harmony again to bodily function. A disease characterized by too much heat and dryness had to be combated with a medicine that imparted cold and moisture (Temkin, 1973). Thus, a Galenic pharmacopœia might be composed of from three to six thousand plants, all of which were classified according to their qualities and the strength of the qualities. Empirically testing all of those classifications in patients was not practical, so the herbalist and the physician more often determined their classification by taste. In addition, if the disease was too severe, the humors were considered to be in too great an excess. Therefore, to speed resolution, the patient was bled and purged to expel the humors and bring the body back to equilibrium.

Paracelsus

The arcana of the alchemists became focused on therapeutics with the work of Theophrastus Bombastus ab Hohenheim (1494–1541), known to history as Paracelsus (Pachter, 1951). A contemporary of Luther, Paracelsus propounded the heretical view that there were specific diseases, and they required specific remedies. His treatments were frequently based on folk medicine, and he encouraged his fellow alchemists to extract the essences and quintessences from plants and minerals to find the right treatment for each disease.

Although his treatments, such as arsenicals for syphilis, failed to displace the Galenic system of his time, his thought and the obvious successes of certain folk remedies, such as the root of the autumn saffron (colchicine) for gout, foxglove (digitalis) for dropsy, Peruvian bark (quinine) for fevers, ipecac (emetine) for dysentery, ergot for childbirth, and opium (morphine) for pain, ultimately undermined the Galenic system.

The medical knowledge of eighteenth-century Europeans was in most cases no more advanced than that of folk medicine. However, there was a major difference in how patients were treated. The folk healer treated individual patients. In the large cities, medical practice required treating large numbers of patients in a hospital or clinic situation. This development set the stage for open, critical observation of disease and its therapy (Ackerkneckt, 1967). The ability to observe numbers of patients with the same disease sharpened diagnosis and provided an opportunity to assess the effectiveness of a treatment. The numerical method of Pierre Charles-Alexander Louis (1787–1872), a Paris physician, was applied to the analysis of blood-letting (Louis, 1835) and other diseases. Although primarily categorization and numerical tabulation of patient results, it helped rationalize therapy by giving a quantitative form to the assessment outcome.

Therapeutics Becomes a Science

The apothecary Frederick Sertürner (1783–1841) (see Figure 1-1) took the first major step toward a rational pharmacology by developing, through trial and error experimentation, a procedure for isolation of pure, crystalline morphine from opium (Schmitz, 1985). He had demonstrated, for the first time, a way to prepare the quintessence of a medicinal plant. Purity made possible the standardization of dose. François Magendie (1783–1855) (see Figure 1-2) and Pierre-Joseph Pelletier (1788–1842) quickly adapted his method to the purification of emetine. Pelletier was joined by Joseph Bienaime Caventou (1795–1877) in the isolation of colchicine, the first pure substance for the treatment of an arthritic disease—gout. However, quinine, also isolated by Pelletier and Caventou, provided the strongest impetus for the early rational development of drugs (Delépine, 1951). The great diseases of the day were the various fevers: yellow fever, scarlet fever, typhoid fever, quartan fever, rheumatic fever, etc. Lowering of fever was considered the way to cure disease, and quinine, the best fever-lowering agent available, was considered a panacea for all fevers (Rageth, 1964), although in fact it only could cure tertian and quartan fevers (malaria).

Figure 1-1. Fredrick Sertürner. An apothecary by training, he was the first to develop a procedure for the isolation of pure morphine. Courtesy American Institute for the History of Pharmacy, Madison, WI.

Magendie first placed the study of drugs on a firm foundation (Olmsted, 1944). He broke with the deistic belief that man is unique and was able to declare that "after 20 years of study, man and animals respond" essentially in the same way to drugs. Detailed studies of drugs could now be done in animals. He was also the first to define the site of action of a drug. Most important, he reduced the Galenic pharmacopoeia from over 6000 substances of unproven therapeutic value to 60 pure substances, each of which he had tested in animals and in patients. From his own experience, Magendie, in 1834, could recommend the best use and dose for each substance. This was the beginning of therapeutics as we know it today: testing of pure substances with detailed evaluation first in animals and later in the clinic.

Figure 1-2. François Magendie. A physician, he can be considered the founder of rational therapeutics. He carefully explored the actions of drugs in both humans and animals. Courtesy of Roget-Viollet, Paris.

The First Synthetic Drug

Herman Kolbe (1818–1884) (Figure 1-3), was one of the more prolific chemists and probably in his day the most successful teacher of chemistry by the experimental method (von Meyer, 1884). He coined the term synthesis (*kunstliche*) for the *de novo* preparation of compounds. As a chemist before chemical structure was understood, he used the cumbersome theory of types to describe molecules. He had prepared a small amount of a substance (he named it salylic acid) and declared it was an isomer of benzoic acid. However, Kekulé had just proposed a structure for benzene which did not permit benzoic acid to have isomers. Reichenbach and Beilstein declared Kolbe to be wrong. Since salylic acid had been prepared from salicylic acid, in order to prove or disprove the possibility of an isomer of benzoic acid, bulk amounts of salicylic acid had to be prepared as a starting product

Figure 1-3. Herman Kolbe. A chemist, he developed a synthesis of salicylic acid while at Leipzig. Through his efforts salicylic acid found many uses in hygiene, as a preservative, and as a drug. Courtesy of the Edgar Fahs Memorial Library of the History of Chemistry, Philadelphia, PA.

for the production of salylic acid. Kolbe developed and patented a synthesis of salicylic acid. While Kolbe never found his isomer of benzoic acid, the new synthesis of salicylic acid was to launch the pharmaceutical industry.

Lister (1827–1912) had just published his method of antiseptic surgery in which phenol (known then as carbolic acid) was used to sterilize the surgical field. While phenol was a success in dramatically reducing the incidence of secondary infection, it was a rather toxic compound presenting a problem for patient and surgeon alike. Since Kolbe's synthesis took carbon dioxide and phenol under heat and pressure to make salicylic acid, Kolbe reasoned that salicylic acid could break down in the body to phenol and carbon dioxide, and thus should be a good antiseptic. Thus began the process in which salicylic acid became a drug. In spite of numerous discoveries and rediscoveries of the therapeutic properties of plant products such as willow bark, which contained derivatives of salicylic acid, it was Kolbe's reasoning about the pure chemical that provided the impetus for commercialization.

Kolbe's Propoal
Salicylic Acid as a prodrug

STRUCTURE 1-A

Kolbe proposed that salicylic acid (see Structure 1-A) could be used not only for surgery, but as a preservative for food, wine, beer, and water, in foot powders and mouth washes, and as an internal antiseptic. As an internal antiseptic, it was tested first by Buss (1849–1878). He examined its effects in many different diseases, but also noted on the basis of four patients that it might have some specific action in arthritis (Buess and Balmer, 1969). Subsequently, Franz Stricker, in 1876, showed systematically that salicylic acid had particular usefulness for treatment of arthritis. Kolbe, through his contacts, convinced Frederick von Heyden (1838–1926) (Figure 1-4), a newly graduated chemist, to begin commercial production of salicylic acid. It was the first pharmaceutical to be prepared in the modern way, i.e., synthetically, with patent protection, and by a single manufacturer.

Drugs by Accident and by Design

ACETANILIDE.   A. Cahn and Paul Hepp, both physicians working in a clinic in Strasbourg, had a patient with worms. They were intending to treat him with naphthalene, but their supply of drug had run out. They sent down to the Koppschen apothecary to obtain more and received a package of drug. After treatment, the patient's fever was lowered and appetite was restored. These effects were clearly not characteristic of naphthalene. The substance was sent to Edward Hepp, Paul's brother, who was working as a chemist in the dye manufacturing firm of Kalle and Company (see Figure 1-5). The substance was identified as acetanilide, a cheap, synthetic chemical. As acetanilide was not a pharmaceutical, no explanation has ever been found for the mix-up or even for the presence of acetanilide in the apothecary's

Figure 1-4. Fredrick von Heyden, the chemist who, with Herman Kolbe's patent and support, established the first commercial production of salicylic acid. Courtesy of von Heyden and Company.

shop. Since acetanilide was a known compound, it could not be patented. Kalle and Company introduced it for sale with the registered trade name Antifebrin. As Cahn and Hepp noted in their 1886 paper, in cases of acute rheumatism, Antifebrin brought prompt relief of joint pain. Cheap and effective, acetanilide became the key ingredient of patent medicines and over-the-counter headache and arthritis remedies. It was used for almost 80 years until it was banned because of methemoglobinemia.

KAIRIN. The production of synthetic quinine was a goal of many academic chemists and chemical manufacturers. Quinine itself has a structure too complex to be understood at that time, so a more practical approach was adapted. Derivatives of breakdown products of the quinine molecule that were simpler and chemically understood (quinolines, for example) were prepared and tested to see if they might share the properties of quinine.

Figure 1-5. The chemical laboratory of Kalle and Company at the time of the clarification of the structure of Antifebrin. While principally a dye manufacturer, they undertook the marketing of Antifebrin. Courtesy of von Kalle and Company.

Wilhelm Filehne (1844–1926) (Figure 1-6), a physician at the University of Erlangen, tested compounds for Hoechst to determine their usefulness. If they possessed good antipyretic activity, Hoechst licensed them. The Hoechst series Kairin M, Kairin A, and Kairoline (see Structure 1-B) were the foretaste of the modern pharmaceutical industry (Squibb, 1888). Each was a patented product that improved on its predecessor. Each was advertised and promoted in succession, and would constitute what could today be called a continually improving product line.

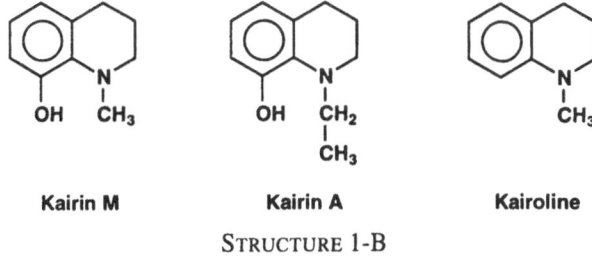

Kairin M   Kairin A   Kairoline

STRUCTURE 1-B

ANTIPYRINE.   Ludwig Knorr (1859–1921) (Figure 1-7), then a student of Emil Fisher, decided to try a new synthetic approach to the quinolines

Figure 1-6. Wilhelm Filehne, a physician cum pharmacologist. He specialized in the study of antipyretic drugs at the University of Erlangen and was intimately involved in the development of many drugs beginning with kairin M on through aminopyrine. Courtesy of Kay Brune, Professor of Pharmacology at the University of Erlangen.

Knorr's structure
Methyloxy quinazine

Antipyrine

STRUCTURE 1-C

Figure 1-7. Ludwig Knorr, a chemist. While a student, L. Knorr synthesized some novel pyrazolones. Antipyrine and its successors opened up an entirely new field of chemistry and a rich source of new drugs. Frontispiece from Berichte der Deutschen Chemischen Gesellschaft 1, 1927.

starting with phenyl hydrazine. He prepared a new quinoline and sent it to Filehne for testing. It was inactive, but Filehne had noted that the addition of a methyl group could often be relied on to improve pharmacologic activity. Accordingly, Knorr methylated his product, and Filehne's testing showed it to be a very good antipyretic (see Structure 1-C). It was patented and was the best antipyretic up to that time. Hoechst licensed it and introduced it as Antipyrine, whereupon it quickly supplanted the Kairin series (Brune, 1985). Knorr continued to study his new chemical and found that the structure was wrong. No matter that it was not a quinolone, but a pyrazolone; the compound had been tested and found effective in humans!

PHENACETIN. With the Kairin series, acetanilide, and antipyrine, it was clear that drugs did not have to be derived from natural products. It appeared that drugs could be systematically designed. At Bayer, there was an extreme oversupply of para-nitro aniline, a side product from the production of certain dyes. Carl Duisberg, the director of chemistry at that time, asked each chemist to try to find a use for it. Oskar Hinsberg suggested a structure that combined features of acetanilide and a quinolone drug called thallin. The new structure was synthesized, tested by an outside pharmacologist (Bayer at that time had no pharmacologists in their employ) and found to be poorly tolerated. Remembering that the ethoxy group (kairin A) was generally better tolerated than the methoxy group in kairin M, the ethoxy substituent was also made (see Structure 1-D). It was active and did not show any toxicity, and so, shortly thereafter, it was introduced into medicine as phenacetin. Soon it became a major drug (Duisberg, 1913). Having apparently proved the point that drugs could be designed, Bayer established a pharmaceutical division.

| Thallin | Acetanilide | Phenacetin |

STRUCTURE 1-D

Improvement on Existing Drugs

It turned out not to be so easy to discover new drugs. However, it was also clear that there were many ways one could modify the structures of previously discovered drugs and seek to improve them.

ASPIRIN. Sodium salicylate was a good drug for arthritis, but it was very sweet and had to be taken in four to five or more gram quantities to have a good effect in arthritis. A taste aversion to the drug often developed. Felix Hoffman tried a number of chemical derivatives of salicylate among which was the acetylated product, acetylsalicylic acid. It was assumed to be a prodrug for salicylate, and Heinrich Dresser introduced it into medicine as such in 1899. It turned out to be a much more effective analgesic than salicylate while maintaining its antiarthritic effects at higher doses.

DIPYRONE. Filehne suggested to Hoechst that antipyrine might be improved by further methylation. Although the specific product he suggested had not been synthesized, F. Stoltz had synthesized an analog with an additional methyl group. It turned out to be effective and was introduced for the treatment of arthritis as aminopyrine. Aminopyrine, however, was rather insoluble so a sulfate group was added to the structure for better solubility. The new compound, dipyrone, turned out to be a good analgesic (see Structure 1-E). For a time, it rivaled aspirin. However, a low incidence of agranulocytosis caused it eventually to be banned from sales.

Aminopyrine   **Dipyrone**

STRUCTURE 1-E

PHENYLBUTAZONE.   Aminopyrine still appeared to be the better product for arthritis so Stenzel at Geigy took another approach to improving it. Since it is a basic compound, he mixed it with various acids in hopes of obtaining a soluble salt that might also be useful for intravenous injection. A good solubilizing acid was found (phenylbutazone), and the new soluble preparation butapyrin (a 1:1 mixture of aminopyrine and phenylbutazone) was put on the market for treatment of arthritis (Wilhelmi, 1949). Clinical studies showed unexpectedly greater therapeutic activity and a much longer half-life of action than that of aminopyrine (see Structure 1-F). Studies soon demonstrated that the solubilizing agent, phenylbutazone, was therapeutically more effective and had a longer half-life than the drug aminopyrine which they had been trying to solubilize. First phenylbutazone, and then its major metabolite, oxyphenylbutazone, were marketed. They became mainstays of arthritis therapy for the next decade.

ACETAMINOPHEN.   Phenacetin meanwhile was coming under criticism. In spite of good tolerance in the majority of patients, hepatic problems were not uncommon among those who used it daily. Von Mering (1893) had tried acetaminophen as an antipyretic in 1893, but it was not accepted at that time because phenacetin was more potent. In studying the metabolites of both acetanilide and phenacetin, Brodie and Axelrod (1948, 1949) noted that acetaminophen (see Structure 1-F) was the major metabolite of

Phenylbutazone   **Acetaminophen**

STRUCTURE 1-F

phenacetin and acetanilide, and contributed to the therapeutic effects of both compounds. When examined in more detail, it was found that the methemoglobinemia and the hepatic side effects were not attributable to acetaminophen, but to other metabolites. Acetaminophen was then introduced as a minor analgesic and still finds extensive use.

Back to Empiricism

The pharmaceutical manufacturers now could make a number of good drugs available to the physician for treating arthritis, but all of them were for relief of pain. Even those that were available were largely discovered by one of two methods: chance discovery, or modification of an already existing compound. However, one could still speculate on the causes of arthritis and then try known agents that would be expected to counteract those causes.

GOLD.   In the 1890s, Robert Koch, who had discovered the tubercle bacillus, wrote a summary of his efforts to find a compound that would destroy or at least halt the growth of the bacillus. Although he had tested many substances, only a few had any activity; gold salts were among the most active agents. However, he noted that the gold salts were completely ineffective when tested in tubercular animals. Others followed up his findings and began testing various gold preparations in animals in hopes of discovering an *in vivo* active agent. Some gold salts were found with modest activity against tuberculosis.

   Landé meanwhile had concluded that gold had nonspecific antiseptic effects and therefore began trials of gold in various nontubercular diseases using aurothioglucose. In 1927 he reported on his clinical trials. He was particularly impressed by the relief of joint pain and reported that gold was "worthwhile in chronically febrile cases of painful arthritis resistant to the usual treatment." Forestier followed up with a systematic study of the effects of gold-thiopropanol sodium sulfonate in rheumatoid arthritis and, in 1935, reported favorable results after treatment of 550 cases. The rationale for his studies: "If gold salts are active in a chronic disease like human tuberculosis, why should they not be active in another chronic disease in which an important infectious factor seems to be present?" Although no infectious agent has been shown to be the cause of rheumatoid arthritis, gold nevertheless continues to be effective therapy.

D-PENICILLAMINE.   D-penicillamine, a metabolite of penicillin, had been found in the urine of patients treated with penicillin. Walshe (1956) thought that the free sulfhydryl group would be bioavailable and could be used to chelate and remove the pathological excess of copper in Wilson's disease, and successfully introduced it for the treatment of Wilson's disease. Later, Jaffe suggested that as a reducing agent, D-penicillamine could also potentially disaggregate rheumatoid factor and render it inactive. His studies (1963) showed that rheumatoid factor could be reduced *in vitro*. Thus he initially studied D-penicillamine in rheumatoid arthritis patients with arteritis and found it was a useful treatment. It was not long before D-penicillamine was also found to have activity in severe rheumatoid arthritis. Its mode of action is, however, on the disease itself and not on rheumatoid factor. Its mechanism remains unknown.

CHLOROQUINE AND HYDROXYCHLOROQUINE.   Since the earliest efforts to prepare a synthetic quinine, quinolone derivatives had been used for rheumatic diseases, based on a presupposed structural similarity with quinine. It should be no surprise that after World War II, some of the synthetic antimalarials were examined as treatments for arthritis. Freedman and Bach (1952) first examined quinacrine, but other antimalarials were rapidly tested, e.g., amodiquine, dapsone, chloroquine, and hydroxychloroquine, and found to have a useful therapeutic effect in arthritis. Chloroquine became the most common antimalarial for treatment of RA, but after retinotoxicity was demonstrated (Hobbs et al., 1959), the use of antimalarials dramatically decreased. Over a decade later, it was found that the ocular changes were reversible if drug was discontinued at an early stage. Thus, when regular ophthalmic testing was instituted, retinopathy was virtually eliminated (Rynes et al., 1979), and the popularity of the antimalarials, hydroxychloroquine in particular, rebounded. These drugs, the antimalarials, D-penicillamine, and gold, have been called DMARDs (disease-modifying antirheumatic drugs) because they show a slow-onset improvement in the patient's disease.

SULFASALAZINE.   Nana Svartz was convinced that rheumatoid arthritis must have an infectious etiology. She therefore began synthesizing derivatives of sulfonamides, the only antibiotics known at the time, and salicylate. She teamed up with chemists from Pharmacia in the effort. One of her colleagues, Helander, suggested that azo compounds had specific affinity for connective tissues, and this led Svartz to focus her interest on sulfasalazine, in which sulfapyridine was linked to 5-aminosalicylic acid by an azo bond. Open trials were conducted with favorable results (Svartz, 1942). However, Sinclair and Duthie (1948) carried out a clinical trial in which they failed to follow the recommendations for dose and duration of treatment and, in a well-publicized study, found the drug inactive. The drug was dropped from further consideration as therapy for arthritis.

It wasn't until 1978 that Brian McConkey and colleagues reexamined the effectiveness of sulfasalazine in arthritis. They were unaware of Svartz's earlier work and selected sulfasalazine because of a perceived structural similarity to the antimalarial dapsone, which they had successfully used earlier for treatment of arthritis. After the success of their initial study, they then followed with a larger study in which they showed it had DMARD-activity as shown by good but slow-onset clinical effects and changes in the acute phase reactant CRP.

METHOTREXATE.   The discovery of the folate antagonist methotrexate as a treatment for leukemia suggested to many investigators that it should be used for the treatment of other proliferative disorders. Gubner et al. (1951) were the first to test methotrexate as a therapeutic in a nonmalignant disease. He studied seven patients with rheumatoid arthritis. They found most of the patients improved, but one patient with psoriatic arthritis had dramatic improvement in his skin condition. More patients with psoriasis were treated, and a similar improvement was seen. Although improvement in arthritis was observed, the hepatic side effects and general toxicity left a pessimistic view of the utility of the drug. Only later, after the development of lower-dose, better-tolerated pulse therapy for psoriasis, was the issue of using methotrexate for arthritis reopened. Hofmeister (1972) examined low-dose pulse therapy, which minimized the side effects, and obtained very good results. Many studies extended his initial observations and made it possible for methotrexate to be approved for use in rheumatoid arthritis. However, the mechanism by which it acts is still not known. The very low dose and leucovorin rescue experiments suggest it is more than simple folic antagonism.

GLUCOCORTICOIDS.   Thomas Addison, a physician at Guy's Hospital in London, found that patients with a severe wasting disease that inevitably progressed on to death were characterized on autopsy by atrophied or absent adrenals. In 1855, Addison published a detailed description of the syndrome (now known as Addison's disease). In the following year, Brown-Séquard (1856) elicited the disease in animals by adrenalectomy. In the 1930s, extracts of the adrenal cortex were shown to reverse the disease, and in the following years, the structures of many of the steroids in adrenal extracts were elucidated by the groups of Reichstein in Switzerland and Kendall in the US. Desoxycortisone was the first cortical hormone to be synthesized, albeit in minute amounts (Steiger and Reichstein, 1937). Thus at the outbreak of the Second World War, the synthetic adrenal corticosteroids were an academic endeavor; adrenal extracts were used to treat Addison's patients, and small amounts of steroids could be produced in the laboratory. All that changed when Allied intelligence reported that the Germans had an adrenal product that would allow their pilots to fly at high (low oxygen) altitudes. The US wartime research program funded an effort to prepare large amounts of the adrenal steroids. With time, it was found that the rumor was false, and, in addition, the oxygen mask was sufficient for high-altitude flight. The program was terminated, but Louis Sarrett at Merck finished the synthesis of a few compounds. They were tested in Addison's disease, and the leftovers placed on the shelf. There they might have stayed except that a colleague of Kendall, Philip Hench at the Mayo Clinic, had noted that rheumatoid arthritic patients got better during pregnancy or during infectious stress, both being times of excess adrenal output. Hench asked Kendall for the drug, and it was obtained from Merck. The drug was tested on a single, totally bedridden patient. Astonishingly, by the third morning of daily treatment with cortisone, the patient was able to get up and walk, and by the end of the week, she went shopping for the first time in years.

More cortisone was synthesized and the results repeated in other severe arthritics (Hench et al., 1949). A film of the before and after treatment presented to the American Medical Association created a nationwide sensation, and a rush was on to prepare the drug at a reasonable cost. Merck succeeded by further improving its chemical synthesis; Upjohn succeeded via bacterial-mediated 11-hydroxylation. It soon became clear that glucocorticoid treatment was a two-edged sword: the patients got better, but the side effects were all too severe for routine treatment.

Design by Rational Empiricism

There was now a new goal in the pharmaceutical industry. A drug with steroid-like anti-inflammatory activity was needed, but without the steroid's side effects: a nonsteroidal anti-inflammatory drug (NSAID). Two prototype agents, aspirin and phenylbutazone, were available and both showed much greater efficacy than any earlier compounds, i.e., the quinolones, antipyrine, phenacetin, dipyrone, or pyramidon. The question was how to improve on those two drugs.

INDOMETHACIN. Charlie Winter and colleagues at Merck had been using the cotton string granuloma test to evaluate the steroidal anti-inflammatory agents being prepared at Merck to see if they could reduce their side effects. With Porter (1957), Winter found that the test would detect aspirin and phenylbutazone but not detect anti-histamine compounds. Although he did not know the mechanism of the two standard compounds, he had found a model that would detect both of them selectively. Because of the large compound requirements, empirical screening was out of the question. Initially, T.-Y. Shen and Winter (1977) thought that compounds that had some structural resemblance to serotonin would be the type in which to find a new anti-inflammatory agent. Serotonin was believed at that time to an important mediator of inflammation, so they started testing with some indoles from a serotonin program. Although the serotonin theory was incorrect, they ended up discovering indomethacin, a totally new NSAID (see Structure 1-G).

Indomethacin   Ibuprofen

STRUCTURE 1-G

PROFENS. Gerhard Wilhelmi (1949) of Geigy (Figure 1-8) was also exploring the properties of aspirin and phenylbutazone. He found that they would inhibit a model of sunburn, i.e., an ultraviolet light induced erythema in guinea pigs. The model was not inhibited by antihistamine or antiserotonin agents. With modifications of this model, Stewart Adams at Boots began examining a series of phenoxyalkanoic acids. Interestingly, both the Merck lead and the Boots lead were plant growth regulators, i.e., indolylacetic and phenoxyalkanoic acids. Ibufenac was the first to be marketed, but hepatic side effects prevented its continued marketing (Nichol-

Figure 1-8. Gerhard Wilhelmi, a pharmacologist. While at Geigy Pharmaceuticals, he developed numerous assays for anti-inflammatory activity. His publication on the ultraviolet light test spawned the profen and fenamic classes of anti-inflammatory compounds. He also was involved in the discovery of phenylbutazone. Photo a gift of G. Wilhelmi.

son, 1982). Eventually, structure-activity studies brought them to a series of phenyl propanoic acids. Ibuprofen (see Structure 1-G) was the first to be marketed. It was safe and sufficiently well tolerated after many years of prescription sales to be approved for over-the-counter sales. It became the first drug to reverse the trend toward physician-controlled drug usage and to become freely available to the self-prescribing public.

Further Developments

Charlie Winter (Figure 1-9) continued his examination of animal models and reported in 1962 that subcutaneously injected carrageenan would cause paw swelling that could be inhibited by aspirin, phenylbutazone, and indomethacin. The method quickly supplanted both the cotton string granuloma and ultraviolet light erythema tests.

Figure 1-9. Charles Winter, a pharmacologist. At Merck he and T.-Y. Shen discovered indomethacin. His carrageenan edema assay provided the method for developing most modern NSAID. Photo a gift of C. Winter.

Winter had shown the empirical way to develop such NSAIDs, and thus the focus became the chemical matter with which to develop new drugs. In a general sense most of the active drugs were aromatic acids. Several companies chose to develop new analogs of propanoic acid, e.g., suprofen, flurbiprofen, and naproxen. Ted Wiseman and Joe Lombardino (1982) at Pfizer chose to take a different route and developed agents using nontraditional acids. Lombardino obtained the acidic hydrogen from a β-diketone instead of a carboxylic acid. Piroxicam was the result (see Structure 1-H).

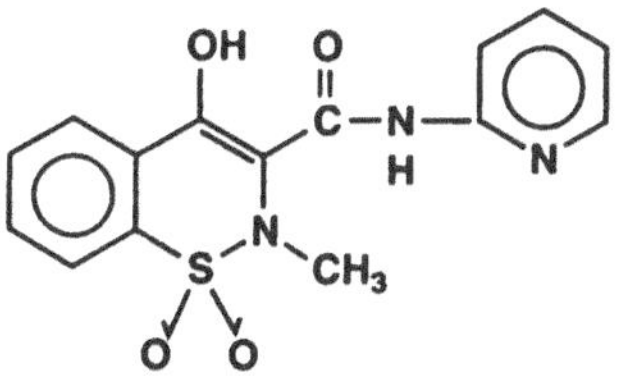

Piroxicam

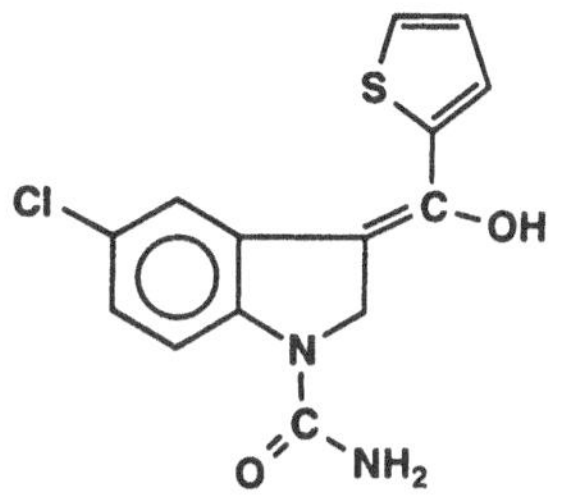

Tenidap

STRUCTURE 1-H

A Mechanism

Studies of the mode of action of aspirin allowed John Vane (1971) (Figure 1-10) to solve the problem of the mechanism of the NSAIDs. They inhibited the enzyme cyclooxygenase, which was responsible for the synthesis of a key intermediate along the pathways to prostaglandins, thromboxanes, and prostacylins. Inhibition of the synthesis of these substances was responsible for most of the good (inhibition of pain, edema and cell infiltration) and bad (ulcerogenicity) effects of the NSAIDs. The discovery of a second inducible cyclooxygenase enzyme (COX II) raises the hope that safer, more effective nonsteroidal anti-inflammatory drugs can be obtained from isozyme selective cyclooxygenase inhibitors.

TENIDAP. The presence of different classes of anti-inflammatory drugs such as the profens, the aryl acetic acids, the fenamic acids, and the oxicams meant that there were many NSAIDs on the market for the treatment of the symptoms of arthritis. The question was whether one could significantly improve on the standard NSAID. One approach was to take the fundamental NSAID activities and to add additional mechanisms. Compounds were sought that inhibited 5-lipoxygenase (5-LO) activity *in vitro* and *in vivo* (Otterness et al, 1989). The cutaneous basophil anaphylaxis reaction was used as a surrogate measure of 5-LO production *in vivo* since the tools to measure 5 LO products *in vivo* were not available in the early 1980s when the work was done. These activities were added to compounds that had the fundamental NSAID profile and resulted in the discovery of tenidap (see Structure 1-H). In clinical trials, tenidap possesses activities characteristic of a DMARD and cytokine modulator rather than a 5-LO inhibitor as reflected by lowered IL-6, CRP and SAA in patients with arthritis. It is more effective than NSAIDs alone and as effective as an NSAID plus a DMARD.

Figure 1-10. John Vane, a pharmacologist. At Burroughs-Wellcome, he established bioassays for detection of minute quantities of pharmacologically active substances. He received the Nobel Prize for the discovery of prostacyclin and the mode of action of aspirin-like drugs. Courtesy of Rolf Adlecreutz/Claes Löfgren, Pressens Bild, Sweden.

The New Rationalism

To the present, there are no marketed drugs for arthritis that have been developed by a strictly *in vitro* design approach based on a known pathological mechanism. There are many potential mediators from immunological and biochemical pathways that offer opportunities to intervene in rheumatoid arthritis, and we do have some effective, empirically derived agents for its treatment. For osteoarthritis, there is no current proven therapy. Anti-inflammatory agents confer some pain relief and functional improvement, but therapy that slows or halts disease progression is not available.

Mechanisms in Abundance

It is easy today to make a list of potentially important mechanisms in the pathology of arthritis. To mention but a few, there are effector molecules—cytokines (TNF, IL-1), T-cell mediators (IL-2, IL-4, perforins), chemokines (IL-8, NAP2, MIPs), leukotrienes (LTB$_4$, lipoxins), NO, several families of adhesion molecules, degradative enzymes (metalloproteinases, cathepsins, etc.), growth factors (IGF, TGFβ, FGF), and so on. In addition, there are signaling pathways—receptor-transducing enzymes and factors, transcriptional regulators, translational regulators, processing enzymes, and secretory pathways. Each step could be individually explored, a pharmacological tool prepared, and the importance of the pathway or mediator in the disease process tested. This is the opposite of the earlier methodology. Formerly, a model of the disease was probed pharmacologically, and an inhibitor of the model derived. Today, a single mechanism is chosen, an inhibitor developed, and the therapeutic importance of the mechanism probed in human arthritis.

The understanding of current pharmaceutical scientists exists like a mountain top towering over the meager knowledge of their predecessors. However, in the richness of scientific knowledge lies the problem of drug discovery today. Drugs are now targeted extremely specifically against a single mechanism. This greatly reduces the chances of undesired side effects. However, the wealth of choices increases the probability that any particular choice will be ineffective. Today the challenge for drug discovery is rarely whether we can develop an inhibitor of a specific mechanism. With the scientific tools available and commitment of sufficient resources, success can usually be attained. The challenge is now reversed. Can we choose a target mechanism well enough that an inhibitor will have a good probability of being therapeutically effective in humans and so justify committing resources to the project.

The Future

There are now many drugs on the market for the treatment of rheumatoid and osteoarthritis. Many are quite good drugs, but they cannot be classified as great drugs. Unlike penicillin for a patient with a streptococcal infection, the antiarthritic drugs do not cure rheumatoid arthritis nor have any been proven to reverse or slow the joint deterioration of osteoarthritis. Largely empirical efforts have given rise to a steady stream of improvements in therapy over the past 100 years. It can be hoped that there is a single fundamental etiology or mechanism whose inhibition could reverse disease progression or whose discovery could guide the targeting of new therapy.

However, arthritis may be a multimechanism-mediated disease. If that were the case, the introduction of new highly specific therapy would lead to further real, but incremental, therapeutic improvements. It might then take a combination of these single agents to corral and tame arthritis.

REFERENCES

Ackerkneckt EH (1967): *Medicine at the Paris Hospital: 1794–1848*. Baltimore: Johns Hopkins Press

Addison T (1855): *On the Constitutional and Local Effects of Disease of the Suprarenal Capsules*. London: Samuel Highley

Brodie BB, Axelrod JR (1948): The fate of acetanilide in man. *J Pharmacol Exp Therap* 94:29–38

Brodie BB, Axelrod JR (1949): Metabolic fate of acetophenetidin in man. *J Pharmacol Exp Therap* 97:58–67

Brown-Séquard CE (1856): Recherches expérmental sur la physiologie et la pathologie des capsules surrenales. *C R Hebd Séances Acad Sci Paris* 43:422–425

Brune K (1985): Knorr und Filehne in Erlangen. In: *100 Jahre Pyrazolone*, Brune K, Lanz R, eds. Munich: Urban & Schwarzenberg

Buess H, Balmer H (1969): Carl Emil Buß (1849-1878) und die Begründung der Salicylsäure-Therapie. *Gesnerus* 19:130–154

Cahn A, Hepp P (1886): Das Antifebrin, ein neues Fiebermittel. *Centralblatt für Klinische Medicin* 33:561–565

Delépine M (1951): Joseph Pelletier and Joseph Caventou. *J Chem Education* 28:454–461

Dresser H (1899): Pharmakologisches über Aspirin (Acetylsalicylsäure). *Pflügers Arch Ges Physiol* 76:306–318

Duisberg C (1913): Zur Geschichte der Entdeckung des Phenacetins. *Angew Chemie* 26:240

Forestier J (1935): Rheumatoid arthritis and its treatment by gold salts. The results of six years experience. *J Lab Clin Med* 20:827–840

Freedman A, Bach R (1952): Mepacrine and rheumatoid arthritis. *Lancet* 2:321

Gubner R, August S, Ginsberg R (1951): Therapeutic suppression of tissue reactivity. II. Affect of aminopterin in rheumatoid arthritis and psoriasis. *Am J Med Sci* 221:176–182

Hench PS, Kendall EC, Slocumb CH, Polley HF (1949): The effect of a hormone of the adrenal cortex (17-hydroxy-11-dehydrocorticosterone; compound E) of the pituitary adrenocorticotropic hormone on rheumatoid arthritis. *Proc Staff Meet Mayo Clin* 24:181–197

Hobbs HE, Sorsby A, Freedman A (1959): Retinopathy following chloroquine therapy. *Lancet* 2:478–480

Hofmeister RT (1972): Methotrexate in rheumatoid arthritis (abstract). *Arthritis Rheum* 15:114

Jaffe IA (1963): Comparison of the effect of plasmapheresis and penicillamine on the level of circulating rheumatoid factor. *Ann Rheum Dis* 22:71–73

Landé K (1927): Die Günstige Beeinflüssung schleichender Dauerinfekte durch Solganal. *Münch Med Wschr* 74:1132–1134

Louis PC-A (1835): *Researches on the Effects of Bloodletting*. Birmingham: The Classics of Medicine Library

Magendie F (1834): *Formulary for the Preparation and Employment of Several New Remedies*. Philadelphia: EL Carey and A Hart

McConkey B, Amos RS, Butler EP, Crockson RA, Crockson AP, Walsh L (1978). Salazopyrin in rheumatoid arthritis. *Agents Actions* 8:438–441

Nicholson JS (1982): Ibuprofen. In: *Chronicles of Drug Discovery*, Bindra JS, Lednicer D, eds. New York: John Wiley & Sons

Olmsted JMD (1944): *François Magendie: Pioneer in Experimental Physiology and Scientific Medicine in XIX Century France*. New York: Schuman

Otterness IG, Carty TJ, Loose LD (1989): Tenidap: A new drug for arthritis. In: *Therapeutic Approaches to Inflammatory Diseases*, Lewis AJ, Doherty NS, Ackerman NR, eds. New York: Elsevier

Pachter HM (1951): *Magic into Science, The Story of Paracelsus*. New York: Schuman

Rageth S (1964): *Die antipyretische Welle in der zweiten Hälfte des 19. Jahrhunderts*. Zürich: Juris-Verlag

Rynes Rl, Krohel G, Falbo A, Reineche RD, Wilfe B, Bartholomew LD (1979): Ophthalmologic safety of long-term hydroxychloroquine treatment. *Arthritis Rheum* 22:832–836

Schmitz R (1985): Friedrich Wilhelm Sertürner and the discovery of morphine. *Pharm Hist* 27:61–74

Shen T-Y, Winter CA (1977): Chemical and biological studies on indomethacin, sulindac and their analogs. In: *Advances in Drug Research*, Harper NJ, Simmonds AB, eds. New York: Academic Press

Sinclair RJG, Duthie JJR (1948): Salazopyrin in the treatment of rheumatoid arthritis. *Ann Rheum Dis* 8:226–231

Squibb ER (1888): Note on antipyretics. *Am J Pharm* 60:361–365

Steiger M, Reichstein T (1937): Desoxy-cortico-steron (21-oxy-progesteron) aus Δ^5-3-oxy-ätio-cholensäure. *Helv Chim Acta* 20:1164–1179

Stricker (1876): Über die Resultate der Behandlung der Polyarthritis rhematica mit Salicylsäure. *Berl Klin Wochenschr* 30:1–2, 16–17, 99–102

Svartz N (1942): Salazopyyrin, a new sulfanilamide preparation. *Act Med Scand* 60:577–598

Temkin O (1973): *Galenism: Rise and Decline of a Medical Philosophy*. Ithaca: Cornell University Press.

Vane J (1971): Inhibition of prostaglandin synthesis as a mechanism of action for aspirin-like drugs. *Nature (New Biology)* 231:232–235

von Mering J (1893): Beiträge zur Kenntniss der Antipyretica. *Therapeut Monsatshefte* 7:577–587 von Meyer E (1884): Zur Erinnerung an Hermann Kolbe. *J Prakt Chem* 138 (30 new series):417–466

Walshe JM (1956): Wilson's disease. New oral therapy. *Lancet* 1:25–26.

Wilhelmi G (1949): Ueber die pharmakologischen Eigenschaften von Irgapyrin, einem neuen Präparat aus der Pyrazolreihe. *Schweiz Med Wochenschr* 79:577–582

Winter CA, Porter CC (1957): Effect of alterations in side chain upon anti-inflammatory and liver glycogen activities of hydrocortisone esters. *J Am Pharm Assoc (Scientific Edition)* 46:515–519

Winter CA, Risley EA, Nuss GW (1962): Carrageenan-induced edema in the hind paw of the rat as an assay for antiinflammatory drugs. *Proc Soc Exp Biol Med* 111:544–547

Wiseman EH, Lombardino JG (1982): Piroxicam. In: *Chronicles of Drug Discovery*, Bindra JS, Lednicer D, eds. New York: John Wiley & Sons

2

The History of the Discovery and Development of Cyclosporine (Sandimmune®)

J.F. Borel, Z.L. Kis, and T. Beveridge

Introduction

Important scientific achievements always build on previous work and, even when they merit being considered a major leap forward, they constitute a logical step in the advancement of science, at least retrospectively. Though a researcher's intuition and the meanderings of his thinking remain unpredictable, the realization of his ideas is largely dependent on preexisting basic methods or on their development and perfection. Nevertheless, it is often difficult to trace the primal cause of a particular discovery, since it may rest on various crucial events and result from the efforts of several people. However, a rather more than less precise reconstruction of the true history is usually possible if it is based on all the remaining experimental protocols, existing publications, and many other records. We the authors, who were deeply involved in the discovery and/or development of cyclosporine, are combining our efforts to recount here as faithfully as we can our personal experiences and to include some of the contributions of our colleagues.

In 1969, no one envisaged how profoundly a handful of earth picked up by a scientist (H.P. Frey from Sandoz Ltd., Basel) vacationing in Norway would change modern medicine. From that soil sample, a fungus was isolated that was to take transplantation surgery from experimental science into the mainstream of twentieth-century medicine. Cyclosporine, extracted in the Microbiology Department of Sandoz in Basel, Switzerland, is the main metabolite of the fungus *Tolypocladium inflatum*. In the past

The Search for Anti-Inflammatory Drugs
Vincent J. Merluzzi and Julian Adams, Editors
© Birkhäuser Boston 1995

20 years the compound has done even more than open up a new branch of medicine—it has promoted understanding of basic mechanisms in immunology, including immunosuppression, and is beginning to light the way towards an understanding of a range of recalcitrant autoimmune diseases such as rheumatoid arthritis, insulin-dependent diabetes mellitus type I, and psoriasis, among others.

Cyclosporine selectively inhibits both the ability of the immune system to reject foreign tissues, as in transplantation, and chronic immune-mediated inflammation, as in autoimmunity. Equally momentous, it is largely free of the life-threatening side effects that had previously been a major problem with other immunosuppressants.

Microbiology and Chemistry (Z.L. Kis)

Microbiology as a Source of Potential Drugs

Since the beginning of the antibiotic screening program in 1958, it was usual for Sandoz employees on business trips or vacation to take with them small plastic bags for collecting soil samples. Such samples, catalogued by date and place of collection, might contain several thousand microorganisms, but since different samples usually had many identical strains, only a restricted number of strains were isolated. These strains were purified to obtain viable monocultures, which were checked for antimicrobial activity, and the active ones were cultivated in Erlenmeyer flasks and sent to the chemists for analysis. Success in antibiotic screening required fast and efficient chemical analysis. At this stage the search was for a new drug, hidden in a complex mixture, that could be differentiated from the many known and ubiquitous metabolites; the large number of strains producing the same metabolites, called multiplicates, confused the picture further. Therefore the ability to evaluate a great number of strains was of paramount importance because there was such a slight chance of discovering a new, active metabolite.

We devised in 1970 a two-stage system that largely solved our problems. The first stage was a microbiologically visualized chromatographic fingerprinting (e.g., thin-layer chromatography, electrophoresis at different pH) combined with computer-aided evaluation. This enabled us to identify the multiplicates and strains producing known compounds and allowed our small chemistry group to concentrate its research work on strains with only few or no multiplicates. These rare and peculiar microorganisms, we expected, would be the ones most likely to produce novel antibiotic substances.

The second stage was to isolate the active metabolites on a micropreparative scale and compare their physicochemical data with the literature. For this we needed more extracts and, therefore, repeated culture batches, either in bigger Erlenmeyer flasks or in glass fermentors. We had already gained some information from the first stage about the properties of the metabolite that were useful in planning the purification. From the beginning we intended to evaluate the data by computer, but because no software existed, H.J. Tobler and J. Gautschi, had to develop a workable system. The database generated from the literature could be searched using different physicochemical criteria, such as melting point, UV- and mass-spectra, as well as antibiotic activity. The database was searched for a particular criterion, such as a melting point with a confidence range; the resulting list was then searched for further characteristics, such as IR- and UV-spectroscopy, etc., so that the novelty of the compound could be rapidly established. This strategy spared us lot of laboratory work and time. At that time we processed the data in batches as that was the contemporary state of the art.

Isolation and Purification of Cyclosporine

Early in 1970, B. Thiele, the specialist for fungi in our Microbiology Department, isolated new strains of fungi imperfecti from samples of soil from the Hardanger Vidda in Norway and from Wisconsin in the USA (*Cylindrocarpon lucidum Booth*). The Norwegian strains were originally classified as *Trichoderma polysporum* (Link ex Pers.) *Rifai*, but the correct taxonomic name is *Tolypocladium inflatum Gams*. These fungi synthesize the family of metabolites that were later called cyclosporins, and a strain from the Norwegian soil sample is the one now used for the large-scale fermentation of cyclosporin A. The cyclosporins are not released in the culture media but have to be extracted from the mycelia. The analytical procedure described above revealed the similarity of the metabolites from the strains (multiplicates) and showed that they were indeed novel, at least for Sandoz.

In Z.L. Kis's laboratory, the metabolites were first purified from the original crude extracts on a micropreparative scale. Initially, we were able to isolate and characterize 80 mg of a mixture of cyclosporins as neutral, lipophilic peptides that were presumably cyclic, since no amino or carboxyl endgroups were detectable. The two components in the mixture had a molecular weight greater than 700 Daltons (the limit of mass spectroscopy in those days) and nuclear magnetic resonance suggested that the peptides contained several N-methylated amino acids. They possessed antifungal activity *in vitro*. The computer evaluation of these data showed that we were dealing with a mixture of a novel family of metabolites. Novel and active compounds merit further *in vivo* investigation, but experiments in animals require larger quantities of material. This meant scaling up the delicate fermentation and extraction procedures and making the transition from culturing in flasks to culturing in stirred tank fermentors, a step often involving major complications.

For the cyclosporins, the first attempt by E. Härri produced a good yield of more than 100 g of a mixture of the two homologous components, thus providing more than sufficient quantities to be used for detailed biological screening. A. Rüegger, who was responsible for preparative extraction of metabolites of fungi, sent 2 g of this mixture, designated as 24-556, to the Sandoz Research Institute in Vienna for *in vivo* antibiotic testing. A few weeks later, the Viennese Group reported the fungistatic activity of 24-556 in animals as restricted to a relatively narrow range of fungal strains. Our colleague there also commented on the low level of toxicity.

Though the antifungal activity did not seem adequate to warrant development at the time, this immediate disappointment was nothing new to us. But the mixture's unusually low toxicity, and the knowledge that microbial metabolites often possess interesting pharmacologic activities other than the antibiotic effects for which they were originally selected, caused us to press on with further pharmacological screening.

For many years it had been our practice at Sandoz to screen fungal products as broadly as possible, not only for antimicrobial but also for cytostatic, antiviral, and immunosuppressive activity. The screening tests included a cell culture assay for inhibition of cell proliferation (cytostatic effects) and an *in vitro* model to detect agents able to neutralize cytotoxic T-cell activity (immunosuppression). In addition, a combined *in vivo* test in mice was used to assess both the immunosuppressive and the anticancer activity. This early chemotherapy screening program was eventually enlarged and integrated in 1970 into a general screening program in which some 50 pharmacological parameters were evaluated. Thus, the metabolite mixture 24-556 entered this general screening program in December 1971 and was found to possess potent immunosuppressive properties. By the first half of 1973, a variety of experimental studies showed that 24-556 suppressed both antibody- and cell-mediated immunity. These interesting findings justified the separation of the single components of the mixture 24-556.

Mapping the Structure of Cyclosporins

The separation was done by absorption-column chromatography and resulted in two pure components: cyclosporin A, the main component, and cyclosporin B, so that elucidation of their chemical structures became feasible (Rüegger et al., 1976). In 1975, using chemical degradation and single-crystal x-ray studies with the J-derivatives, our colleagues, A. Rüegger et al. (1976) and T. Petcher et al. (1976), succeeded in establishing the chemical structure: the two cyclosporins were cyclic peptides consisting of eleven amino acids, seven of which were N-methylated and one of which had never been seen before. The peptides were neutral, and rich in hydrophobic amino acids. The correct structure was later confirmed by the x-ray analysis of crystalline cyclosporin A by H.P. Weber (Petcher et al., 1976) (see Figures 2-1, 2-2, and 2-3).

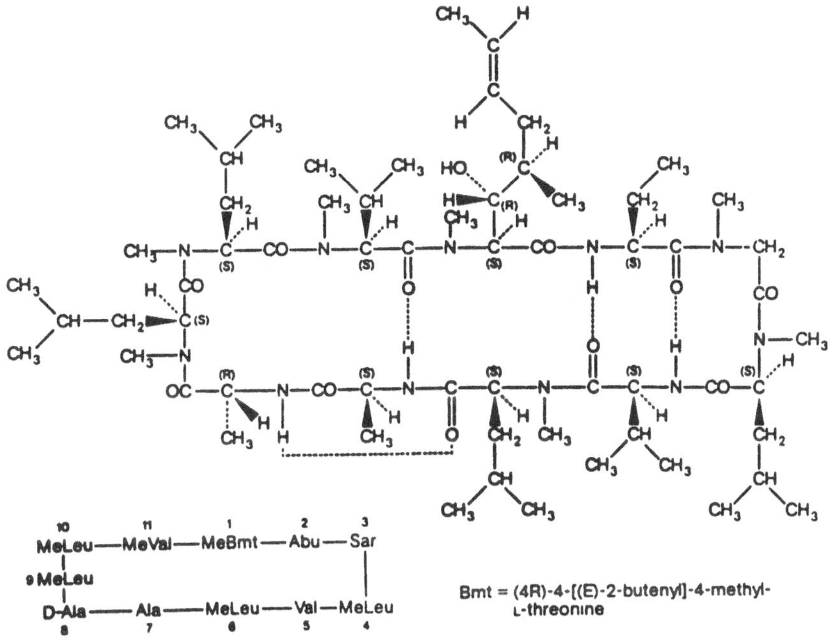

Figure 2-1. The structure of cyclosporine was established by chemical degradation and x-ray analysis of an iodo-derivative. Having a molecular weight of 1202, $C_{62}H_{111}N_{11}O_{12}$, cyclosporine is a neutral, hydrophobic, cyclic peptide with eleven amino acid residues, all having the L-configuration of the natural amino acids, except for the D-alanine in position 8 and the achiral sarcosine in position 3. One amino acid had never been seen before. Located in position 1, it is abbreviated as MeBmt, for (4R)-4-[(E)-2-butenyl]-4, N-dimethyl-L-threonine.

Knowledge of structure helps scientists to construct hypotheses about the structure-activity relationship. Generally, the larger the molecule, the more correlations between structure and activity are needed to corrobo-

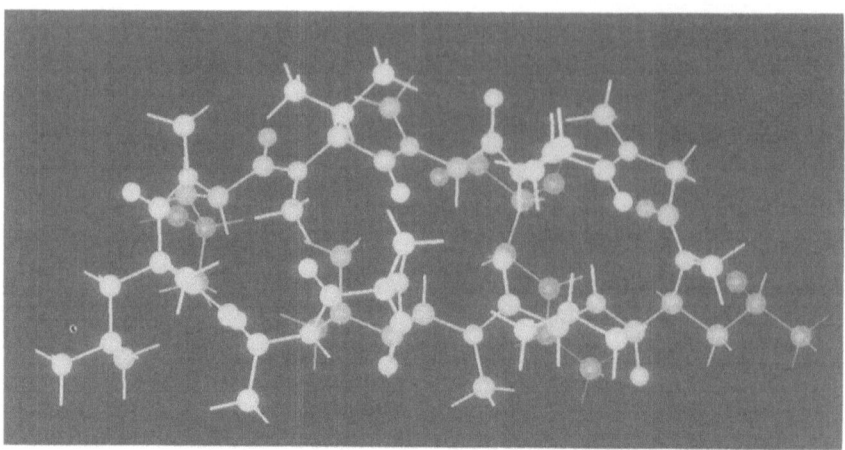

Figure 2-2. This is an orthogonal view on the crystalline conformation of cyclosporin A, as a stick and ball model. The hydrogen atoms are indicated with sticks only.

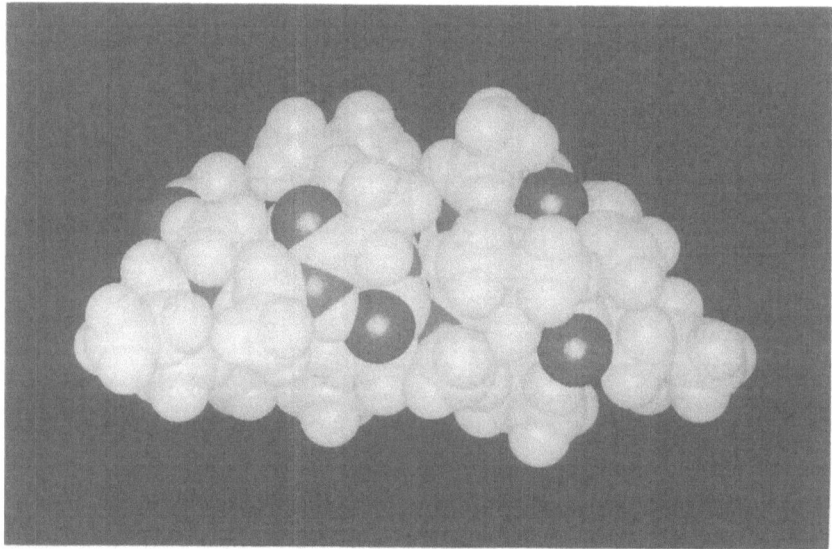

Figure 2-3. Space-filling model of cyclosporin A in the same view as in Figure 2-2. Here the butterfly shape can be seen.

rate a hypothesis. In addition to cyclosporin A, the fungus *Tolypocladium inflatum* produces large numbers of minor metabolites of the same cyclic structure but different amino acids. This building of a superfamily is the consequence of the nonribosomal biosynthesis of cyclosporin A, by a 1.4-MDa multienzyme polypeptide, compared with the much more reliable ribosomal synthesis of proteins (Kleinkauf et al., 1992; Lawen and Traber, 1993). At least 25 of these natural cyclosporins have been isolated, and all have been found to be composed of eleven amino acids, seldom differing by more than two amino acids (Dreyfuss et al., 1976; von Wartburg and Traber, 1986). The amino acids sarcosine in position 3 and the alanines in positions 7 and 8 are highly conserved. The reason for this is still unknown. From 1980 onwards, the systematic study of the structure-activity relationship has been made possible by total synthesis of the compound—a difficult task achieved by R. Wenger (Wenger 1982, 1986). Since the molecule could now be modified in every possible way, about 2000 semisynthetic or synthetic analogues were produced and tested *in vitro*, but only a restricted number of them were available in sufficient quantity for *in vivo* characterization. Although our knowledge of the pharmacological effects of all these cyclosporins remains incomplete, it is surprising that so far none of them has proved to possess significantly greater potency than cyclosporin A in either *in vitro* tests or *in vivo* models.

Pharmacology (J.F. Borel)

Early Pharmacologic Development

The first stage in the development of pharmacologic immunosuppressive agents began around 1960 and was aimed at the destruction of all rapidly dividing cells with cytostatic drugs such as azathioprine (Schwartz and Dameshek, 1959; Calne, 1961). Later, more selective drugs or procedures were tested and were mostly restricted to the elimination of the immunocompetent cells, namely the lymphocytes. Thus, the lymphocytotoxic effects of steroids, antilymphocyte serum, and total lymphoid irradiation were used (CIBA, 1967; Bell, 1981; Strober and Weissman, 1981; Brent and Sells, 1989).

It was in this setting that, in the mid-1960s, M. Täschler, the Head of our Pharmacology Research at Sandoz Ltd., Basel, had the foresight to make immunology a specific area of research. A laboratory for immunology was established in 1966 by a young immunologist, S. Lazáry, under the supervision of H. Stähelin, Chief of the Molecular Pharmacology Division. Once the relevant test systems had been set up, several compounds previously isolated by the Microbiology Department were investigated for their potential immunosuppressive effects.

When I joined the Medical and Biology Research Division at Sandoz Ltd. in Basel in the early spring of 1970, I brought with me eleven years of academic research experience in immunogenetic (mainly blood group

serology in animals [Borel, 1962] and humans [Borel, 1967]) and inflammation (antibody-mediated response and chemotaxis [Borel and Sorkin, 1967;
Borel, 1970]) research. I had the good fortune to take over from S. Lazáry a
well-equipped laboratory with excellent experience in assessing immunosuppressive agents. During the next few months, until his departure for
academia, my predecessor introduced me to this new field of research in a
most helpful and efficient way. I therefore met with very favorable conditions for a good start and I am still very grateful to him and to all my other
colleagues who were so supportive during those early days.

Together with many technicians I began to investigate in depth the
methodology of selected immunologic assays used in our screening program (Borel, 1974, 1976), and we soon found that, although generally
reliable, on several occasions they indicated false negative results with
some reference compounds known to be immunosuppressive. We made
modifications that were later to be crucial, for the detection of the immunosuppressive property of cyclosporine (cyclosporin A or CS) would
not have been possible with the old version of the hemagglutination test
used for screening. These modifications were reported in an internal mem-
orandum (October 28,1970) addressed to my then chief, H. Stähelin, who
accepted them immediately (see Figure 2-4).

When the compound 24-556 was fed into the general screening program
in December 1971, it fell to my laboratory to test it for its immunosuppressive potential. The *in vivo* screening model in which the immunosuppressive effect of CS was discovered started with the intravenous immunization
of mice with sheep erythrocytes on day 0. The compounds to be tested
were administered intraperitoneally on four consecutive days, blood was
drawn at day 7, and a hemagglutination assay performed; if the drug prevented hemagglutination, it possessed immunosuppressive activity. The
intraperitoneal injection of 24-556 resulted in an almost complete absence
of hemagglutination, demonstrating the potent immunosuppressive action
of the drug mixture. Unexpectedly, though, when the drug was given orally,
the results of this first experiment could not be reproduced. Further experiments with 24-556, using the more sensitive plaque-forming cell assay,
demonstrated a rather mediocre immunosuppressive effect. However, by
increasing the dose and the number of treatments, it eventually became
possible to reproduce unequivocally the initial potent suppressive activity
of 24-556 (Borel, 1982). It was only years later that it was discovered why
the first test was not confirmed by the second: higher doses had been given
for oral administration, and a procedure using a different galenic formulation for solubilizing the highly water-repellent compound had been chosen
the second and subsequent times, and this partial solution was poorly ab-

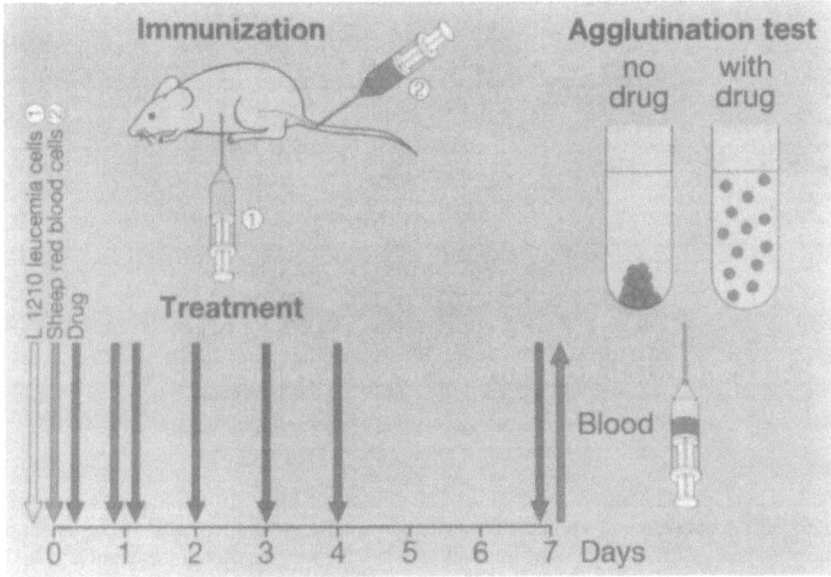

Figure 2-4. The hemagglutination test in which the immunosuppressive effect of cyclosporine was first discovered in January 1972. Mice were immunized on day 0 with antigen (sheep red blood cells) intravenously. In addition, they were injected intraperitoneally with L 1210 leukemia cells for testing the cytostatic effect of drugs. Then they were treated with the compound every day starting on day 0 up to day 3 inclusive (as modified by Borel). In the original screening model treatment occurred on days 1, 4, and 7, a protocol that would not have detected cyclosporine. At day 7 blood was collected. The serum of the blood contains antibodies against sheep red blood cells in solvent-treated control mice, i.e., these antibodies will agglutinate sheep red cells in a test tube. Animals treated with an immunosuppressive compound, e.g., cyclosporine, will not produce antibodies, and the sheep red cells do not agglutinate in the test tube.

sorbed by the mice. The various results of the very first tests are shown in Table 2-1. However, in the laboratory of H. Stähelin, another and most remarkable finding emerged: 24-556 had no effect on murine tumor cells (P-815) *in vitro* or on the survival time of leukemic mice inoculated with the murine leukemia cell line L-1210, indicating that immunosuppression was not linked with general cytostatic activity (Borel, 1982). Indeed, it should be remembered in this context that most of the immunosuppressive drugs used before CS act indiscriminately by blocking all cells in mitosis. After an incompatible transplant, an immune response is elicited that triggers active proliferation of lymphoid cells, but other cells replicate too, and

Table 2-1. Summary of early results (in 1972) on humoral immunity with compound 24-556. (Reprinted with permission from Sandorama 1983/II.)

Experiment number	Dose (mg/kg)	Route	Result
1 HA*	4 × 37	IP[†]	SI[•] <0.1
	4 × 112	IP	LD-100[◁]
2 HA	4 × 40	IP	SI 0.9
	4 × 200	PO[‡]	SI 0.8 (0.7)
3 PFC**	75	IP	73% inhibition
	150	PO	73% inhibition
4 HA	5 × 200	PO	SI 0.35
	5 × 400	PO	SI <0.2; LD-50[◁]
5 PFC	150	PO	73% inhibition
	3 × 150	PO	95% inhibition
	3 × 300	PO	<99% inhibition

*HA = hemagglutination test; **PFC = plaque-forming cells; [†]IP = intraperitoneal injection; [‡]PO = oral administration; [•]SI = suppressive index, which represents the quotient of the average titer (in $-\log_2$) of the treated group and the average titer of the controls; [◁]LD-100 = lethal dose inducing 100% (LD-50 = 50%) death.

the transplant is saved at the expense of other body systems such as bone marrow, gut, and other compartments with rapid cell turnover.

When efforts in the area of immunology were methodically extended at Sandoz, it turned out that a fungal metabolite, ovalicin, was able to suppress considerably the immune response of animals (Lazáry and Stähelin, 1968, 1969; Sigg and Weber, 1968; Bollinger et al., 1973). Ovalicin is a sesquiterpene with two epoxy groups, which had been isolated from the broth of the fungus *Pseudorotium ovalis* because of its high cytostatic activity against P-815 mastocytoma cells of the mouse *in vitro* (Stähelin, 1986). It had, however, no effect on other cells, e.g., human tumor cell lines, except at extremely high concentrations (Stähelin, 1986). Proliferation of cells other than lymphocytes seemed to be little affected in animals and, in particular, no significant leukopenia was observed at immunosuppressive doses. At higher doses, however, there was an early, transient reduction of leukocyte counts and, in particular, thrombocytopenia was observed in rodents and squirrel monkeys (Borel et al., 1974). This apparent selectivity of ovalicin for lymphocytes was confirmed by *in vitro* studies that, furthermore, produced evidence in favor of the hypothesis that the primary target of ovalicin action was the synthesis of ribosomes (Zimmermann and Hartmann, 1981). In clinical trials, however, ovalicin produced thrombocytopenia and central nervous system disturbances, and these side effects necessitated abandonment of any further use of the compound in humans (Stähelin, 1986).

This was happening around the time I joined the company. We were, therefore, soon involved in assessing the suppressive and the toxic effects of a derivative, ovalicin-semicarbazone (Borel et al., 1974), and in carrying out combination studies with other drugs, in particular some podophyllotoxin analogues (Borel and Stähelin, 1972, unpublished results). Although these endeavors did not lead to the expected breakthrough, it is evident that the work with ovalicin in particular had both prepared the minds of the group for a drug with selective biological properties and ensured that test models for the immune response were maintained in our screening programs (Stähelin, 1986). Nevertheless, after the numerous failures not only in our laboratory but worldwide, it was difficult to realize that with 24-556 we had stumbled on that rare compound that was able to inhibit very selectively an unknown step unique to the proliferation process of lymphocytes, while apparently sparing the proliferation of other somatic cells.

In the first half of 1973, a variety of experimental studies showed that 24-556 suppressed both antibody- and cell-mediated immunity, but that it was not effective in acute inflammatory reaction, and that it did not induce leukopenia. These interesting findings justified the separation of the single components of mixture 24-556, as mentioned earlier.

At this point, the stock of cyclosporins was nearly depleted, and we needed to have larger amounts from the chemistry group in order to pursue our animal experiments. This, however, was impossible without microbiologic fermentation on a larger scale.

Reevaluation of Research Goals

The next hurdle for CS appeared not in the laboratory but in the executive offices of Sandoz, where it was decided to integrate the company's very limited involvement in immunology into another major field of research. It should be recalled that during the 1960s and early 1970s, immunology had developed rapidly. Lymphocytes were recognized as belonging to either the T- (thymus derived) or B- (bone marrow derived) cell lineage, and the discovery of new cell-surface markers had made it possible to distinguish their subpopulations. The structure and isotypes of antibodies had been determined, and it became evident that the B-cell receptor was the antibody itself. Eventually, the concept of an immune response as the result of a series of cell interactions was reached. This increase in basic knowledge had given rise to great hopes, but offered too little in the way of clinical application. The field of clinical organ transplantation was largely restricted to kidney allografts, and most immunosuppressive drugs (aza-thioprine, methotrexate, cyclophosphamide, steroids) were quite cheap, the exception being antithymocyte globulin. Some of these drugs were also used for treating autoimmune diseases. Finally, there were strong arguments for abandoning immunology because of the failure of ovalicin and because huge sums of money would be needed to pursue the development of 24-556. From the standpoint of research planners, this meant investing in what was then a small, unattractive market—transplantation—with the added risk that the compound might have no clinical value and the company's outlay would never be recouped. It was estimated that about $ 250 million was needed to take a new chemical entity through development to FDA approval. Ovalicin, the previous immunosuppressant of interest, had just failed the clinical trials for toxic side effects. Management believed that prospects for a new immunosuppressant were less promising than other avenues of research and proposed abandoning 24-556.

Cyclosporine Promoted as an Anti-Inflammatory Drug

For the moment, it seemed we had been brought to a halt. Although discouraged, we few champions of CS knew our only hope was to find an application for 24-556 in an approved area of research. We found it in inflammation. We had earlier shown that 24-556, when given to rats, either preventively (developing disease) or therapeutically (established disease), markedly reduced experimental allergic encephalomyelitis, an autoimmune disease model. Because of this important finding, we had suspected that this immunosuppressive compound would also show inhibitory activity in a model of chronic inflammation, such as adjuvant arthritis, since chronicity is immune-mediated.

Although the mixture 24-556 showed no effect at all in preliminary acute inflammation testing (carrageenan-induced edema), which is not immune-mediated, we suggested to our colleague H.U. Gubler, from the Sandoz Research Institute in Berne, that this compound be retested in the more time-consuming, adjuvant arthritis model in the rat. In his 1973 report he described 24-556's strong inhibition of symptoms in this immune-mediated inflammatory reaction, when administered either preventively or therapeutically. It showed a further benefit: in contrast to other antiphlogistic drugs, CS did not induce ulcers. Because inflammation was among Sandoz research priorities, Gubler's crucial report enabled us now to propose the project as an official goal.

Management accepted our proposal. Microbiologists set about producing larger quantities of the mixture, and the chemists were able to supply the two metabolites. Further biologic testing with the single components soon revealed that cyclosporin A (CS) was the major active metabolite; cyclosporin B showed much weaker activity, and was not further pursued. Consequently, CS was initially promoted to the first formal development phase in the indication of rheumatoid arthritis, even though its remarkable immunosuppressive properties in the transplantation models were unquestionable. However, it might well have been realized by the experts that the field of autoimmune diseases, including arthritis, was unquestionably a larger indication than that of transplantation.

Preclinical Pharmacology at the Sandoz Laboratories

Early basic work in our immunology laboratory clearly revealed that, in contrast to all previous agents, the purified CS was the very first compound to inhibit the immunocompetent lymphocytes specifically and reversibly, and that it might be considered the prototype of a new generation of immunosuppressive drugs. Indeed, two studies were framed to exclude possible nonspecific cytostatic effects of CS on cells other than lymphocytes.

The first study investigated the cytostatic activity of CS on spleen and mastocytoma cells (cell line P-815) in comparison with that of other immunosuppressive drugs. *In vitro*, CS demonstrated a very selective activity on the proliferation of spleen lymphoid cells, which was 300 times more potent than on the nonlymphoid mastocytoma cells. This selective activity on lymphocytes was not shared by the other immunosuppressants (except hydrocortisone and antilymphocyte serum), which proved equally cytotoxic to both cell types (Borel and Wiesinger, 1977).

In the second study, we analyzed the effect of CS on bone marrow cell counts and stem cell proliferation in mice (Figure 2-5). In contrast to azathioprine, the *in vivo* effects of CS on bone marrow cells were minimal. Repeated treatment, even with high doses of CS, reduced the number of bone marrow cells only marginally. It never appeared to impair the proliferative capacity of the hematopoietic myeloid stem cells (Borel et al., 1977). This was an important finding that decisively increased CS's clinical potential, since the leading immunosuppressants at the time, azathioprine and methotrexate, were causing serious myelotoxicity in patients.

The effect of CS was also assessed in the tuberculin-type hypersensitivity reaction in guinea pigs. In contrast to the procedure for chemical-induced, delayed skin reactions, the animals were not treated with drugs during the entire sensitization phase, the first drug dose being given only at the time of tuberculin challenge, which followed the sensitizing antigen dose after about six weeks. CS injected just before and just after

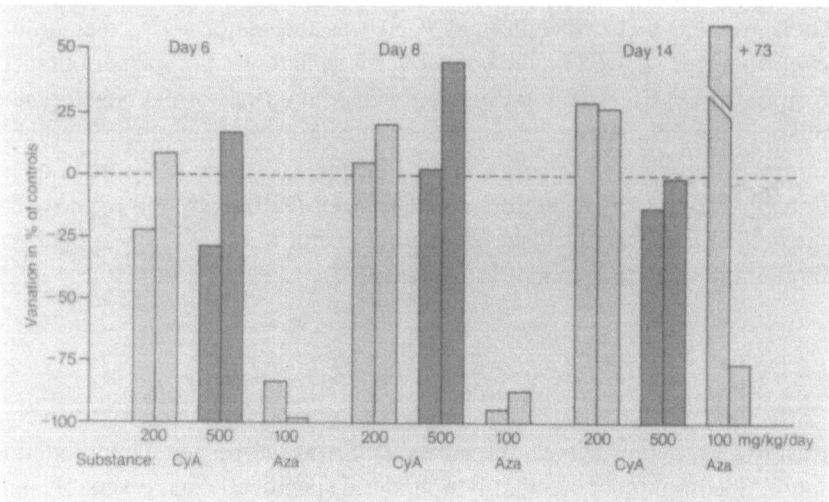

Figure 2-5. Comparison of the effect of cyclosporine (CyA) and azathioprine (Aza) on bone marrow cell counts and stem cell proliferation in mice. (Reprinted with permission from Sandorama 1983/II).

C57BL/6 male mice were treated orally on 6 consecutive days (0 to 5) with the indicated doses of drug. Two animals of each group were killed on days 6, 8, and 14, and the nucleated bone marrow cells counted. Counts from treated mice are expressed in percent of the corresponding control counts and shown in the first column of each pair. An assay for colony-forming stem cells was also performed on the same days using a pool of 10^6 bone marrow cells per group. After 7 days of culture the colonies were counted. The number of colony-forming cells derived from treated mice is expressed in percent of the respective control and indicated in the second column of each pair. The immunosuppressive potency of 200 mg/kg/day of cyclosporine corresponds to 100 mg/kg/day of azathioprine.

antigenic challenge considerably impaired the hypersensitivity reaction to tuberculin (Borel et al., 1977). This observation made us suspect one possible mechanism of action of CS, namely the suppression of T-helper cell function by inhibition of lymphokine release. It is well known that in this model the presence of specifically sensitized T cells is required, but that the swelling reaction is caused by invading phagocytes which are attracted by lymphokines locally released from these sensitized T lymphocytes.

From 1972 to 1976, only those within Sandoz knew about CS. Now it was time to share this important development. The first publication of our results in 1976 was entitled, "Biological effects of cyclosporin A: A new antilymphocytic agent" (Borel et al., 1976). This article has become a citation classic, according to *Current Contents* (*Current Contents*, 1984). This classic study demonstrated the following characteristics:

1. CS shows selectivity for lymphocytes, mainly for T-helper cells but sometimes for T-effector cells, depending on the test model used. It is clearly not myelotoxic at immunosuppressive doses.

2. CS exerts an immunosuppressive action on antibody- and cell-mediated immunity and in chronic, but not in acute, inflammatory reactions.

3. CS inhibits the induction phase of lymphoid cell proliferation. It affects the early mitogenic triggering, but not mitosis.

4. CS is not lymphocytotoxic, because the reversibility of the effect can be demonstrated.

5. CS is effective in all species tested, i.e., in the mouse, rat, guinea pig, rabbit, and monkey. (Because an inappropriate test system was used, it was long thought that it lacked efficacy in the dog, but this later proved to be erroneous.)

We also reported that CS had no other pharmacologic activity (cardiovascular, psychotropic, or other) that would limit its usefulness in humans. Furthermore, extensive toxicity studies in rats, dogs, and monkeys performed at Sandoz had confirmed the drug's selective effect on lymphocytes and lack of effect on hematopoiesis (Matter et al., 1982; Ryffel, 1982). If, at that time, a drug that immunologists and transplant surgeons were dreaming about had been designed, it would have been similar to CS.

However, it is often forgotten today that the development of CS was at the time a marginal project which was maintained by the special interest of only a few persons. Our major official project at that time was the investigation of lymphocyte chalones, i.e., factors involved in the control of lymphoid cell proliferation and differentiation. Both our *in vitro* and *in vivo* results were published (Hiestand et al., 1977; Borel et al., 1978), but one major impact of these studies was that, especially as a result of the collaboration of my highly competent research assistant, Camille Feurer, we enlarged our armamentarium with several important immunological models. We also remained quite active in the field of chemotaxis, as part of inflammation research, where we used *in vitro* tests (Borel, 1973; Borel and Feurer, 1975) and developed an *in vivo* cell migration technique using plastic collection chambers in the rabbit ear (Feurer and Borel, 1974; Borel and Feurer, 1978). Besides further screening work, we also studied numerous other compounds with suspected immunomodulating properties that were routinely sent to our immunology laboratory. This probably explains our expertise in the field and also why we were so confident of the pharmacologic potential of CS.

First Outside Animal Tests

At the April 1976 meeting of the British Society for Immunology, the main characteristics of CS were presented. In the audience was D.J.G. White from Cambridge, a co-worker of Sir Roy Y. Calne, who had been involved in experimental transplantation research, particularly in the development of azathioprine in the early 1960s (Calne, 1961). He immediately expressed great interest in the new fungal metabolite, and a supply of CS was shipped to Cambridge where the first animal tests outside Sandoz were performed. Initial results were impressive, in heterotopic heart allografts in the rat (Kostakis et al., 1977), in renal allografts in the rabbit (Dunn et al., 1978), and in the dog (Calne and White, 1977), and especially in orthotopic heart grafts in the pig (Calne et al., 1978a). The median survival of the grafted pig hearts of over 68 days with CS compared very favorably with only six days when the classical treatment with azathioprine and steroids was used (Table 2-2). Calne said of the outcome, "Such clear-cut results are truly meaningful and do not require the help of statistics to make them significant!" From this experiment, he concluded that "Cyclosporin A is sufficiently non-toxic and powerful as an immunosuppressant to make it an attractive candidate for clinical investigation in patients receiving organ grafts" (Calne et al., 1978a). Rapid publication of these findings boosted worldwide interest.

Cyclosporine Saved by Self-experimentation

The next step was to study the absorption, distribution, metabolism, and excretion of the drug in normal human volunteers. Generally, studies done in animals must be repeated in humans, in both healthy volunteers and patients.

Serious troubles arose at the very beginning of this phase in late 1976. The first time a single oral dose of CS was given to humans (pure, undissolved drug administered in gelatin capsules), it was not absorbed; that is, no pharmacologically active levels of CS were detected in their blood. Yet we remained convinced from our considerable experience with various galenic forms in animal studies that improvement of absorption was only a technical problem. Earlier, our colleague H. Wagner, who had formerly worked on lipid absorption in animals, had suggested dissolving the compound in pure olive oil for two reasons: because of its lipophilicity and because, during absorption in the gut, the resulting emulsion of the olive

Table 2-2. Prolonged survival of serologically mismatched orthotopic pig heart transplants.

Immunosuppression (dosage in mg/kg/day)	Survival (days)	Median survival (days)
No drug	5, 6, 6	6
Azathioprine (5) + methylprednisolone (5)	4, 5, 6, 6, 16, 51	6
Cyclosporine (15) on days 0, 2, and 4	19, 21, 22, 33, 33	22
Cyclosporine (25)	22, 43, >48, >88, >95, >111	>68

Five pigs (weighing 20 kg) given cyclosporine at 15 mg/kg by intramuscular injection on the day of operation and on the second and fourth postoperative days had a median survival of 22 days. Histological evidence of rejection varied from minimal to severe. In a second group of six animals given 25 mg/kg/day of cyclosporine (intramuscularly for the first two days and subsequently orally), two died of heart-failure with moderate histological features of rejection at 22 and 43 days. The remaining four animals were still alive at the time of publication. The median survival time in these six animals is > 68 days. The surviving animals were in good health and were gaining weight and eating normally. They showed no evidence of abnormal liver or renal function, and their hemoglobin, white blood cell, differential, and platelet counts were all normal.

oil might act as a vehicle to transport the dissolved CS across the intestinal wall. Experimental evidence in animals supported the accuracy of this crucial suggestion (J.F. Borel, 1977, unpublished results).

In March 1977, I and two other colleagues (B. von Graffenried and H. Stähelin) volunteered to swallow the drug in three different forms. I myself took the highly hydrophobic compound mixed in a new but efficiently absorbed vehicle, using pure ethanol, water, and some solvent (Tween 80). It was a distasteful concoction that made me feel intoxicated, but two hours later, using two different bioassays, a blood level of 1 μg/mL was measured. Drug dissolved in olive oil produced lower but still significant blood levels, but in gelatin capsules it produced no detectable level. Later, better vehicles, including olive oil, would prove acceptable to patients, and blood levels may now be accurately determined by radioimmunoassay (Quesniaux, 1989).

From a marketing perspective, there were still skeptics. Production of the drug proved extremely difficult and required an expensive and elaborate purification process. Because resolution was faltering, someone had to take the lead, create enthusiasm, and encourage further work. Someone also had to correlate our experimental findings with trials outside Sandoz. Consequently, at the end of 1977, I arranged to bring Calne and White's outstanding animal work to the attention of Sandoz management. Our results and theirs at Cambridge were consonant: CS was a novel, safer, and more selective immunosuppressant than any other seen in a laboratory. We had found the prototype of a new generation of immunosuppressive drugs, the first compound to inhibit the immunocompetent lymphocytes specifically and reversibly. At the end of that memorable meeting in Basel, management was convinced. Even more vital, it agreed to commit research and development funds and staff to continue testing the pharmacological potential of CS, a decision that opened the way for pilot clinical trials.

How Cyclosporine Works

Before moving on to the clinical trial phase, it should be recalled that CS is not only a successful drug in clinical indications, such as organ transplantation and autoimmunity, but it is also widely employed as an experimental tool for basic research. To achieve optimal control in the use of a drug, it is crucial to understand its mechanism of action.

It was demonstrated very early that CS spares the stem cells in the bone marrow and their maturation (Borel et al., 1977; Wiesinger and Borel, 1979). The compound affects resting and immunocompetent lymphocytes, which are blocked into their resting stage and cannot respond to an antigen by mounting an immune response. Present-day knowledge of the mechanism of action of CS has been extensively reviewed (Baumann, 1992). It is now evident that the compound inhibits lymphokine expression (interleukin-2, γ-interferon, and others) at the level of gene transcription. Its action *in vivo* is, however, much more complex (Borel, 1989). Interestingly, CS has been demonstrated to control an ongoing immune reaction effectively in a number of experimental and clinical situations, which implies that it can also inhibit the function of already activated lymphocytes and, therefore, be useful in helping to overcome a rejection crisis or to induce a remission in autoimmune diseases. There is, moreover, another important aspect to be considered: the complete cascade of biochemical events occurring within a lymphocyte during its activation process remains partly hypothetical. CS is proving to be an invaluable tool in this type of basic immunologic research. Conversely, increased knowledge of the activation steps of lymphocytes helps to increase knowledge of the mode of action of CS. This is leading into the exploration of novel possibilities for pharmacologic intervention (Walkinshaw et al., 1992).

Clinical Research Aspects (T. Beveridge)

Cyclosporine (CS, Sandimmune®) has been, in scientific, medical, and financial terms, a tremendous success story with clinical research playing a vital but modest part in the total development of this new immunosuppressive drug. Although successful in this case, one is mindful of the fact that, as R.B. Smith said, "in the development of drugs there is no guarantee of success and when success is obtained it is usually the result of effective planning, great amounts of expended nervous energy and, of course, old fashioned good luck" (Smith, 1985). Just think of it. A doctor working in the pharmaceutical industry might spend an entire working career without ever being directly responsible for the successful clinical development of a new drug for widespread use in humans. One might compare the daily routine of the industry, which involves the discarding of many thousands of compounds until finding one that will succeed as a new drug in humans, with that of a person existing for years on a monotonous diet of beans on toast and then suddenly being offered caviar. For me, personally, CS was certainly the equivalent of caviar, and the ensuing 15 years have been, without doubt, the most challenging, enjoyable, busy, and satisfying that I have experienced in over 22 years at Sandoz. The clinical development of CS was unique and its importance recognized by the award of the Galien Prize in France and Belgium and the Claudius Galenus Prize in Germany. Each of these was for an outstanding achievement in drug research and for the most important therapeutic progress achieved in that year. CS became generally available in 1983, and by 1984 it had heralded a new era in organ transplantation and made a worldwide impact (Beveridge, 1992). What follows will add some more details to what has been previously published (Borel and Kis, 1991; Beveridge, 1992) and provide some additional information concerning the clinical testing of CS.

In June 1978 when CS was introduced into the clinic, I was not directly involved in the project. It was my colleague, Dr. B. von Graffenried, who inaugurated the studies with Prof. Sir Roy Y. Calne at Addenbrooke's Hospital, Cambridge (Calne et al., 1978b) and Dr. R.L. Powles, Royal Marsden Hospital, Sutton (Powles et al., 1978) in renal transplantation and bone marrow transplant patients, respectively. It was in January, 1979 when Dr. von Graffenried changed his position within the firm (he was later to return to take charge of the CS autoimmune diseases program) that my then boss, Dr. R. Schmidt, Head of the Experimental Therapeutics Department, passed on to me the responsibility for clinical research relating to CS. This remit embraced the whole world outside of the USA. It was a date to remember. My initial direct involvement with CS related to both transplantation and autoimmune diseases. However, because of the sheer size of the clinical program for CS in transplantation and because the transplantation and autoimmune disease areas required essentially different approaches, in particular in terms of time-frame and clinical management, they became wholly differentiated as programs. Thus, an important early decision had been reached; namely, to concentrate on transplantation, since this was a life-saving indication and rapid feed-back from trials could be anticipated. Although I shall touch on the early clinical work done in autoimmune diseases, it is with transplantation that I shall be primarily concerned in this review.

Clinical development of CS followed, in principle, the usual four phases, but it was necessary to undertake radical and original departures from this basic plan and planning had to be exceptionally flexible. Phase I trials, for example, had many characteristics more common to Phase II or Phase III trials. The efforts required and the difficulties that had to be met were of major proportions. I believe it is true to say that the clinical program for CS trod unexplored ground and set new precedents. Why was this so? Well, first, there was the ever-present anxiety in the early years concerning drug supply, which must have caused Dr. E. Wiskott (then project coordinator) many sleepless nights. Second, this was a new therapeutic area for Sandoz and our first contact with the life and death considerations pertaining to transplantation. Third, at the time we entered the clinic, there was no universal scheme of drug therapy established for transplant patients. Each center tended to favor their own particular regimen, which meant that some were prepared to accept CS initially as sole therapy with the addition of steroids and/or azathioprine as seemed justified, while others favored the use of CS with steroids from the beginning. Fourth, in the early trials there was no method generally available for monitoring CS blood concentrations. A high-performance liquid chromatography method had been developed but this was complex and of only limited availability. Therefore, we had little guidance as to what dosages would actually be required to achieve effective and safe immunosuppression in humans. Earlier attempts at answering the dose-response question on the basis of some functional *ex vivo* assays involving the immunosuppressive effect had proven futile due to the large variability inherent in such tests. These results were reported in November 1976 and predated my involvement. Suffice to say that, understandably for an immunosuppressive drug, only a very limited number of normal volunteers could be recruited for this study, which also involved patients with rheumatoid arthritis. Dose-finding had therefore to be done empirically in patients, with the result that the 25 mg/kg per day given to the first renal transplant patients (Calne et al., 1978b) was that which had successfully prolonged survival of renal allografts in mongrel dogs (Calne and White, 1977). Fifth, there was the considerable problem of finding a suitable pharmaceutical formulation for the very lipophilic CS from which the drug could be reasonably well absorbed. First attempts using encapsulated powder were a failure; then, in March 1977 (Borel and Kis, 1991), came the first glimmer of hope when olive oil was used as a vehicle and provided the first clear evidence that CS could induce immunosuppression in humans. It was to be March 1979 before a more refined oral solution could be given to patients. By mid-July 1979 this form was being used in all clinical studies and remained unchanged for the remaining clinical program. The injectable form of CS in 1979, and up until mid-1981, had to be given intramuscularly but was replaced by a concentrate for intravenous infusion on the above date. Still later, soft gelatin capsules became available as an alternative oral form. Sixth, there was the fact that in transplantation we could not, of course, conduct placebo-controlled studies and even found it necessary in some cases (bone marrow, liver, and heart or heart-lung transplantation) to rely solely on historical controls.

If a man will begin with certainties, he shall end in doubts; but if he will be content to begin with doubts, he shall end in certainties.
                                                                    – Francis Bacon

It was against this background that the first published clinical results were viewed with cautious optimism. They were encouraging but not problem free. Nephrotoxicity, although previously found in rats, was only highlighted when CS was first given to humans (Calne et al., 1978b). This was particularly unfortunate in renal transplant patients in whom it might easily be confused with a sign of acute rejection, thus leading to additional immunosuppressive agents being given and the (already too high) dose of CS being maintained. Work continued at both of these centers (Calne et al., 1978b; Powles et al., 1978), and the initial publications inevitably stimulated major interest in other transplant centers as well as amongst clinicians working in the autoimmune disease area. At this stage, although the galenical difficulties were approaching resolution, we were still hampered in extending our work by the limited supplies of CS available. By July 1979, in addition to the studies ongoing at Cambridge and Sutton, trials had also been arranged in renal transplantation at the Royal Free Hospital, London and in various autoimmune diseases including rheumatoid arthritis at Basel, Goodpasture's syndrome and myasthenia gravis at the Hammersmith Hospital, London, and psoriasis at St. John's Hospital in London. These latter trials in autoimmune diseases were particularly important for the investigation of galenical forms and as dose-finding studies. Preparations for many other trials were under way, including plans for transplant trials at centers in the USA, in renal transplantation at centers in Boston and Denver in August and September 1979, respectively, and in BMT at Seattle and Baltimore in December 1979 and January 1980, respectively. At Cambridge, transplant experience was extended into other areas including liver and pancreas transplantation.

Prof. W. Müller, Basel, who was among the very first to use CS in patients, published his first results in August 1979, which described the effect of CS on rheumatoid arthritis and on psoriasis associated with arthritis (Herrmann and Müller, 1979). The fact that this was in a somewhat obscure journal may have meant that not many noted the pictures depicting a dramatic disappearance of long-standing psoriatic lesions in a 45-year-old patient after four weeks of CS therapy at 900 mg daily for one week, then 450 mg daily (reduced because of nausea). However, not long afterwards the same authors published a letter (Müller and Herrmann, 1979) which described four cases of psoriasis treated with CS and which would later assume greater significance when CS was formally studied in psoriasis. These early observations suggested that the skin lesions of psoriasis could be beneficially affected by CS. Even relatively small doses were found to act rapidly, but the effect quickly disappeared when the drug was discontinued.

Then, in the autumn of 1979, we ran into apparently severe difficulty. Of the first 34 recipients of cadaveric organs in Cambridge, three were found to have developed lymphoma. We were immediately advised of these findings in October and they were published in November (Calne et al., 1979). If CS had a general tendency to induce lymphoma, this would mean that the drug would have to be withdrawn. As soon as possible after receiving this information, a crisis meeting was arranged between representatives of the Cambridge and Sutton groups and of Sandoz. I also attended the meeting, which was held at Sutton on November 6, 1979. It was the belief of all attending that CS was indeed a valuable drug with a real contribution to make to the advancement of transplantation. However, it was equally clear to all present that the effects seen, if they continued, could result in the drug's abandonment. However, it was important to realize that CS was the sole immunosuppressant in only one of these three cases, the others having received additional immunosuppression. The possibility of overdosing and over-immunosuppression was seen as a likely explanation for the events that had occurred. Lack of experience with CS had led to the false interpretation that the renal dysfunction caused by CS represented signs of rejection requiring increased immunosuppression. It was agreed to lower the CS dosage from 25 mg/kg per day to 17–20 mg/kg per day initially and to reduce this still further when circumstances allowed. In addition, it was planned to take regular blood samples from individual patients at intervals during therapy. These would be stored and subjected to later analysis using high-performance liquid chromatography, the method being still restricted. Looking back, the decision to reduce the dose of CS was correct. The effects of over-immunosuppression per se had been responsible and not CS in particular (Calne et al., 1981). This meeting effectively saved the day for CS and also meant that CS could go on to later become the most important drug at Sandoz. Incidentally, only one lymphoma was reported in Europe (Crawford et al., 1980) during the $1\frac{1}{2}$ years following the meeting. Nevertheless, continuing extreme caution was required.

By now few doubted that CS was a powerful and effective immuno-suppressant, but as always tends to occur in drug development, whenever the efficacy is no longer newsworthy, it becomes a matter of focusing on adverse events. In the case of CS, it was first the perceived nephrotoxicity and then the occurrence of lymphomas which produced many headlines. At this stage there was also a desire for a better knowledge of the kinetics, metabolism, absorption, distribution, and excretion of CS, although there was still no widely available chemical analytical method to measure plasma or blood concentrations of CS. Dosing of patients was still done very much empirically. During this period, in particular, the task facing Sandoz was indeed massive. It required close cooperation with all centers involved and the monitoring of all possible relevant data. All this information had to be analyzed and the accumulated experience, for example in relation to dosing, monitoring, and occurrence of side effects, had to be exchanged between the various groups involved. The program entailed maximal efforts in information gathering, evaluation, and, crucially, exchange.

In view of the particular difficulties mentioned earlier, discussion with registration authorities in the UK and US had already been opened in early 1979 with a view to establishing what might be an appropriate formula for registration of CS. In both cases, the difficulties of carrying out trials towards registration and the potential value of CS were easily appreciated. With the UK authorities, it was agreed that it would be appropriate to compare accumulated results for transplantation under CS therapy with historical data, e.g., records of graft failure, patient survival, graft rejection and so forth, recorded pre-CS. In the US, the IND (Investigational New Drug) approval, allowing the early US trials to which I have referred above, was filed in May 1979. From that point on, there was a close collaboration with our US colleagues. My counterpart in the US was Dr. David Winter who ran the Sandoz clinical trials program for CS in that country. This proved to be a very pleasant cooperation, and we were able, as will be clear later, to exchange our data when it came to preparing registration dossiers.

By February 1980 some twenty studies were already approved or ongoing in Europe, the US, and Canada, with four in bone marrow transplantation (including the Marsden group), eight in transplantation (including the Cambridge group), seven in renal autoimmune diseases, and a bioavailability/pharmacokinetic trial in Basel under my direct supervision. At this time sufficient data were available to complete Phase I clinical trials. Specifically, 52 patients had received CS for bone marrow transplantation, 59 for cadaveric kidney transplantation, and three receiving liver and pancreas transplants had recently started. A further 67 patients had received CS for the treatment of various autoimmune diseases. On the basis of the studies completed by this time, it was possible to conclude that, in humans, CS was the most powerful immunosuppressant so far used in the treatment of renal transplant and bone marrow transplant patients. Results in both these indications had proven to be very encouraging; no kidney had been lost due to rejection, and only one patient had died of acute graft-versus-host disease in the only large series of trials performed. CS was not myelotoxic and, since it could be given alone, side effects of steroids could be avoided. Therefore, diabetics in end-stage renal failure and children could also be transplanted. Bacterial infection was not a problem. Viral infection had only occurred sporadically and had not been a problem either. The three lymphomas, to which I have already referred, were increasingly seen as being more likely an indication of the potency of CS than an inherent problem with the drug. A dose-dependent and reversible increase in serum creatinine and urea had been the most commonly encountered side effect. Abnormalities in liver function tests were also observed in transplant and some other patients. Hypertrichosis had also been fairly common, as well as anorexia and tremor. Gum hypertrophy had been reported only in renal transplant patients. All side effects had been shown to be reversible on stopping the drug. An exact dosage schedule could still not be defined, but there was every indication that about 17 mg/kg per day intramuscularly initially, decreasing to about 10 mg/kg per day orally was effective in renal transplantation. In bone marrow transplantation for the prophylaxis of graft-versus-host disease, a maintenance dose of 12.5 mg/kg per day on average had been effective, and CS could be stopped after 4–6 months. Every effort was being made to reduce still further the maintenance dose in renal transplant patients. It had also been established that by using a new hydration protocol (i.e., paying special attention to an adequate state of hydration of the patient at the time of transplantation) problems with renal dysfunction due to CS could be greatly reduced or eliminated (Calne et al., 1981). An 86% predicted graft survival at one year was considered unusual and much better than what had been obtained before. Such information was to be invaluable in initiating trials at new centers. Not surprisingly, on the basis of these results, Sandoz Research Management was able to take the decision to enter Phase II trials.

It was on April 1, 1980 that a meeting took place in the Britannia Hotel on London's Grosvenor Square that was to have far-reaching consequences. It was attended, on a very wet afternoon, by representatives from Cambridge (Prof. Sir Roy Calne and Dr. David White) and from Sandoz (Dr. E. Wiskott and Dr. T. Beveridge). A somewhat bedraggled Prof. Calne was last to arrive (having first gone to an even more prestigious hotel just around the corner) and then, over afternoon tea with toasted muffins, the decision was taken, in principle, to conduct a European multicenter, controlled, randomized trial in renal transplantation. We, on the Sandoz side, were overjoyed at this decision in view of the difficulties until then of conducting anything other than open studies. However, with an apparent 20% improvement in one-year graft survival when using CS, the time had clearly come to compare CS with the best local therapy in a formal setting. Within a few months it was possible for Prof. Calne to convene a special meeting during the period of the Boston International Congress of the Transplantation Society and eight possible centers were identified. The first meeting of the group was held in Basel in September 1980, and the European Multicenter Trial Group became reality. As there had never been anything comparable until then in the field of transplantation, it is worthwhile recording the names of the principal investigators and those from Sandoz responsible for trial coordination and data analysis. They were F. Harder (Basel), R.Y. Calne (Cambridge), R. Pichlmayr (Hannover), R. Margreiter (Innsbruck), R.A. Sells (Liverpool), R.W.G. Johnson (Manchester), W. Land (Munich), M. Slapak (Portsmouth), and T. Beveridge, W. Maurer, T.W. Poole, E. Wiskott, and A.J. Wood (Sandoz).

Further meetings took place in November 1980 to finalize the study protocol, and in December 1980 there were meetings with the local coordinators (one was designated for each center) and those centers that till then had not used CS were allowed to gain their first experience with the drug before the trial began on January 1, 1981. It had been calculated that we would require about 200 patients in order to be able to show a significant difference and by December 31, 1981, 232 patients (CS, 117; azathioprine + steroids, 115) had entered, and entry was complete. When all patients had been followed up for at least one year, one-year graft survival was 72% in the CS group and 52% in the control group (European Multicentre Trial Group, 1983). Thus we had obtained the perfect result in this pivotal trial. The story, however, does not end there because it was decided to continue the follow-up for ten years. So it was that on June 26, 1992 the group came together in Vitznau, Switzerland, to discuss the ten-year results (European Multicentre Trial Group, 1995), which still showed a statistically significant advantage in graft survival for patients treated with CS, thus confirming the three-year and five-year analyses that had previously been published (Calne and Wood, 1985; Calne, 1987). The coordinators, whose names can be found on the publications cited, deserve special credit for doggedly sticking to their task and sometimes going to extraordinary lengths to ensure that very few patients were lost to follow-up. From the outset, however, this group exhibited a fine team spirit. Regular contact had the effect of cementing the relationships, reinforcing the need for continued discipline in the conduct of the trial and of offering mutual support. Added to that was the presence of a fine sense of humor which, fortunately, was abundantly evident throughout the entire trial. For all the participants this was an exhilarating and unforgettable experience.

By April 1981, a little over one year after commencing Phase II trials, a total of more than 450 patients had received CS for periods of up to two years with doses ranging from 5 to 25 mg/kg per day in the various indications. Of these there were about 230 renal transplant and 130 bone marrow transplant patients, with the remainder having received CS for other transplants (liver, pancreas, heart) and, relatively few, for autoimmune diseases. It was also by this time becoming possible to carry out some comparative trials. In addition to the European multicenter trial there was a Canadian multicenter trial, organized in liaison with Dr. C. Stiller, London, Ontario, involving twelve centers and more than 200 patients. Whereas the European trial used CS alone, the Canadian trial combined CS with low-dose steroids. The latter trial also planned for a five-year follow-up period. Further controlled studies were also being performed by groups in Australia; Helsinki, Finland; Oxford, UK; Denmark; and Huddinge, Sweden. In the USA a further four monitored trials were ongoing at Boston, Denver/Pittsburgh, Minneapolis, and Houston. These entailed both open and randomized protocols involving the concomitant use of low-dose steroids along with CS. The pattern for the use of CS in Europe had been set by Calne (Calne et al., 1981), who avoided steroids whenever possible, whereas Prof. T.E. Starzl (Starzl et al., 1981c) advocated the use of CS with steroids, thus setting the pattern for the USA. In pancreas transplantation four additional studies were planned to follow the pilot study at Cambridge: two in the UK at Cambridge and Birmingham, one at Lyon in France, and one at Minneapolis, USA. The first time CS was used in heart transplantation was at Stanford, California, in December 1980 (Jamieson et al., 1981; Oyer et al., 1982). The excellent results obtained at Stanford reawakened interest in this operation throughout the world.

In March 1981, the first successful heart-lung transplant was carried out at Stanford using CS along with conventional immunosuppressants (Reitz et al., 1982a, 1982b). The patient, a former newspaper executive in Arizona, had to rely on her newspaper colleague and coauthor of the book she later wrote (Gohlke and Jennings, 1985) to use all of his political contacts and persistence to finally obtain the drug's approval by the FDA in time for her operation, which had undergone enforced delay pending this decision. I am proud to have a signed copy of this book in my possession. In liver transplantation, early results from the use of CS with (Starzl et al., 1981a, 1981b, 1982) or without steroids (Calne et al., 1979, 1981) were most encouraging and had the effect of revitalizing the whole field. In bone marrow transplantation five trials were ongoing: with the original Marsden group (R.L. Powles) at Sutton, UK, and with groups in Basel (B. Speck), Paris (E. Gluckman), Hammersmith (E.C. Gordon-Smith), and Westminster (A.J. Barrett) Hospitals, London, UK. A further pilot trial had been opened in South Africa. For ethical reasons there was no possibility of conducting any randomized comparative trials in bone marrow transplantation.

Results obtained had been extremely promising, with actuarial one-year kidney graft survival rates as high as around 85% in some centers. This was substantially better than the previous average of some 50%–60%. Initial doses of CS in kidney transplant patients of 17 mg/kg/per day, tapering down to about 8 mg/kg per day had proved effective and, in the liver transplant patients, a starting dose of 15 mg/kg per day had proved effective. In bone marrow transplantation, initial doses had been as high as 20 mg/kg per day, which was then reduced to a maintenance dose of 12.5 mg/kg per day. This could then be gradually tapered down and stopped altogether after 6–12 months. About 75% of patients had survived, and out of 109 patients, only ten deaths associated with graft-versus-host disease had been recorded. The results of bone marrow transplantation in severe aplastic anemia appeared particularly gratifying, with a survival rate of the order of 70% as compared with only 40%–50% on conventional therapy. No significant additional side effects had become apparent, and the reversibility of renal dysfunction and other side effects previously recorded was confirmed. On the basis of the results already at hand, it was possible to conclude the Phase II program and enter Phase III studies.

Knowledge of these results had spread quickly throughout the transplant community and generated a tremendous amount of interest coupled with the desire to obtain CS for clinical studies. Considerable time and effort went into the careful consideration of all these requests and proposals for initiating studies. By October 1981, we had a total of 99 investigators on our files who had expressed their interest in CS. They represented 24 countries from all over the world, from Norway to New Zealand and from Brazil to China.

At this time there was still a severe restraint on our ability to involve further centers due to the CS supply situation. However, by the autumn of 1981, thanks to the efforts of those involved in upscaling of fermentation and supply, the position was improving. It was estimated that, in addition to amounts of CS required to maintain ongoing studies and follow-up studies in 1982, about 100 kg of material would be available. This could allow for the treatment of about 300 additional patients in organ transplantation and a further 100 or 200 patients in bone marrow transplantation. This was seen as being vital to the expansion and earliest possible completion of Phase III studies.

In mid-1981, not long after the start of Phase III trials, the concentrate for intravenous infusion became available and was entering preliminary clinical testing as a potential, and later actual, replacement for the intramuscularly injectable form hitherto in use. The increased availability of high-performance liquid chromatography (Niederberger et al., 1980; Nussbaumer et al., 1982) and radioimmunoassay (Donatsch et al., 1981) techniques for measuring CS concentrations in blood and either plasma or serum enabled further evaluation to be made of the pharmacokinetics of CS (Beveridge et al., 1981).

It was in December 1981 that it became possible to prepare a clinical synopsis containing data that supported the claim that CS is of benefit in the prevention of graft rejection following bone marrow transplantation, and in the prophylaxis of graft-versus-host disease. From September 1978 to May 1981, CS was given prophylactically to a series of 153 patients at five centers. There were 81 patients with acute myeloid leukemia (AML), 27 with acute lymphoblastic leukemia (ALL), 36 with severe aplastic anemia (SAA), seven with chronic myeloid leukemia (CML), and one patient each with Hurler's syndrome and Gaucher's disease. Twenty of the 153 patients received mismatched grafts. A group of 25 patients previously treated with methotrexate served as historical controls. One-year survival for all CS-treated patients with matched grafts was 68%; for leukemics, 65%; SAA, 79%. AML (67%) was better than ALL (58%), and leukemic patients transplanted in first remission did better (71%) than those in later remission or relapse (49%). For historical controls treated with methotrexate the one-year survival was 52%. Death associated with graft-versus-host disease occurred in ten out of 132 patients (8%). CS had proven to be a most powerful immunosuppressant and superior to methotrexate in its ability to reduce the severity of graft-versus-host disease. While graft-versus-host disease still occurred in 73% of CS-treated patients, morbidity and mortality were considerably less. In the CS series 86% of the episodes of graft-versus-host disease were mild in severity (56% being only a rash or of Grade I severity and 30% Grade II). The other advantages of CS were: no myelotoxicity, prompt and sustained engraftment, quicker hematological recovery, and less time spent in the hospital. These results were presented in March 1982 at the Sixth European Bone Marrow Transplant Meeting held in Courmayeur, Italy (Beveridge, 1982).

During the course of 1982, despite the logistic difficulties and the sheer volume of material to be sorted through, it was possible to begin compiling results towards the preparation of a clinical synopsis dealing with organ transplantation for submission to registration authorities. This work was completed in December 1982.

The clinical data in this file incorporated the results of five studies in Europe and Australia involving 296 patients (Part A) and of ten studies in North America involving 411 patients (Part B). In Part A there were 279 control patients and in Part B 266 control patients (all in renal transplanta-tion). Also in Part B were 170 historical controls for liver transplantation and 70 for heart transplantation. Apart from patients who had received CS for renal transplantation, there was data for seven liver and ten pancreas transplants in Part A, and in Part B there were 40 patients who underwent a total of 44 liver transplant procedures, 40 patients who received 41 heart transplants, and six who received a combined heart-lung transplant, all of whom received CS. This brought together information and details of the use of CS in organ transplantation during the period June 1978 until August 1982.

CS either given alone (Part A) or along with steroids (Part B) was able to provide excellent results. In the two largest randomized controlled studies the predicted one-year graft survival figures were (CS first) 73% versus 53% (European Multicentre Trial Group, 1982) and 80% versus 64% (Canadian Multicentre Transplant Study Group, 1983). In each of the studies in Part B of the file a common feature was the decrease in the number of rejection episodes seen in CS-treated patients. In the European multicenter trial (European Multicentre Trial Group, 1982), graft failure due to rejection was markedly less common in the CS group than in the control group (23 versus 40, respectively) and there were many more patients in the CS group than in the control group who did not experience any rejection episodes (20% versus 11%). In liver transplantation, one-year patient survival rates of 71% and 64% were recorded (versus 32% for historical controls). One-year survival for heart transplant patients was 76% (versus 62% for historical controls) at one center and 67% at another. Four out of the six heart-lung transplant patients were alive and well.

Regarding infections, the general tendency was for CS patients to have fewer and less severe infections than the control patients. The side effects recorded were as previously described. Among laboratory parameters the serum potassium was consistently higher in CS-treated patients than in control patients. Serum total bilirubin and alkaline phosphatase were also consistently higher in CS-treated patients, as were blood urea and serum creatinine values. In renal transplant patients, many of whom by then had had CS for up to two years, there was no evidence of any substantial problems accompanying long-term therapy. It was clear that, on balance, the benefits to be gained from CS in organ transplantation far outweighed the risks. Dosage recommendations could be made including instructions on the use of the intravenous concentrate for infusion, and a consensus was obtained that routine use of the radioimmunoassay to monitor the blood CS concentration could be of value in providing good patient care even in the absence, at that time, of definite therapeutic or toxic ranges for CS concentrations. CS was no panacea but had been shown to be the most potent and specific immunosuppressant so far discovered and had an acceptable incidence of side effects. CS (Sandimmune®) became generally available in 1983.

> *Now this is not the end. It is not even the beginning of the end. But it is, perhaps, the end of the beginning.*
>
> — Winston S. Churchill

Reflections

When CS came along in 1978, it was the first new immunosuppressant to be used in transplantation in over 15 years. Naturally, CS generated tremendous interest and excitement, and the result was a powerful pressure exerted on Sandoz as the outside investigators tried to obtain some drug to satisfy their various interests. In one sense this was good, in that we could more or less choose our partners, but bad in that we could never hope to satisfy everyone. The policy of being as liberal as possible with CS and allowing many leading figures to use the drug and publish their findings was, in my view, a wise one and was certainly appreciated by the transplant community. There can be no doubt that the Sandoz policy at that time, and the general way in which we conducted the clinical trials with CS as well as the manner in which we dealt with the investigators, created a good track record and was the foundation of an enormous amount of goodwill and respect for the company. The reputation of Sandoz today in the transplant area is largely based on that vital early experience.

In the early 1980s it was unusual for someone in the Experimental Therapeutics Dept to continue to be responsible for Phase II and III projects.

It was my great good fortune (and perhaps a good company decision) that I was allowed to remain as Medical Expert for CS in transplantation, a position I still hold today. Starting off with two dosage strategies, CS with or without steroids initially, might have been seen as creating problems yet this was able to be turned to our advantage when it became clear that CS could be used in several different regimens. Today one speaks of CS-based immunosuppressive regimens, and many patients who have had a kidney transplant profit from the absence of steroids in the maintenance phase. Indeed there are still a few centers who continue to use CS alone as initial immunosuppression. The importance of Cambridge in much of our early work with CS cannot be overestimated. Theirs was a pioneer role, and by example, other centers were helped and encouraged to follow. It was a successful pattern which was repeated over and over again in other countries as the use of CS was gradually expanded. Since it is widely accepted that CS is not an easy drug to use, it seemed to make sense to impart as much knowledge as possible in one center and let them guide other centers, with our help, as development continued. Perhaps it was our good fortune with CS to work mostly with surgeons who, when required to make a tricky decision, tend to favor a direct and decisive approach rather than trust the committee method of deliberating while the situation deteriorates. It was certainly the innovativeness of CS itself which was mainly responsible for the rapid and unique development, but another factor was the rapport enjoyed with investigators from all parts of the world coupled with the fine team spirit in Basel among those most closely involved with the project. With a time of just five years from first clinical studies to registration, one wonders how our flexible program of the early 1980s might fare today if we had to conform with our current sets of rules and guidelines for clinical trials.

What happened with the two main problems we encountered along the way, nephrotoxicity and lymphomas? Investigators in transplantation quickly learned how to cope with the nephrotoxicity, mainly by judicious dosage reduction aided by the monitoring of blood CS concentrations as assay methods became available but also aided by an increased understanding of the morphological findings in kidneys that had been exposed to CS (Mihatsch et al., 1983). As for lymphomas, one had concluded by 1983 that CS, if used with care, presented no more risk to the patient than conventional immunosuppression (*Lancet* editorial, 1983). This view was underlined (Beveridge et al., 1984) shortly following the finding that lymphomas and lymphoproliferative disorders developing under CS and steroid therapy are relatively innocuous and reversible if appropriately treated (Starzl et al., 1984).

With such a drug, not only patients stood to gain something. Registration authorities also recognized that with CS they had much to gain and little to lose by helping to make it available to as many patients as possible and as soon as possible.

It did not take long for CS to establish itself as the immunosuppressive agent of first choice in transplantation and claim its place in history. As of December 1993, there were more than 20,000 publications on CS and about 9,000 of these related to clinical transplantation. CS was the most significant factor in the expansion not only of renal transplantation but also of liver and heart transplantation (Gordon et al., 1988). In 1983, liver transplantation was accepted as a therapeutic modality for end-stage liver disease (National Institutes of Health, 1983) and no longer considered experimental. Only 17 centers were active in heart transplantation in December 1980 (Kriett and Kaye, 1990) when CS was used at Stanford in their first patient (Jamieson et al., 1981; Oyer et al., 1982). Now there are over 200. Before 1980, the annual total of such operations was less than 100; since 1986 more than 2000 per year have been performed with one-year survival generally about 85%.

As with other great discoveries, for example penicillin, there was an element of fortuitousness surrounding it. However, one still has to recognize good fortune when it is met. In the discovery and development of penicillin, three people were essential to the project: Fleming, Chain, and Florey (Macfarlane, 1979). With CS the number three again features. It was thanks to the continuity provided for more than twelve years by Borel, Wiskott, and Beveridge that the CS development was smoother than might otherwise have been possible.

REFERENCES

Baumann G (1992): Molecular mechanism of immunosuppressive agents. *Transplant Proc* 24(Suppl 2):4–7

Bell PA (1981): Steroids and the cells of the immune system. In: *Mechanisms of Steroid Action*, Lewis GP, Ginsburg M eds. London: Macmillan

Beveridge T (1982): Combined European experience with Cyclosporin A in bone marrow transplantation. *Exp Haematol* 10(Suppl 10):88–91

Beveridge T (1992): Clinical development of cyclosporine. *Transplant Proc* 24 (suppl 2):64–66

Beveridge T, Calne RY for the European Multicentre Trial Group (1995): Cyclosporin A (Sandimmune®) in cadaveric renal transplantation. Ten-year follow-up of a multicentre trial. *Transplantation* 59:1–3

Beveridge T, Gratwohl A, Michot F, Niederberger W, Nüesch E, Nussbaumer K, Schaub P, Speck B (1981): Cyclosporin A: Pharmacokinetics after a single dose in man and serum levels after multiple dosing in recipients of allogeneic bone-marrow grafts. *Curr Ther Res* 30:5–18

Beveridge T, Krupp P, McKibbin C (1984): Lymphomas and lymphoproliferative lesions developing under cyclosporin therapy. *Lancet* i:788

Bollinger P, Sigg HP, Weber HP (1973): Die Struktur von Ovalicin. *Helv Chim Acta* 56:819–830

Borel JF (1962): The usefulness of lectins for detecting individual differences among chickens. *Vox Sang* 7:632–637

Borel JF (1967): Serological analysis of anti-A antibodies in relation to the ABO Morbus haemolyticus neonatorum. *Zschr ImmunForsch klin Immunol* 132:72–92

Borel JF (1970): Effect of subcellular leukocyte fractions on neutrophils and macrophages. *Int Arch Allerg* 39:247–271

Borel JF (1973): Effect of some drugs on the chemotaxis of rabbit neutrophils in vitro. *Experientia* 29:676–678

Borel JF (1974): Comparison of the immune response to sheep erythrocytes, tetanus toxoid and endotoxin in different strains of mice. *Agents Actions* 4:277–285

Borel JF (1976): Comparative study of in vitro and in vivo drug effects on cell-mediated cytotoxicity. *Immunology* 31:631–641

Borel JF (1982): History of Cyclosporin A and its significance in immunology. In: *Cyclosporin A*, DJG White ed. Amsterdam: Elsevier

Borel JF (1989): Pharmacology of Cyclosporine (Sandimmune). IV. Pharmacological properties in vivo. *Pharmacol Rev* 41:259–371

Borel JF, Feurer C (1975): Chemotaxis of rabbit macrophages in vitro: Inhibition by drugs. *Experientia* 31:1437–1439

Borel JF, Feurer C (1978): In vivo effects of anti-inflammatory and other drugs on granulocyte emigration in the rabbit skin collection chamber. *J Path* 124: 85–93

Borel JF, Feurer C, Gubler HU, Stähelin H (1976): Biological effects of cyclosporin A: A new antilymphocytic agent. *Agents Actions* 6:468–475

Borel JF, Feurer C, Hiestand PC, Stähelin H (1978): The effects of franctions (chalones) obtained from lymphoid organs on the immune response in vivo. *Agents Actions* 8:523–531

Borel JF, Feurer C, Magnée C, Stähelin H (1977): Effects of the new anti-lymphocytic peptide cyclosporin A in animals. *Immunology* 32:1017–1025

Borel JF, Kis ZL (1991): The discovery and development of cyclosporine (Sandimmun). *Transplant Proc* 23:1867–1874

Borel JF, Lazáry S, Stähelin H (1974): Immunosuppressive effects of ovalicin-semicarbazone. *Agents Actions* 4:357–363

Borel JF, Sorkin E (1967): Antikörperbildung gegen Schaferythrozyten durch Milzzellen und Blutleukozyten bei der Ratte. *Zschr ImmunForsch klin Immunol* 133:207–220

Borel JF, Wiesinger D (1977): Effect of cyclosporin A on murine lymphoid cells. In: *Regulatory Mechanisms in Lymphocyte Activation*. Lucas DO ed. New York: Academic Press

Brent L, Sells RA (1989): Notes on the history of tissue and organ transplantation. In: *Organ Transplantation*, Brent L, Sells, RA eds. London: Ballière Tindall

Calne RY (1961): Inhibition of the rejection of renal homografts in dogs by purine analogues. *Transplant Bull* 28:445–461

Calne RY for the European Multicentre Trial Group. (1987): Cyclosporin in cadaveric renal transplantation: 5-year follow-up of a Multicentre Trial. *Lancet* ii: 506–507

Calne RY, Rolles K, White DJG, Thiru S, Evans DB, McMaster P, Dunn DC, Craddock GN, Henderson RR, Aziz S, Lewis P (1979): Cyclosporin A initially as the only immunosuppressant in 34 recipients of cadaveric organs: 32 kidneys, 2 pancreases and 2 livers. *Lancet* ii:1033–1036

Calne RY, White DJG (1977): Cyclosporin A – a powerful immunosuppressant in dogs with renal allografts. *IRCS Med Sci* 5:595

Calne RY, White DJG, Evans DB, Thiru S, Henderson RG, Hamilton DV, Rolles K, McMaster P, Duffy TJ, MacDougall BRD, Williams R (1981): Cyclosporin A in cadaveric organ transplantation. *Brit Med J* 282:934–936

Calne RY, White DJG, Rolles K, Smith DP, Herbertson BM (1978a): Prolonged survival of pig orthotopic heart grafts treated with cyclosporin A. *Lancet* i:1183–1185

Calne RY, White DJG, Thiru S, Evans DB, McMaster P, Dunn DC, Craddock GN, Pentlow BD, Rolles K (1978b): Cyclosporin A in patients receiving renal allografts from cadaver donors. *Lancet* ii: 1323–1327

Calne RY, Wood AJ for the European Multicentre Trial Group (1985): Cyclosporin in cadaveric renal transplantation: 3-year follow-up of a European Multicentre Trial. *Lancet* ii:549

Canadian Multicentre Transplant Study Group (1983): A randomized clinical trial of cyclosporine in cadaveric renal transplantation. *New Engl J Med* 309:809–815

CIBA Foundation Study Group No. 29 (1967): In *Antilymphocytic Serum*, Wolstenholme GEW, O'Connor M eds. London: J. & A. Churchill

Crawford DH, Thomas JA, Janossy G, Sweny P, Fernando ON, Moorhead JF, Thompson JH (1980): Epstein Barr virus nuclear antigen positive lymphoma after cyclosporin A treatment in patient with renal allograft. *Lancet* i:1355–1356

Current Contents (1984): 27(6):16

Donatsch P, Abisch E, Homberger M, Traber R, Trapp M, Voges R (1981): A radioimmunoassay to measure cyclosporin A in plasma and serum samples. *J Immunoassay* 2:19–32

Dreyfuss M, Härri E, Hofmann H, Kobel H, Pache W, Tscherter H (1976): Cyclosporin A and C. New metabolites from *Trichoderma polysporum* (Link ex Pers.) *Rifai. Europ J Appl Microbiol* 3:125–133

Dunn DC, White DJG, Wade J (1978): Survival of first and second kidney allografts after withdrawal of cyclosporin A therapy. *IRCS Med Sci* 6:464

European Multicentre Trial Group (1982): Cyclosporin A as sole immunosuppressive agent in recipients of kidney allografts from cadaver donors: Preliminary results. *Lancet* ii:57–60

European Multicentre Trial Group (1983): Cyclosporin in cadaveric renal transplantation: One-year follow-up of a Multicentre Trial. *Lancet* ii:986–989

Feurer C, Borel JF (1974): Localised leukocyte mobilisation in the rabbit ear. An in vivo cell migration technique using plastic collection chambers. *Antibiot Chemother* 19:161–178

Gohlke M, Jennings M (1985): *I'll Take Tomorrow.* New York: M. Evans

Gordon RD, Iwatsuki S, Esquivel CO, Makowka L, Tzakis AG, Todo S, Starzl TE (1988): Liver transplantation. In: *Organ Transplantation and Replacement*, Cerilli GJ, ed. Philadelphia: J.B. Lippincott

Herrmann B, Müller W (1979): Die Therapie der chronischen Polyarthritis mit Cyclosporin A, einem neuen Immunsuppressivum. *Akt rheumatol* 4:173–186

Hiestand PC, Borel JF, Bauer W, Kis ZL, Magnée C, Stähelin H (1977): The effects of franctions (chalones) obtained from lymphoid organs on lymphocyte proliferation in vitro. *Agents Actions* 7:327–335

Jamieson SW, Oyer PE, Reitz BA, Baumgartner WA, Bieber CP, Stinson EB, Shumway NE (1981): Cardiac transplantation at Stanford. *Heart Transplant* 1:86–91

Kleinkauf H, van Liempt H, Palissa H, von Döhren H (1992): Biosynthese von Peptiden: Ein nichtribosomales System. *Naturwissenschaften* 79:153–162

Kostakis AJ, White DJG, Calne RY (1977): Prolongation of the rat heart allograft survival by cyclosporin A. *IRCS Med Sci* 5:280

Kriett JM, Kaye MP (1990): The registry of the international society for heart transplantation: Seventh official report – 1990. *J Heart Transplant* 9:323–330

Lancet (1983): Editorial: Cyclosporin and neoplasia. i:1083

Lawen A, Traber R (1993): Substrate specificities of cyclosporin synthetase and peptolide SDZ 214-103 synthetase. *J Biol Chem* 268:20452–20465

Lazáry S, Stähelin H (1968): Immunosuppressive and specific antimitotic effects of ovalicin. *Experientia* 24:1171–1173

Lazáry S, Stähelin H (1969): Immunosuppressive effect of a new antibiotic: Ovalicin. *Antibiot Chemother* 15:177–181

Macfarlane G (1979): *Howard Florey: The Making of a Great Scientist.* Oxford: Oxford University Press

Matter BE, Donatsch P, Racine RR, Schmid B, Suter W (1982): Genotoxicity evaluation of cyclosporin A, a new immunosuppressive agent. *Mutation Res* 105:257–264

Mihatsch MJ, Thiel G, Spichtin HP, Oberholzer M, Brunner FP, Harder F, Olivieri V, Bremer R, Ryffel B, Stocklin E, Torhorst J, Gudat F, Zollinger HU, Loertscher R (1983): Morphological findings in kidney transplants after treatment with cyclosporine. *Transplant Proc* 15(Suppl 1):2821–2835

Müller W, Herrmann B (1979): Cyclosporin A for psoriasis. *New Engl J Med* 301:555

National Institutes of Health Consensus Conference (1983): *JAMA* 250:2961

Niederberger W, Schaub P, Beveridge T (1980): High-performance liquid chromatographic determination of cyclosporin A in human plasma and urine. *J Chromatogr, Biomed Appl* 182:454–458

Nussbaumer K, Niederberger W, Keller HP (1982): Determination of cyclosporin A in blood and plasma by column-switching HPLC after rapid sample preparation. *J High Resol Chromatography & Chromatography Commun* 5: 424–427

Oyer PE, Stinson EB, Reitz BA, Jamieson SW, Hunt SA, Schroeder JS, Billingham ME, Wallwork JL, Bieber CP, Baumgartner WA, Gamberg PL, Miller JL, Shumway NE (1982): Preliminary results with cyclosporin A in clinical cardiac transplantation. In: *Cyclosporin A*, White DJG, ed. Amsterdam: Elsevier

Petcher TJ, Weber HP, Rüegger A (1976): Crystal and molecular structure of an iodo-derivative of the cyclic undecapeptide cyclosporin A. *Helv Chim Acta* 59:1480–1488

Powles RL, Barrett AJ, Clink H, Kay HEM, Sloane J, McElwain TJ (1978): Cyclosporin A for the treatment of graft-versus-host disease in man. *Lancet* ii:1327–1331

Quesniaux VFJ (1989): Pharmacology of Cyclosporine (Sandimmun). III. Immunochemistry and Monitoring. *Pharmacol Rev* 41:249–258

Reitz BA, Wallwork JL, Hunt SA, Pennock JL, Billingham ME, Oyer PE, Stinson EB, Shumway NE (1982): Heart-lung transplantation: Successful therapy for patients with pulmonary vascular disease. *New Engl J Med* 306:557–564

Reitz BA, Wallwork JL, Hunt SA, Pennock JL, Oyer PE, Stinson EB, Shumway NE (1982): Cyclosporin A for combined heart-lung transplantation. In: *Cyclosporin A*, White DJG, ed. Amsterdam: Elsevier

Ryffel B (1982): Experimental toxicological studies with cyclosporin A. In: *Cyclosporin A*, White DJG, ed. Amsterdam: Elsevier

Rüegger A, Kuhn M, Lichti H, Loosli HR, Huguenin R, Quiquerez C, von Wartburg A (1976): Cyclosporin A, ein immunsuppressiv wirksamer Peptidmetabolit aus *Trichoderma polysporum* (Link ex Pers.)*Rifai. Helv Chim Acta* 59:1075–1092

Schwartz R, Dameshek W (1959): Drug-induced immunological tolerance. *Nature (Lond)* 183:1682–1683

Sigg HP, Weber HP (1968): Isolierung und Strukturaufklarung von Ovalicin. *Helv chim Acta* 51:1395–1408

Smith RB (1985): *The Development of a Medicine*. Hong Kong: The MacMillan Press

Starzl TE, Nalesnik MA, Porter KA, Ho M, Iwatsuki S, Griffith BP, Rosenthal JT, Hakala TR, Shaw BW Jr., Hardesty RL, Atchison RW, Jaffe R, Bahnson HT (1984): Reversibility of lymphomas and lymphoproliferative lesions developing under cyclosporin-steroid therapy. *Lancet* i:583–587

Starzl TE, Iwatsuki S, Bahnson HB, van Thiel DH, Hardesty R, Griffith B, Shaw BWJr., Klintmalm GBG, Porter KA (1982): Cyclosporin A and steroids for liver and heart transplantation. In: *Cyclosporin A*, White DJG, ed. Amsterdam: Elsevier

Starzl TE, Iwatsuki S, Klintmalm G, Schröter GPJ, Weil R, Koep LJ, Porter KA (1981): Liver transplantation, 1980, with particular reference to cyclosporin A. *Transplant Proc* 13:281–285

Starzl TE, Klintmalm GBG, Porter KA, Iwatsuki S, Schröter GPJ (1981): Liver transplantation with use of cyclosporin A and prednisone. *New Engl J Med* 305: 266–269

Starzl TE, Klintmalm GBG, Weil R, Porter KA, Iwatsuki S, Schröter GPJ, Fernandez-Bueno C, MacHugh N (1981): Cyclosporin A and steroid therapy in sixty-six cadaver kidney recipients. *Surgery Gynec Obst* 153:486–494

Strober S, Weissman IL (1981): Immunosuppressive and tolerogenic effects of whole-body, total lymphoid, and regional irradiation. In: *Immunosuppressive Therapy*, Salaman JR, ed. Lancaster: MTP Press

Stähelin H (1986): Historical background. *Prog Allergy* 38:19–27

Von Wartburg A, Traber R (1986): Chemistry of the Natural Cyclosporin Metabolites. In: *Ciclosporin*. Borel JF, ed. Basel: Karger

Walkinshaw MD, Kallen J, Weber HP, Widmer A, Widmer H, Zurini M (1992): Immunophilin structure: A template for immunosuppressive drug design? *Transplant Proc* 24 (suppl 2):8–13

Wenger RM (1982): Chemistry of cyclosporin A. In: *Cyclosporin A*, White DJG, ed. Amsterdam: Elsevier

Wenger RM (1986): Synthesis of ciclosporin and analogues, structural and conformational requirements for immunosuppressive activity. In: *Ciclosporin*. Borel JF, ed. Basel: Karger

Wiesinger D, Borel JF (1979): Studies on the mechanism of action of cyclosporin A. *Immunobiol* 156:454–463

Zimmermann WA, Hartmann GR (1981): On the mode of action of the immunosuppressive sesquiterpene ovalicin. *Eur J Biochem* 118:143–150

3

Discovery and Development of FK506 (Tacrolimus), A Potent Immunosuppressant of Microbial Origin

Michihisa Nishiyama, Shizue Izumi, and Masakuni Okuhara

Introduction

A little over a decade has passed since the discovery of the potent immunosuppressant FK506 (tacrolimus). In April 1983, Fujisawa expanded the research arm of the company by setting up new research laboratories at Tsukuba science city, Japan's comprehensive research complex for business and industry. Fujisawa then began a year of screening for new immunosuppressants at the Tsukuba laboratories, which introduced us to the field of immunosuppressive drugs.

The story of our discovery and development of FK506 begins with a description of the major factors that contributed to the success of our research program. Looking back, we can say that there were several factors, besides luck, that led us to the discovery of FK506. One was the accumulation of a wealth of expertise in our laboratories on natural product research; another was an increased understanding of the basic mechanisms of immunological diseases, such as graft rejection after transplantation and the development of autoimmune diseases.

Over the past three decades, Fujisawa scientists, searching for natural products, have undertaken comprehensive programs to discover new drugs of microbial origin. Antibiotic research, begun early in this century, has demonstrated that microbial products have amazing diversity and complexity in their chemical structure as well as in their biological activities. By early 1980, we succeeded in finding numerous, diverse agents

The Search for Anti-Inflammatory Drugs
Vincent J. Merluzzi and Julian Adams, Editors
© Birkhäuser Boston 1995

which included antibiotics, anticancer agents, and a variety of pharmacologically active substances such as vasodilators, PAF antagonists, aldose reductase inhibitors, an angiotensin-converting enzyme inhibitor, and, in the immunology field, the potent immunostimulant FK156 (Okuhara and Kino, 1994). Through this experience, we became proficient at detecting various pharmacological actions of microbial products and, therefore, started our screening program for immunosuppressants.

For many decades, immunity was not well understood due to the complexity of the reaction mechanism, although physicians used it to control certain diseases. In recent years, however, our understanding of the immune system has increased enormously. Since the early 1960s, when survival of allogeneic skin grafts was observed to be prolonged in neonatal thymectomized mice, we have considered T cells to play a central role in graft rejection mechanisms. However, the actual T-cell subpopulations involved in the phenomenon remained obscure (Miller, 1962). In the late 1970s, functionally active cytotoxic T lymphocytes were cloned and could be cultured indefinitely in the presence of exogenous T-cell growth factor, interleukin 2 (IL-2), a critical protein in T-cell immune responses (Morgan et al., 1976; Gillis and Smith, 1977). In 1982, using cloned, activated T cells, Engers and co-workers demonstrated that intravenously injected T-cell clones can induce allogeneic tumor cell destruction within the peritoneal cavity of immunosuppressed mice (Engers et al., 1982), thus providing direct evidence of activated T-cell involvement in organ allograft rejection.

In the autoimmune disease field, considerable evidence had accumulated demonstrating that activated T cells participated in the manifestation of the diseases. For example, lymphocyte depletion by thoracic duct drainage or blood leukapheresis led to short-term improvement in rheumatoid arthritis (Paulus et al., 1977; Karsh et al., 1981). In 1981, Ben-Nun and co-workers provided direct evidence that activated T cells are involved in provoking autoimmune diseases (Ben-Nun et al., 1981). They established activated T-cell lines derived from rats with autoimmune encephalomyelitis, and showed that when the cells were injected intravenously clinical paralysis could be induced in syngeneic rats.

The Screening System

Against this background of advances in basic immunology and better understanding of microbial versatility, we initiated our screening program. Generally, in the search for novel bioactive microbial products, we have to screen thousands of fermentation samples for the desired activity. According to many researchers, there are several crucial factors to consider when designing assay methods. Among them are simplicity, reproducibility, high throughput, and specificity. Fujisawa's in vitro assay system for immunosuppressive activity satisfied all of these factors.

Fujisawa's screening program was directed at blocking an abnormal immune response through suppressing the activity of activated T cells. One of the approaches for this objective was inhibition of IL-2 production. Selective inhibition of IL-2 production could inhibit the growth of activated T cells, which in turn would result in suppression of an abnormal immune response. Among the available methods to measure IL-2, we used mixed lymphocyte culture (MLC)-induced or concanavalin A (ConA)-induced lymphocyte blastogenesis.

Mixed lymphocyte cultures were prepared from spleen cells from two strains of mice with different major histocompatibility complex (MHC) antigens. Splenocytes from these two strains were co-cultured in multi-well microtiter plates in the presence of serial diluted screening samples (Figure 3-1). In ConA-induced blastogenesis, ConA was added to a spleen cell culture system prepared from a mouse strain. The degree of suppression of lymphocyte blastogenesis in the screening procedure can be determined by measuring [^{3}H]-thymidine incorporation or through direct microscopic observation of lymphocyte proliferation.

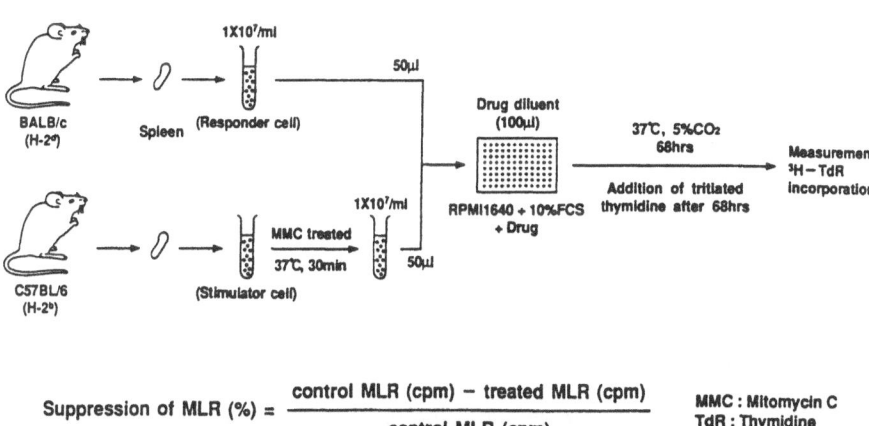

$$\text{Suppression of MLR (\%)} = \frac{\text{control MLR (cpm)} - \text{treated MLR (cpm)}}{\text{control MLR (cpm)}}$$

MMC : Mitomycin C
TdR : Thymidine

Figure 3-1. Mixed lymphocyte reaction screening assay for immunosuppressants.

Since fermentation samples often contain cytotoxic substances that inhibit the growth of lymphocytes, we needed a method that could exclude nonspecific interference. For this specificity screen, we prepared a set of two microtiter plates. One plate was used to test for the suppressive effect on T-cell blastogenesis, and the other was used to test for the nonspecific cytotoxicity. Mouse lymphoma EL-4 was found to be the most appropriate cell line to examine nonspecific cytotoxicity because of its constitutive proliferative properties. This set of two plates, one containing the splenocytes and the other containing EL-4, were cultured in the presence of serial twofold dilutions of the screening samples in a CO_2 incubator. After three days of incubation, the plates were studied under a microscope, and the samples showing suppression of lymphocyte proliferation at a lower concentration than necessary for the growth of EL-4 were selected as positive. The use of MLC in conjunction with EL-4 assays provided the specificity necessary to differentiate cytotoxic compounds from promising agents that were immunosuppressive but not cytotoxic.

The Discovery of FK506

We screened 250 samples per week, prepared from fermentation broths. After screening about 3000 samples, activity was detected that inhibited the mixed lymphocyte reaction (MLR). The producing fungus strain *Penicillium jensenii* was found to produce two active principles: one was fumagillol, a known carcinolytic compound, and the other was its new demethyl derivative FR65814 (Figure 3-2) (Hatanaka et al., 1988b). Both compounds demonstrated a potent suppressive effect on MLR. However, a careful microscopic observation of both lymphocyte blastogenesis in MLR and EL-4 growth showed that, at the range of concentration required for MLR inhibition, the EL-4 growth was also found to be weakly inhibited. Therefore, we did not pursue development of these compounds and continued the screening program. In March 1984, after screening about 10,000 samples, potent suppression of MLR was detected from a cultured broth of an actinomycete, strain No. 9993 (Kino et al., 1987a). The culture filtrate inhibited the lymphocyte proliferation at a 1000-fold dilution, but did not inhibit the growth of EL-4 lymphoma cell at a tenfold dilution (Figure 3-3). Another 1000 samples later, broth from another producing strain (No. 7238) was found to have MLR-suppressing activity (Hatanaka et al., 1988a).

We tried to isolate the active principles from the cultured broth; however, during the isolation process, several difficulties preventing complete purification were found. Although isolation and purification techniques have improved markedly in recent years, one of the most laborious and time-consuming steps of screening remains the isolation of pure compounds from the fermentation broth. Both actinomycete strains produced only small amounts of their respective active components (less than 10 μg per ml of filtrate). Furthermore, the active substance from strain No. 9993 possesses

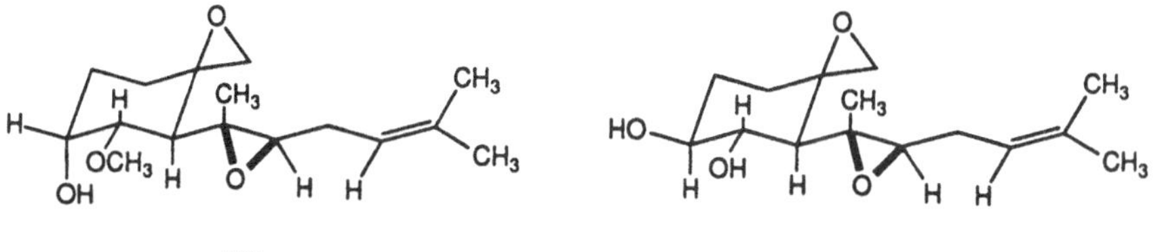

Figure 3-2. Fumagillol (A) and FR65814 (B).

several complicated physicochemical properties, such as lack of stability in methanol and tautomeric equilibration in solution. Minute amounts of derivatives were also produced along with the main active component. Contamination by these minor derivatives, which had similar physicochemical properties, made crystallization of the main substance difficult. Tremendous effort was spent on the isolation process. In May 1984, we obtained 5 mg of white powder that was chromatographically homogeneous, but we had to wait until the following December to obtain it in the completely purified crystalline form. Diaion HP-20 adsorption resin (Mitsubishi Chemical Industrial, Ltd.), ethyl acetate extraction, and silica gel were the key steps employed during the isolation procedures. The active principle was crystallized from acetonitrile as colorless prisms. The crystallized main component was named FR900506, subsequently FK506, and then tacrolimus (Figure 3-4a).

In the course of FK506 mass production for use in animal study evaluations, several minor components with analogous chemical structure were isolated from the culture broths. These were FR900525, FR901154, FR901155, FR901156, and FR900520 (Hatanaka et al., 1989; Okuhara et al., 1990). FR900520 and another new analogue, FR900523, were also isolated from the broth of strain No. 7238 (Hatanaka et al., 1988c) (Figure 3-4b). As shown in Table 3-1, Rf values for several of the analogues were almost identical to that of FK506 on a thin-layer chromatogram using silica gel, which made it difficult to separate them from each other. FR900525, FR901154, and FR901155 could be separated from FK506 by chromatography on silica gel plates using a hexane-ethyl acetate solvent system. FR901156 and FR900520 were impossible to separate from FK506 by chromatography; however, the introduction of Ag^{+}-adsorbed silica gel chromatography, which is often used for separation of compounds with unsaturated chemical bonds, enabled us to separate these physicochemically

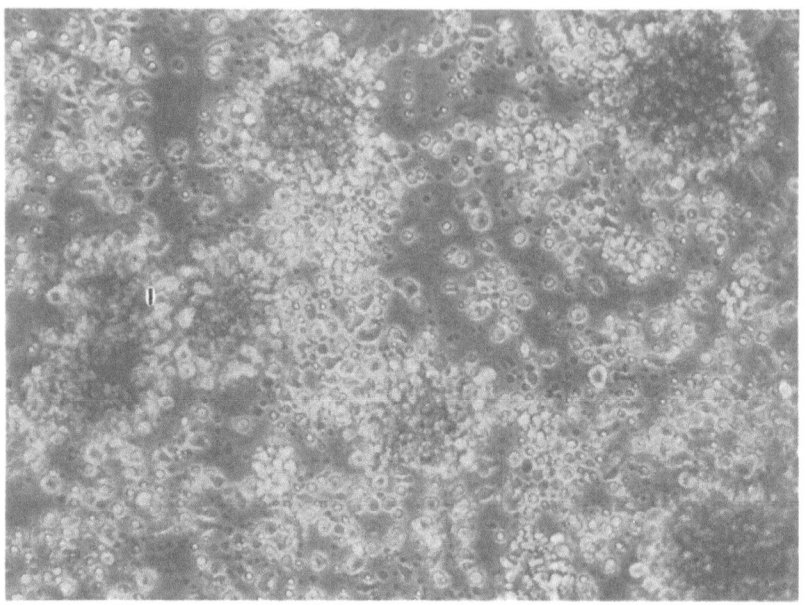

a

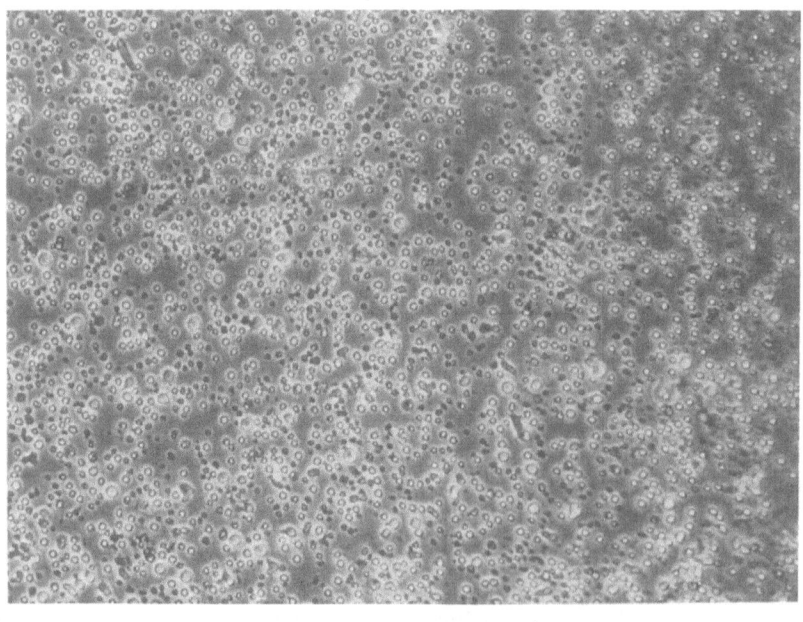

b

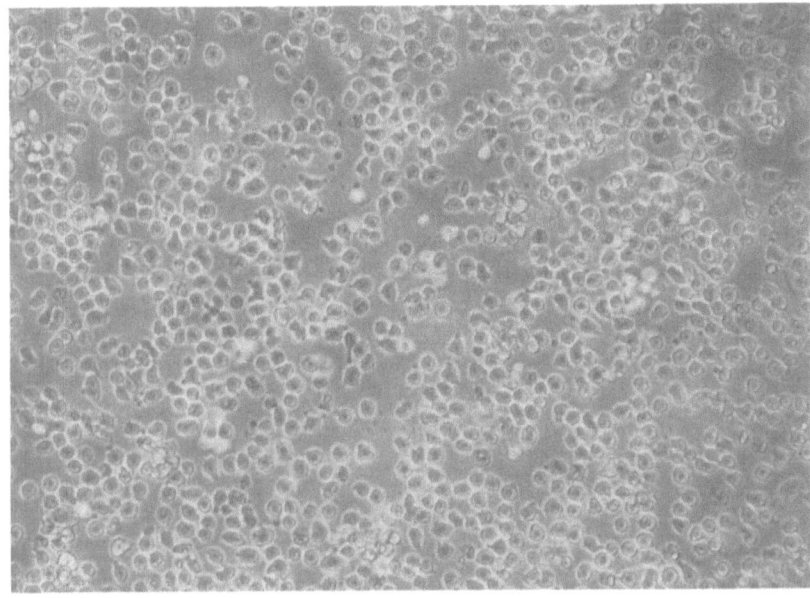

c

Figure 3-3. Effect of FK506 on the MLC-induced lymphocyte blastogenesis ([a]: control, [b]: FK506 1 ng/ml; *see figures on opposite page*) and on the growth of EL-4 ([c]: FK506 1000 ng/ml).

Table 3-1. Rf values of FK506 and its derivatives.*

Compound	Rf
FK506	0.52
FR900525	0.34
FR901154	0.22
FR901155	0.64
FR901156	0.51
FR900520	0.51
FR900523	0.51

* The values were obtained on silica gel TLC, ethyl acetate solvent system.

related minor components. The separation of FR900520 from FR900523 was achieved by reverse-phase chromatography.

Concurrently with the isolation study of FK506, mycological studies on the actinomycete strains were performed in our microbiology section, which was also responsible for strain discovery and improvement. The cultures were carefully examined morphologically and physiologically, and charac-

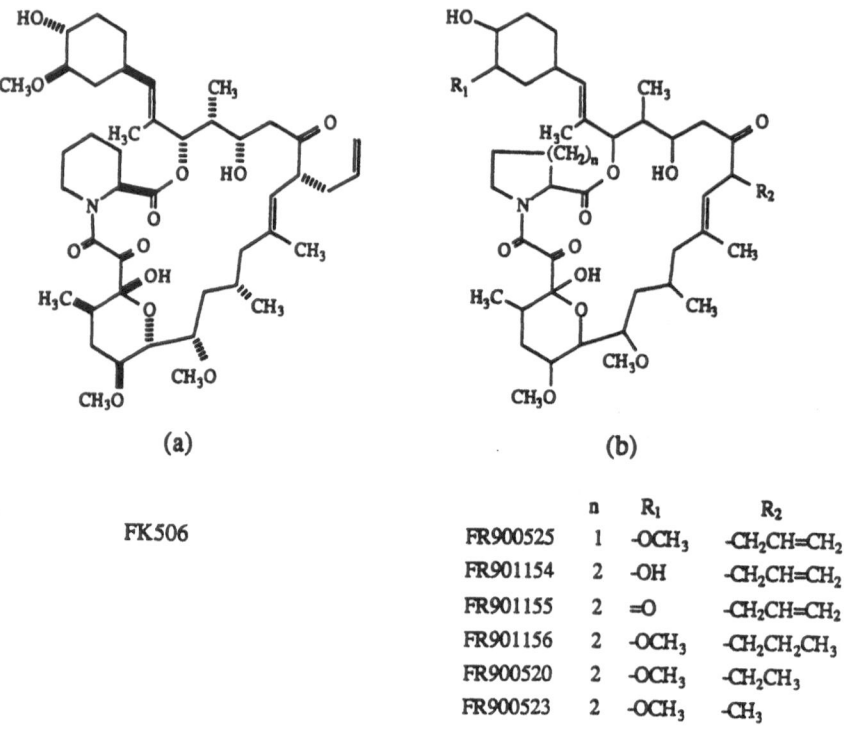

	n	R_1	R_2
FK506			
FR900525	1	-OCH$_3$	-CH$_2$CH=CH$_2$
FR901154	2	-OH	-CH$_2$CH=CH$_2$
FR901155	2	=O	-CH$_2$CH=CH$_2$
FR901156	2	-OCH$_3$	-CH$_2$CH$_2$CH$_3$
FR900520	2	-OCH$_3$	-CH$_2$CH$_3$
FR900523	2	-OCH$_3$	-CH$_3$

Figure 3-4. Structure of FK506 and related compounds.

terized using classical taxonomic methods. Both strains were identified as *Streptomyces* species. The first isolated strain was designated as *Streptomyces tsukubaensis* No. 9993, because the strain was isolated from a soil sample collected near Mt. Tsukuba, close to our laboratories (Figure 3-5). The second isolated strain producing FR900520 and FR900523 was isolated from a soil sample that originated from Yaku-shima, Japan, and was named after the island: *Streptomyces hygroscopicus* subsp. *yakushimaensis* No. 7238 (Table 3-2).

Since fermentation products are frequently present in minute quantities, the yield improvement process, which is also a very labor- and time-consuming step, is essential for the development of microbial products as medicines. A variety of approaches have been carried out for a long period of time. Medium improvement, mutation, and selection of producing strains, as well as optimization of fermentation conditions, such as the size of seed culture volume, term and timing of both the seed culture and the production culture, temperature control, etc. were investigated extensively. As a result, successful yield improvement of more than 100-fold was attained.

Figure 3-5. *Streptomyces tsukubaensis* No. 9993.

Table 3-2. Producing strains of FK506 and its derivatives.

Compound	Strain
FK506	
FR900525	*Streptomyces*
FR901154	*tsukubaensis*
FR901155	No. 9993
FR901156	
FR900520	
FR900520	*Streptomyces hygroscopicus*
FR900523	subsp. *yakushimaensis* No. 7238

When sufficient quantities of purified FK506 became available, its physicochemical properties were investigated further. Studies revealed that FK506 has a molecular weight of 803 and the molecular formula $C_{44}H_{69}NO_{12}$. The compound is soluble in methanol, ethanol, acetone, and chloroform and is insoluble in water. Determination of the chemical structures of FK506 and its biological derivatives was provided by the chemistry division of our laboratories (Tanaka et al., 1987). The plane structure was assigned by a combination of chemical degradation methods and spectroscopic evidence. The result of x-ray crystallographic analysis confirmed the structure and established the relative stereochemistry (Figure 3-6). The structure and absolute stereochemistry of FK506 was determined by acid hydrolysis, which yielded L-pipecolic acid. The results of these experiments indicated that FK506 is an uncommon 23-membered macrolide lactone with a hemiketal-masked triketone functionality.

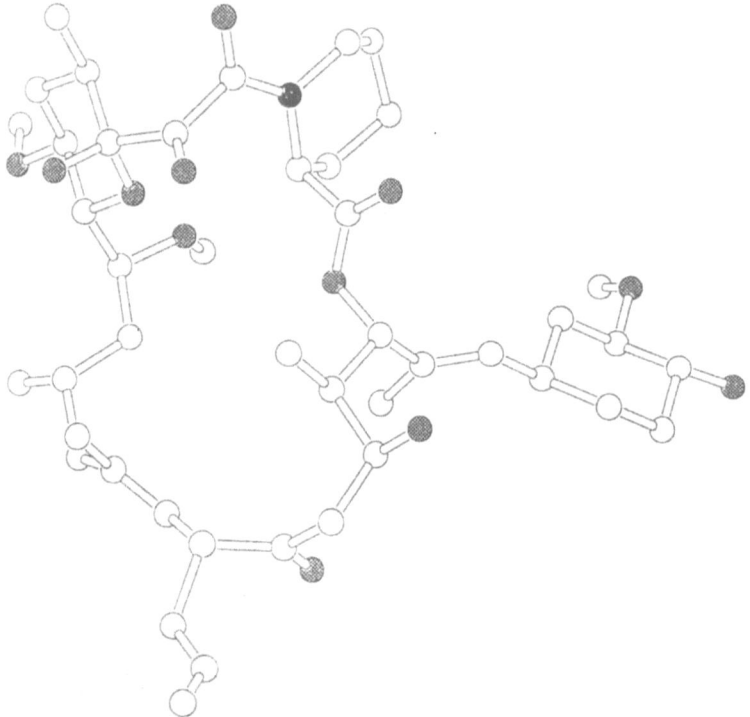

Figure 3-6. OLTEP drawing of FK506.

Immunosuppressive Action of FK506 In Vitro

Originally discovered as an agent that suppressed the proliferation of alloreactive T cells in mixed lymphocyte reactions (MLR), FK506 also inhibits T-cell proliferation induced by various types of stimuli, including specific antigens, mitogenic lectins such as ConA and phytohemagglutinin A (PHA), monoclonal antibodies against CD3 T-cell surface molecules, and phorbol ester plus calcium ionophore. Cyclosporin A (CsA), a well-known immunosuppressant produced by a fungus, *Tolypocladium inflatum*, has similar activities. However, FK506 is 30 to 100 times more potent than CsA in inhibiting T-cell proliferative responses in both murine and human MLR, exhibiting an IC_{50} of approximately $0.2\,nM$ (Kino et al., 1987a) (Table 3-3). Moreover, in a concentration range similar to that which inhibits cell division, FK506 also inhibits the generation of cytotoxic T cells (CTLs) specific for allogeneic targets. CTLs are thought to have the major role in rejection of transplanted tissues. In contrast, proliferation of IL-2-independent mouse lymphoma lines EL4 and BW5147, or IL-2- or IL-4-dependent T-cell clones are unaffected by FK506, even at micromolar concentrations where complete suppression of T-cell activation is obtained. FK506 also shows no toxic effect on murine bone marrow colony formation (Kino et al., 1987b). The inhibitory action of FK506 on T-cell proliferation is partially reversed by exogenously added IL-2, suggesting that FK506 is effective only when endogenous IL-2 is involved in T-cell growth.

Table 3-3. Effects of FK506 and CsA on in vitro immune responses.

Test system	Species	$IC_{50}(nM)$ FK506	CsA
Mixed lymphocyte reaction	human	0.2	14
Induction of cytotoxic T lymphocyte	mouse	0.3	24
IL-2 production	human	0.068	7
IL-3 production	mouse	0.3	32
IFN-γ production	human	0.05	6.2
IL-2 receptor expression	human	0.14	9.9
Bone marrow cell colony formation	mouse	1400	800

T-cell activation is initiated by the recognition of foreign antigen via the T-cell receptor (TCR)/CD3 complex. The TCR/CD3-mediated signaling pathway simultaneously requires the help of cytokines such as IL-1 and IL-6 that are produced by antigen-presenting cells (Weaver and Unanue, 1990). These signals produced at the cell membrane were transferred into the nucleus of T cells through a complex series of biochemical reactions. FK506 (Kay et al., 1989), like CsA (Bijsterbosch and Klaus, 1985) does not affect the early events in these processes including phosphoinositide turnover, generation of the intracellular second messengers such as inositol triphosphate (IP_3) and diacylglycerol (DAG), which cause the increase in intracellular concentration of Ca^{2+} and activation of protein kinase C (PKC), respectively, and subsequent phosphorylation of some cytoplasmic proteins. Following these events, an expression of a set of gene products crucial for T-cell activation and proliferation is eventually induced. FK506 does inhibit these more distal components of the T-cell activation pathway. Thus, FK506 inhibits the transcription of mRNA for several lymphokines produced by $CD4^{+}$ T helper cells, including IL-2, IL-3, IL-4, GM-CSF, TNF-a, and IFN-γ, with 10 to 100 times the potency of CsA (Tocci et al., 1989).

Since IL-2 plays a critical role as a growth and regulatory factor for T cells, inhibition of IL-2 production is thought to be one of the principal mechanisms of immunosuppression by FK506. The effect of FK506 on gene transcription is selective, since FK506 does not affect expression of mRNAs for IL-2 receptors or transferrin receptors that are induced following T-cell activation. Other evidence showing this selectivity is the finding that FK506 does not inhibit expression of IL-1α and IL-1β genes in lipopolysaccharide-stimulated human monocytes.

The IL-2 gene is silent in resting T cells but is activated after antigen recognition by the specific TCR/CD3 complex and the subsequent signal transduction into nucleus. The expression of the IL-2 gene is controlled by the interaction of specific nuclear transcription factors with their corresponding DNA binding sites on the IL-2 gene enhancer/promoter (Fujita et al., 1986; Durand et al., 1988; Crabtree et al., 1989). Several transcription factors, including NF-AT, NF-κB, AP-1, and Oct-1, are involved in the regulation of IL-2 gene expression (Figure 3-7). Among them, NF-AT (nuclear factor of activated T cells) is quite unique because its expression is restricted to activated T cells (Shaw et al., 1988), while the others are commonly found in almost all cells and tissues and are responsible for regulation of the transcription of various genes. In common with CsA, FK506 specifically inhibits the binding of NF-AT to its binding site on the IL-2 enhancer, whereas those of NF-κB, AP-1, and Oct-1 are either marginally or only mildly affected (Emmel et al., 1989; Mattila et al., 1990).

Structurally, functional NF-AT is a complex protein made of two subunits. One subunit (NF-AT$_c$) is a preexisting protein located exclusively in T cells, while the other (NF-AT$_n$) is a ubiquitous protein rapidly induced in nuclei of any cells by activation of protein kinase C. The latter NF-AT$_n$ was recently identified as AP-1 (Jain et al., 1992). Once the T-cell activation signal is delivered, an increase in intracellular concentration of Ca^{2+} drives NF-AT$_c$ to translocate to nucleus and induces the assembly of two subunits to give a functional transcription factor, NF-AT. FK506 inhibits the assembly of the two NF-AT subunits by blocking the translocation of NF-AT$_c$ to

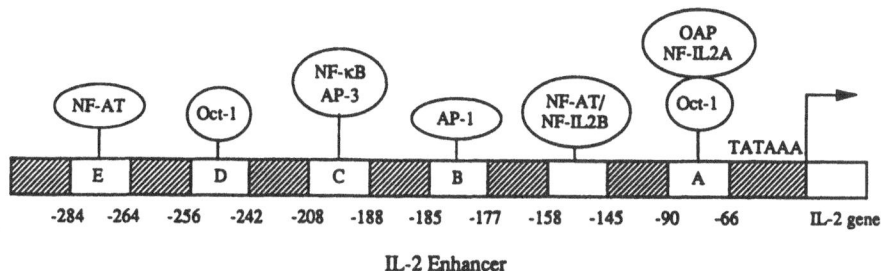

Figure 3-7. Putative binding sites of transcription factors in IL-2 gene.

the nucleus (Flanagan et al., 1991). The highly restricted distribution of NF-AT might well account for the selective action of FK506 on T cells.

However, FK506 (Bierer et al., 1991) and CsA (Thompson et al., 1989) do not inhibit IL-2 production in the Ca^{2+}-independent CD28 pathway, an alternative pathway for T-cell activation. This pathway is triggered by the interaction of the CD28 molecule on T cells with its specific ligand B7/BB1 expressed on monocytes/macrophages or activated B cells, and serves as a co-stimulatory signal for the TCR/CD3 pathway resulting in remarkably enhanced IL-2 production (Fraser et al., 1991). Thus, of particular importance in immunosuppression by FK506 and CsA is that these agents inhibit only such activation events as cause a noticeable rise in intracellular concentration of Ca^{2+}. Thus, a subset of Ca^{2+}-associated signal transduction pathways could be a target of FK506 and CsA, in which transcriptional regulation is not always involved. Good examples of the FK506-sensitive phenomena that do not involve transcriptional regulation are various types of exocytosis-related events in both lymphocytes and non-lymphocytes, including TCR-mediated degranulation of cytotoxic T cells (Dutz et al., 1993), calcium ionophore-induced degranulation of neutrophils and basophils (Forrest et al., 1991; De Paulis et al., 1992), and IgE receptor-mediated histamine and serotonin release from mast cells and basophils (Hultsch et al., 1991; De Paulis et al., 1992).

FK506, like CsA, also inhibits the proliferative response of purified murine and human B cells induced by such stimuli that cause a rapid increase in intracellular calcium levels, including anti-IgM antibody, *Staphylococcus aureus* Cowan strain I (SAC), or phorbol ester plus calcium ionophore at concentrations that inhibit T-cell responses (Wicker et al., 1990; Morikawa et al., 1992), whereas Ca^{2+}-independent lipopolysaccharide-induced B-cell proliferation is not affected by FK506 (Walliser et al., 1989). Furthermore, it was shown that in vitro FK506 inhibited the generation of antibody-producing cells in the presence of T cells but not T-cell-derived helper factors (Suzuki et al., 1990), suggesting that these effects may be due to the inhibition of lymphokine production by activated T cells. However, the fact that FK506 inhibits B-cell immune responses such as proliferation and the expression of TNF-α gene induced by anti-IgM antibody, a T-cell-independent polyclonal B-cell mitogen (Goldfeld et al., 1992), may imply that FK506 is capable of acting directly on B cells.

Immunosuppressive Action of FK506 In Vivo

Effect on Models of Organ Transplantation

At a scientific meeting in mid-1984, Dr. Takenori Ochiai (Chiba University School of Medicine) heard the story of FK506 from Dr. Hatsuo Aoki (Director of Fujisawa Tsukuba Research Laboratories at that time). He was very interested in this potent immunosuppressant and decided to evaluate its capability for preventing graft rejection in organ transplantation models. Since then, FK506 has been extensively tested by many research groups in a wide variety of experimental allograft models including allotransplants of liver, kidney, heart, small bowel, lung, pancreas, pancreatic islet, bone marrow, skin, limb, cornea, and trachea. Animal models used during these studies were mouse, rat, dog, rabbit, pig, monkey, and baboon. In general, FK506 was shown to be effective in preventing allograft rejection in a number of species at doses 10 to 100 times lower than CsA required to exert similar effects in the same models (Table 3-4).

Most of the animal studies were performed by administration of FK506, beginning immediately after transplantation, intramuscularly (i.m.) or orally (p.o.). FK506 is capable of prolonging survival time for hepatic, cardiac, and skin allografts in rats at doses of 0.1–0.32 mg/kg i.m. or 1.0 mg/kg p.o. In dogs, beneficial effects of FK506 on hepatic and renal transplantation were achieved as well and at similar doses. However, severe side-effects were observed more frequently in dogs than in rats, resulting in a narrower therapeutic window. In order to obtain maximal efficacy, administration of FK506 should begin immediately after transplant surgery. Interestingly, FK506 was shown to prolong hepatic allograft survival in dogs at 1.0 mg/kg (i.m.) (Ueda et al., 1990) and cardiac allograft survival in rats at 1.28 mg/kg (i.m.) (Murase et al., 1987; Murase et al., 1990a), even when given as late as four and five days after transplantation, respectively, suggesting that FK506 may be useful not only as a prophylactic but as a therapeutic drug.

FK506 was also tested in experimental heart xenotransplantation models, including mouse to rat (Ochiai et al., 1989), hamster to rat (Gudas

Table 3-4. Overview of FK506 in allotransplantation.

Species	Graft	Route[1]	Dose (mg/kg/day)[2] FK506	CsA	Reference
Mouse	Heart	p.o.	4	21	Morris et al., 1990
		i.p.	0.7	10	Morris et al., 1990
	Skin	i.p.	0.1	–	Lagodzinski et al., 1990
Rat	Liver	i.m.	0.1	20	Tsuchimoto et al., 1989
		p.o.	1.0	–	Isai et al., 1990
	Heart	i.m.	0.1	>3.2	Ochiai et al., 1987b
		p.o.	1.0	10	Ochiai et al., 1987b
	Skin	i.m.	0.32	32	Inamura et al., 1988
	Limb	i.m.	0.32	15	Kuroki et al., 1989
	Pancreatic islets	s.c.	0.32	–	Yasunami et al., 1989 Yasunami et al., 1990
	Small bowel	i.m.	0.32	30	Iga et al., 1990
	Lung	i.m.	0.3	–	Katayama et al., 1991
	Pancreas	i.m.	0.32	–	Yamashita et al., 1988
Rabbit	Cornea	local	0.1	–	Kobayashi et al., 1989
Pig	Liver	i.v.	0.03	10	Lautenschlager et al., 1991
Dog	Liver	i.m.	0.1	–	Yokota et al., 1989
		p.o.	1.0	–	Todo et al., 1987
	Kidney	i.m.	0.08	–	Ochiai et al., 1987a
		p.o.	1.0	–	Ochiai et al., 1987a
	Lung		0.1 (i.m.)	20 (p.o.)	Saitoh et al., 1990
	Pancreas	i.m.	0.2	–	Kenmochi et al., 1991
	Pancreas/ duodenum	i.m.	0.1–0.3	20	Imai et al., 1991
	Trachea	i.m.	0.1	–	Moriyama et al., 1989
Monkey	Liver	i.m. to p.o.	1.0 to 10	–	Monden et al., 1990
	Heart	i.m. to p.o.	1.0 to 10	–	Flavin et al., 1991
	Pancreas	i.m. to p.o.	1.0 to 10	–	Ericzon et al., 1991
Baboon	Kidney	i.m.	0.5	–	Imventarza et al., 1990
	Heart	i.m.	1.0	–	Hildebrandt et al., 1991
		i.v.	0.5	–	Hildebrandt et al., 1991

[1] p.o.: orally; i.p.: intraperitoneally; i.m.: intramuscularly; s.c.: subcutaneously; i.v.: intravenously; local: subconjunctically.
[2] It represents minimal effective dose to prevent allograft rejection except murine cardiac allograft model where ED_{50} are shown.

et al., 1989; Nakajima et al., 1989), and lamb to goat (Kawauchi et al., 1991). Long-term survival of xenografts was not achieved in these models; however, transplantation across a small histoincompatibility barrier, e.g., lamb to goat treated with 0.16 mg/kg FK506 (i.m.), caused significantly prolonged graft survival.

Effect on Animal Models of Autoimmune Diseases

Although the pathogenesis of autoimmune diseases is still unknown, autoreactive T cells are now proven to be extensively involved in the development of some types of autoimmune disease. The effect of FK506 has been examined in models of experimental autoimmune diseases, including spontaneous diabetes mellitus in NOD mice and BB rats, spontaneous glomerulonephritis in (NZBxNZW)F_1 mice (a model of systemic lupus erythematosus), experimental allergic encephalomyelitis in rats (a model of multiple sclerosis), experimental autoimmune uveoretinitis in rats, and collagen-induced arthritis (a model of rheumatoid arthritis) in mice and rats. FK506 shows beneficial effects on all these models, as shown in Table 3-5.

Table 3-5. Overview of FK506 in autoimmune diseases.

Diseases	Species/Strain	Route[1]	Dose (mg/kg/day)[2]		Reference
			FK506	CsA	
Diabetes	Mouse: NOD	s.c.	1	–	Carroll et al., 1991
melitus	Rat: BB	p.o.	1	–	Murase et al., 1990b
Glomerulo-	Mouse:	s.c.	1.5	30	Carrieri et al., 1991
nephritis	(NZBxNZW)F_1				
Encephalo-	Rat: Lewis	p.o.	3.2	–	Deguchi et al., 1991
myelitis					
Uveoretinitis	Rat: Lewis	i.p.	1	–	Kawashima et al., 1990
Arthritis	Rat: SD	s.c.	2.5	100	Miyahara et al., 1991

[1] p.o.: orally; i.p.: intraperitoneally; s.c.: subcutaneously.
[2] It represents minimal effective dose.

Mechanism of Action of FK506

Because of the functional similarities, FK506 was initially thought to bind to the cytoplasmic protein to which CsA binds. However, these two agents exert their potent immunosuppressive activity to interacting with a distinct set of structurally unrelated cytosolic proteins. These cytosolic receptor proteins for FK506 and CsA are named FKBP and cyclophilin (CyP), respectively. Moreover, it has been shown that FKBP and CyP belong to a family of cytosolic immunosuppressant binding proteins called immunophilins. FK506 (Siekierka et al., 1989) and CsA (Handschumacher et al., 1984) bind to their corresponding immunophilins, FKBP and CyP, with K_d values of 0.4 and 200 nM, respectively. The common characteristics of immunophilins are ubiquitous and highly conserved throughout most species, from lower eukaryotes to primates. Five or more FKBPs have been discovered so far, including FKBP12, FKBP13, FKBP25, FKBP59, and FKBP80, called so on the basis of their respective molecular weights. FKBP12, a 12kDa protein, was the first one discovered in 1989 and is the most abundant (Harding et al., 1989; Siekierka et al., 1989). The relative ratio of these five FKBPs in Jurkat cells and calf thymus homogenates is reported to be 42 (FKBP12); < 1 (13kDa); 4.2 (30kDa); 9.2 (60kDa); and 1.0 (80kDa) (Fretz et al., 1991).

Despite the fact that FKBP and CyP have no similarities in amino acid sequence, both proteins were shown to possess peptidyl-prolyl cis-trans isomerase (PPIase, also known as rotamase) activity which catalyzes cis-trans isomerization of peptidyl X-prolyl bonds in oligopeptides (where X is any amino acid), a rate-limiting step in the folding of several proteins (Fischer et al., 1989; Harding, 1989; Siekierka et al., 1989; Takahashi et al., 1989). Surprisingly, the enzymatic activities of FKBP and CyP are inhibited by their respective immunosuppressive ligands, FK506 (Harrison and Stein, 1990) and CsA (Fischer et al., 1989), with K_i values of 1.7 and 2.6 nM, respectively, and no cross-inhibition occurs at all. Further study revealed that FKBP12 has a single ligand-binding site, which is also the site of PPIase catalysis.

Does isomerization of some proteins by PPIase play a critical role in T-cell activation? A number of factors suggest that inhibition of PPIase activity does not explain the immunosuppressive activity of FK506 and CsA. First, the immunophilins are so ubiquitous that inhibition of PPIase activity of these proteins hardly explains the highly selective action of FK506 and CsA on certain cell types. The second factor is related to the abundance of immunophilins. The amount of cytosolic FKBP is reported to be 5 nM; however, inhibition of IL-2 production by FK506 occurs at as low as 1 nM, well below the concentration required to saturate binding to FKBP. Third, FK506 or CsA binds to and inhibits only one of the two PPIases while the other one remains functional; nevertheless, each drug is able to induce complete suppression of IL-2 synthesis. Given the fact that PPIases do not have strict substrate specificity, it might be expected that the other intact PPIase would compensate for the one blocked by FK506 or CsA. These results questioned the biological role of PPIase in T-cell signal transduction and led to another hypothesis that the complex formed between each immunosuppressant and the cognate immunophilin functions as an actual immunosuppressive molecule.

The first evidence supporting this hypothesis has come from studies on the yeast *Saccharomyces cervisiae*, which expresses abundant immunophilins that are highly homologous to human immunophilins. The
proliferation of *Saccharomyces cervisiae* is blocked at much higher concentrations than those required for immunosuppression by both FK506 and
CsA. However, there are some mutants which are resistant to these agents.
The analysis of these mutants revealed that they lacked the cognate immunophilins for FK506 (Heitman et al., 1991) and CsA (Tropschung et al.,
1989), or they made immunophilins that were incapable of binding these
drugs. More importantly, since cell division of yeast lacking immunophilins
is normal, the enzyme activity of the immunophilins is unlikely to play a
critical role in cell proliferation, at least in yeast. These results indicate that
the binding of FK506 and CsA to the cognate immunophilin is a prerequisite step for their growth-inhibitory action on *Saccharomyces cervisiae*, but
the suppression of the enzymatic activity does not account for drug action.

Other evidence supporting the functional importance of the immunophilin drug complex has emerged from studies using mammalian cells and
FK506 analogues, 506BD and rapamycin (RAP) (Figure 3-8). FK506

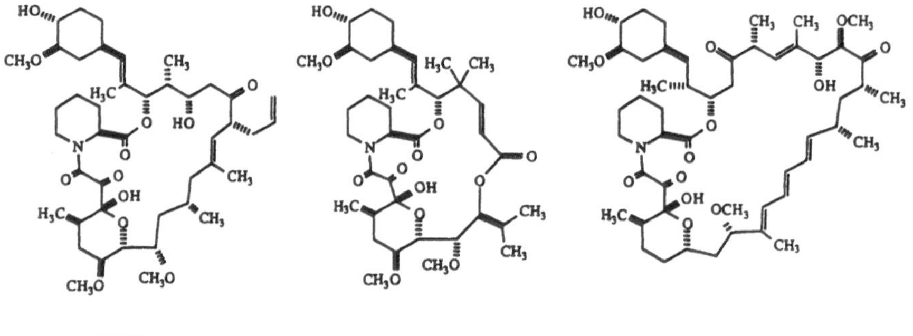

Figure 3-8. Structure of FK-BP ligands.

analogue, 506BD, is a synthetic compound containing only the structural
elements of FK506 necessary for binding to FKBP (Somers et al., 1991).
Therefore, 506BD binds to FKBP with a high affinity and potently inhibits
PPIase activity with a K_i of 5 nM. Nevertheless, it is incapable of inhibition
T-cell activation at all; it antagonizes FK506 action (Bierer et al., 1990b).
These results indicate that the binding of FK506 to FKBP is essential for
exerting immunosuppression, but inhibition of PPIase activity by FKBP is
not linked to FK506 action. Another example of inhibition of FK506 bind-
ing to FKBP is RAP, a microbial product of *Streptomyces hygroscopicus*,

with a very similar structure to FK506. RAP was originally discovered as an antifungal substance in 1975 (Sehgal et al., 1975; Vezina et al., 1975) and was later shown to have potent immunosuppressive activities both in vitro and in vivo (Dumont et al., 1990). However, the RAP's mode of immunosuppressive action is clearly different from that of FK506, although RAP specifically and equipotently binds to FKBP ($K_d = 0.2$ nM), inhibits PPIase activity ($K_i = 0.2$ nM) and acts within the first hours of T-cell activation cascade (Bierer et al., 1990a). Unlike FK506, RAP does not inhibit the expression of IL-2 mRNA but blocks the more downstream events in the T-cell activation cascade such as IL-2-driven expansion of T cells that have already expressed IL-2 receptors. Furthermore, RAP inhibits both Ca^{2+}-dependent and Ca^{2+}-independent reactions, and this profile makes RAP a less selective immunosuppressant (Morris, 1991). Surprisingly, it has been shown that FK506 and RAP act as reciprocal antagonists (Bierer et al., 1990a; Dumont et al., 1990).

These three structurally related compounds, FK506, 506BD, and RAP, bind to FKBP's common binding site and inhibit PPIase activity of FKBP. However, these compounds differ from each other in regard to their immunosuppressive properties (Table 3-6). The inhibition of PPIase activity seemed to be an attractive target for immunosuppression, but it is now evident that inhibition of PPIase activity is not relevant to blockading of the Ca^{2+}-dependent signal transduction pathway leading to IL-2 production.

Table 3-6. Summary of the biological profiles of FK506, CsA, 506BD, and RAP.

Compound	Immunophilin binding Cognate immunophilin	PPIase activity	Effect on T-cell activation Early phase (IL-2 production)	Late phase (proliferation)	Effect on graft rejection
FK506	FKBP (antagonized by RAP or 506BD)	+	+	−	+
CsA	Cyclophilin	+	+	−	+
506BD	FKBP	+	−	−	−
RAP	FKBP (antagonized by FK506 or 506BD)	+	−	+	+

+ suppression   − no effect

All these factors led to the prediction of a second cellular target to which FK506-FKBP and CsA-cyclophilin bind, and which plays a key role in T-cell activation. In 1991, the second target molecule was identified as the protein phosphatase calcineurin (Liu et al., 1991). Calcineurin, also referred to as protein phosphatase 2B (PP2B), is a Ca^{2+}- and calmodulin-dependent serine/threonine phosphatase and exists as a heterodimer composed of A (61kDa) and B (15kDa) subunits. Subunit A contains the calmodulin-binding and catalytic sites, while subunit B has four binding sites for Ca^{2+} (Klee et al., 1989). Although FK506 and CsA fail to bind to calcineurin, in the absence of their cognate immunophilins, FK506-FKBP and CsA-CyP are capable of binding specifically to and inhibiting phosphatase activity of calcineurin with IC_{50} values of 0.4 nM and 7 nM, respectively (Liu et al., 1991). Of significant importance is the fact that both the FK506-FKBP and CsA-CyP complexes bind to calcineurin in a competitive fashion, implying that they share a common binding site on the molecule. Interestingly, overexpression of the calcineurin gene in Jurkat cells reduces the immunosuppressive activity of FK506 and CsA, which suggests that calcineurin is a common downstream biochemical target for both agents (Clipstone and Crabtree, 1992; O'Keefe et al., 1992). The ability of FK506 and CsA to specifically inhibit phosphatase activity of calcineurin was also shown in vivo at such concentrations to sufficiently abrogate IL-2 production by activated T cells (Fruman et al., 1992). Moreover, neither RAP-FKBP nor 506BD-FKBP binds to or inhibits calcineurin, demonstrating a correlation between inhibition of calcineurin and IL-2 production.

A recent report compares the effects of FKBP12, FKBP13, and FKBP25 on calcineurin inhibition. The report shows that unlike FK506-FKBP12 complex, which binds of calcineurin ($K_i = 34$ nM), none of the FK506 complexes with FKBP13 and/or FKBP25 bind to calcineurin ($K_i > 1000$ nM), suggesting that FKBP13 and FKBP25, at least, are not directly involved in immunosuppression by FK506 (Bram et al., 1993).

To understand the relevance of calcineurin to the T-cell signaling pathway, it is necessary to identify the substrate(s) for calcineurin. Quite recently, it has been shown that $NF\text{-}AT_c$, the cytoplasmic subunit of NF-AT, may be a substrate for calcineurin in vitro (Jain et al., 1993). This, and the fact that translocation of $NF\text{-}AT_c$ is sensitive to FK506, has led to a new hypothesis concerning the role of calcineurin and the inhibition mechanism of FK506 in TCR/CD3-mediated IL-2 production (Schreiber and Crabtree, 1992): dephosphorylation of preexisting $NF\text{-}AT_c$ ($NF\text{-}AT_p$) enables it to translocate from the cytoplasm to the nucleus where it binds to newly synthesized $NF\text{-}AT_n$, resulting in formation of functional NF-AT which holds a pivotal role in IL-2 gene expression. Either FK506-FKBP or CsA-CyP complex can bind to calcineurin and inhibit dephosphorylation of NF-AT$_c$, which subsequently prevents the assembly of two subunits of NF-AT (Figure 3-9). Thus, looking at the molecular mechanism underlying the selective immunosuppression of FK506 and CsA, in a sense, these agents are merely prodrugs, and the FK506-FKBP and CsA-CyP complexes function as the actual immunosuppression.

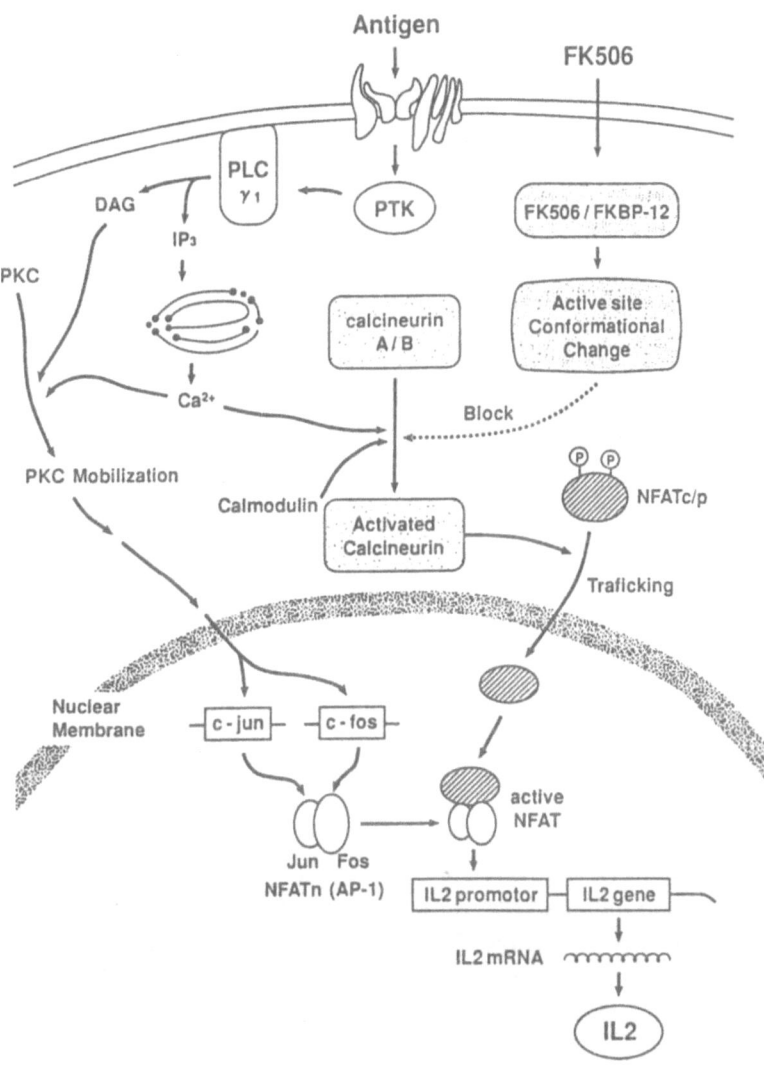

Figure 3-9. FK506-mediated inhibition of T-cell activation.

Toxicology, Pharmacokinetics, and Drug Monitoring Systems

Before a New Drug Application can be submitted, the drug must be determined to be safe for humans. To achieve this, a variety of acute and chronic toxicity studies are first performed in animals to clarify the drug's toxicological properties. Since the early stages of development, the future of FK506 depended on the balance between its potent immunosuppressive activity and its toxicity. In May 1987, we began toxicologic studies in our laboratories. Acute (Ohara et al., 1990), subacute (13-week) (Ohara et al., 1990), and chronic (52-week) (Ohara et al., 1992a, 1992b) toxicity studies, using both rats and baboons, demonstrated that, at high doses, major toxicity results were: (1) decrease in body weight gain resulting from decreased food consumption, and (2) pancreatic and renal dysfunction that were histologically characterized by vacuolation in the islets of Langerhans and mineralization of the corticomedullary junction of the kidney, respectively (Hirano et al., 1992a). Thus, the pancreas and kidney were concluded to be the target organs of toxicity.

In addition to the acute and chronic toxicity studies, mutagenicity tests (Hirai et al., 1992), antigenicity and local irritation studies (Hirano et al., 1992b), and reproductive and developmental studies (Saegusa et al., 1992) were conducted. FK506 did not demonstrate either mutagenicity in the Ames test nor antigenicity in any tested animal. During the reproductive and developmental toxicity studies in rats, some abnormal changes were observed, mainly in the test groups using a high dose (3.2 mg/kg, p.o.).

The pharmacokinetics of FK506 were defined during nonclinical studies conducted in the rat (Iwasaki et al., 1991; Kagayama et al., 1993), dog, monkey, and baboon (Noguchi et al., 1993, unpublished data). Animal studies with ^{14}C-FK506 have demonstrated that the drug is incompletely absorbed and is subject to first-pass metabolism after oral administration. Absorption occurs mainly in the upper intestine, is decreased by the presence of food, and is not dependent on the presence of bile. The latter phenomenon indicates that a patient can safely convert from intravenous (i.v.) dosing to oral dosing with FK506 after liver transplantation and T-tube placement (Jain et al., 1990), whereas CsA absorption requires that bile be present in the intestine.

FK506 is highly lipophilic and undergoes extensive distribution in tissues. Radioactive FK506 administered to rats and baboons by either the i.v. or oral route was detected mainly in the pancreas, kidney, and alimentary tract, and disappeared within 24 hours. Radioactivity was not detected in fetal tissue or breast milk.

FK506 undergoes extensive hepatic metabolism via the P-450 oxidase system, with demethylation and hydroxylation being the primary biotransformational processes. Following incubation of FK506 with NADPH and human or rat liver microsomes, the 13-, 31-, and 15-O-demethylated and the 12-hydroxylated metabolites were identified (Iwasaki et al., 1993).

Only a very small amount of FK506 is excreted unchanged. Metabolized FK506 is primarily excreted in the feces, with only a small amount excreted in urine (Venkataramanan et al., 1990).

Since treatment of animals with the high doses of FK506 during toxicity studies caused renal and pancreatic damage, establishing sensitive assay methods to monitor blood levels was crucial for FK506 therapy in humans. Usually, high-performance liquid chromatography with an ultraviolet detector is precise and has a high throughput. However, it was inappropriate in this case, due to the lack of a strong chromophore in the structure of FK506.

An enzyme-linked immunosorbent assay (ELISA) using a monoclonal antibody to FK506 and a horseradish peroxidase-FK506 conjugate (FK-POD) was established (Tamura et al., 1987; Kobayashi et al., 1991). The capture layer in this assay consisted of immobilized anti-mouse IgG which binds the monoclonal antibody. Samples containing FK506 were allowed to compete with FK-POD for binding to the antibody. An increase in FK506 levels produced a decrease in absorption. This assay system provided a sensitivity of 50 pg/ml. We collaborated with Abbott Laboratories to establish an automated assay system IMx, which enabled us to measure blood FK506 concentrations in patient samples more conveniently (Grenier et al., 1991).

FK506 Clinical Development

The first patient was treated with FK506 on March 1, 1989 at the University of Pittsburgh. This patient was experiencing ongoing, serious rejection after her third liver transplant, which was performed through the transplantation program headed by Dr. Thomas Starzl (Starzl et al., 1989). Since that time, over 4000 transplant patients worldwide have been treated with FK506.

Exploratory Clinical Study

The first FK506 symposium was held in Gothenburg, Sweden in 1987, as a satellite symposium of that year's European Society of Organ Transplantation (ESOT) Conference. At the symposium, FK506 was called a "potent, breakthrough immunosuppressant," and many indications for the drug were

considered in addition to organ transplantation. These included rheumatoid arthritis, Behcet's disease, nephritis syndrome, and multiple sclerosis.
During the animal studies, FK506 demonstrated the most striking effects in
organ transplantation models; therefore, initial clinical testing of the drug
was considered for solid organ transplantation, such as liver, kidney, and
heart.

Among transplant populations, kidney transplant recipients constitute
the largest group. However, patients awaiting kidney transplantation are
often not in life-threatening situations because dialysis is available. The
second largest population, liver transplant patients, have no such alternative
when their grafts are rejected. Because of the high incidence of rejection
with current immunosuppressant therapy, it was decided to target the liver
transplant population for the initial clinical trial of FK506.

As the first step in determining the appropriate FK506 dose for humans,
animal dosing data was evaluated very carefully. In animal models of liver
allograft rejection, FK506 as a single agent was capable of preventing
rejection for $\geq$ 30 days (Todo et al., 1987; Murase et al., 1990a). The
effective doses varied widely between species, however. For example, a
daily oral dose of 1 mg/kg was effective in preventing rejection in dogs
(Todo et al., 1987), but baboons required 12 mg/kg/day p.o. for the same
effect (Imventarza et al., 1990). To assess the safety of FK506, various
pharmacology and toxicology studies also were performed in animals. As
a result, a target therapeutic dose of 3 mg/kg/day p.o. was estimated for
humans.

After discussions with the US Food and Drug Administration (FDA),
the agency recommended that the first dosing in humans begin at one-tenth
the target oral dose, or 0.3 mg/kg/day, so as to obtain a pharmacokinetic
profile of FK506. Therefore, the strategy was to gradually increase the
dose until the target was reached. As it turned out, the lowest dose in the
escalation study was effective in the treatment of refractory rejection. For
that reason, no other dose-finding studies were conducted before the start
of the large-scale, Fujisawa-sponsored multicenter clinical trials.

The first patient ever to receive FK506 was a 26-year-old woman who
was experiencing ongoing, serious rejection of her third liver transplant.
Before FK506 dosing began, she received a detailed explanation from
Dr. Starzl about the potential benefits and risks of FK506. The patient
decided to undertake aggressive treatment with FK506 since she had lost
her first two liver transplants to rejection that seemed refractory to conventional therapies. During the animal studies, FK506 had demonstrated
a synergistic action with CsA; therefore, this first trial utilized coadministration of FK506 and CsA. The woman received FK506 intravenously

at a dose of 0.15 mg/kg/day, and oral CsA at 50% of standard baseline dose (Starzl et al., 1989). After treatment had begun, however, pharmacokinetic data suggested that the CsA blood concentration was increasing. Furthermore, she had an elevation of serum creatinine, indicative of renal dysfunction. Dr. Starzl decided to discontinue the coadministration of CsA with the result that the woman's renal function improved. At that time, Dr. Starzl reasoned that if FK506 dosing was prematurely terminated at the beginning of its first clinical trial, demonstration of the efficacy of FK506 in humans might be delayed indefinitely, especially the potential benefits to suffering patients.

Usually when a new investigative compound is being coadministered with a familiar drug, investigators will choose to discontinue the new compound when complications arise. In fact, the patient's attending physician wanted to terminate FK506 administration. However, Dr. Starzl was abreast of the numerous nonclinical data regarding the efficacy of FK506, and was well aware that the side effects being experienced were typical of CsA toxicity. It is a tribute to Dr. Starzl's sixth sense, developed over many years of clinical experience, that he decided to discontinue CsA instead of FK506. Since the patient's renal function and liver function improved after discontinuation of CsA, Dr. Starzl's judgment proved to be correct. Rejection episodes also ceased, which was confirmed by biopsy.

In total, eleven liver transplant patients with refractory rejection were treated with coadministration of FK506 and CsA. Despite improved hepatic function, several patients experienced episodes of renal dysfunction (Fung et al., 1990). Therefore, Dr. Starzl decided to convert to FK506 alone, with small amounts of corticosteroids, for subsequent patients.

An estimated 13% of liver transplant patients will experience graft rejection that is resistant to therapy with conventional immunosuppressants (Klintmalm et al., 1989). The initial results from the University of Pittsburgh's pilot study of FK506 in the treatment of refractory rejection, therefore, had an immediate impact on transplantation investigators around the world. Many investigators wanted to try FK506 to treat their patients who had refractory rejection and limited chances for survival without retransplantation.

Rescue Studies in Liver Transplantation

The purpose of the first Fujisawa-sponsored study was to evaluate the efficacy and safety of FK506 in the rescue of liver allografts in patients undergoing refractory rejection during treatment with conventional immunosuppressive agents. Due to the high rate of death or retransplantation among these patients, active control arms were not feasible, and it was unethical to perform placebo-controlled studies. As recommended by the FDA, patients were enrolled in the open-label study on a case-by-case basis, subject to review by the FDA and Fujisawa. Later, patients were enrolled on a case-by-case basis subject to review by Fujisawa. The first rescue study patient was treated with FK506 in April 1990. Historical controls were used to compare the efficacy of FK506 with that of conventional treatment.

Results from the participating sites were presented at the First International FK506 Symposium, held in August 1991 in Pittsburgh, Pennsylvania, USA (Lewis et al., 1991; McDiarmid et al., 1991; Shaw et al., 1991), and at the 14th International Transplantation Congress held in August 1992 in Paris, France (US Multicenter FK506 Study Group, 1993). A summarization of the results from the first 125 rescue study patients follows.

All 125 patients in the rescue study had biopsy-confirmed, refractory, acute or chronic rejection which was unresponsive to several courses of steroid and OKT3 therapy. Kaplan-Meier estimates of patient and graft survival at one year were 70.8% and 55.6%, respectively. These results were compared with three historical control groups: the clinical sites control group consisted of 66 liver transplant patients with acute or chronic drug-resistant rejection who remained on CsA-based therapy at four of the eleven centers participating in the study. The United Network for Organ Sharing (UNOS) and University of Pittsburgh controls consisted of 233 and 80 patients, respectively, who underwent retransplantation at least 14 days after the first transplant and continued to be treated with CsA-based therapy.

FK506 was statistically superior to all three historical controls in one-year overall graft survival rates (55.6% versus 43.9%, 53.1%, and 33.1%). FK506 was also statistically superior to the UNOS and University of Pittsburgh historical controls in one-year patient survival rates (70.8% versus 56.7% and 38.1%). The graft survival in the clinical site historical control (43.9%) was considerably lower than patient survival (77.5%), indicating a large number of retransplants in this population. This study and numerous other published rescue trial data indicated that FK506 is an effective agent in the rescue of patients experiencing refractory rejection following liver transplants.

Primary Studies in Liver Transplantation

The University of Pittsburgh expanded the use of FK506 to include treatment of primary liver transplant patients, who were to receive FK506 immediately after transplant surgery. Dr. Starzl and his transplantation program staff, who are the world's most experienced in liver transplantation, were amazed to discover that acute rejection episodes within the first two weeks after transplantation significantly decreased with the use of FK506. At that time, over 60% of patients usually experienced acute rejection episodes within that crucial two-week time frame (Klintmalm et al., 1989; Sher et al., 1991). Additionally, patients taking FK506 recovered liver function more quickly, required lower steroid dosing, and were discharged from the hospital earlier than patients taking CsA.

These encouraging results were presented at the October 1989 ESOT Symposium in Barcelona, Spain (Todo et al., 1990). Many interesting study results were presented at this symposium, including animal toxicological study results, FK506 clinical pharmacokinetics, pathological studies, and other transplantation studies. In fact, many famous transplantation investigators worldwide were very excited by the various results presented at this symposium.

Fujisawa sponsored two clinical trials of FK506 in primary liver transplantation because the US regulatory guidelines require two well-controlled studies for NDA approval. These studies were designed to compare prophylactic efficacy and safety of FK506-based therapy with CsA-based therapy in primary liver transplant patients in the US and Europe. These clinical trials were conducted as randomized, well controlled, multicenter, open-label studies.

The standard regimen for prophylactic immunosuppressant therapy after liver transplantation was CsA plus corticosteroids, with the possible addition of azathioprine and/or antilymphocyte globulin (ALG). However, primary liver transplant patients receiving the standard treatment had 60%–70% incidence of acute rejection post-transplantation, and an estimated 80% one-year graft survival. During the University of Pittsburgh's exploratory rescue study, FK506 had demonstrated its superiority over standard CsA-based therapy in the treatment of this serious problem (Todo et al., 1990). It was hoped that a reduction in the incidence and severity of rejection would improve graft and patient survival after primary liver transplantation.

The US study initially was conducted at five sites, and later was increased to twelve sites. The European study included eight sites in four countries, including the UK, Germany, France, and Sweden. Both studies were almost identical in design, with the efficacy evaluation based on one-year patient and graft survival rates, incidence of rejection episodes, liver function, incidence of infection, and safety evaluations, for at least one year after transplantation. The studies each included over 500 patients, and sample size was approximately equal in both treatment arms.

In the US, ten sites used CsA, steroids, and azathioprine; one site used CsA and steroids; and one site used CsA, steroids, azathioprine, and ALG. Since the participating European sites were principal centers for worldwide transplantation treatment, they had their own standard regimens for CsA-based immunosuppressant therapy (CBIR). All eight European sites used CsA, steroids, and azathioprine; all three German sites added ALG to their regimens.

The same FK506 dose as that used at the University of Pittsburgh was selected for the US and European trials: 0.075 mg/kg infused over four hours twice daily, and 0.15 mg/kg p.o. twice daily. Approximately two months after beginning the studies, some investigators reported that patients had renal dysfunction immediately following transplant surgery. It was suspected that these episodes of renal dysfunction were caused by an overdose of FK506. Thereafter, the initial dose was lowered to 0.05 mg/kg infused over twelve hours, and then stabilized at 0.1 mg/kg/day as a continuous infusion. Daily oral dosing began at 0.3 mg/kg in divided doses.

These study transplants were presented in May 1993 at the American Society of Transplant Surgeons (ASTS) meeting in Houston, Texas. The US trial results demonstrated equivalent patient survival, and a trend towards higher graft survival in the FK506 treatment group (Klintmalm, 1993). In the European study, there was also a trend towards higher patient and graft survival among FK506-treated patients (European FK506 Multicenter Liver Study Group, 1993). Results are shown in Figures 3-10, 3-11, 3-12, and 3-13. Although patient and graft survival rates in the European study tended to be lower than those in the US study, this appears to be caused by the inclusion of more high-risk patients in the European study.

Table 3-7. Comparison of USA and European clinical trial results.

Parameter	USA		Europe	
	FK506	CBIR	FK506	CBIR
One-year patient survival	88%	88%	81%	75%
One-year graft survival	82%	79%	76%	70%
Acute rejection	68%	76%	42%	55%
Refractory rejection	3%	15%	3%	10%

The incidence of acute rejection was significantly lower in the FK506 group than in the CBIR group in both studies (Table 3-7). Also, the incidence of treatment failure/refractory rejection was clinically and statistically lower in the FK506 group than in the CBIR group in both studies.

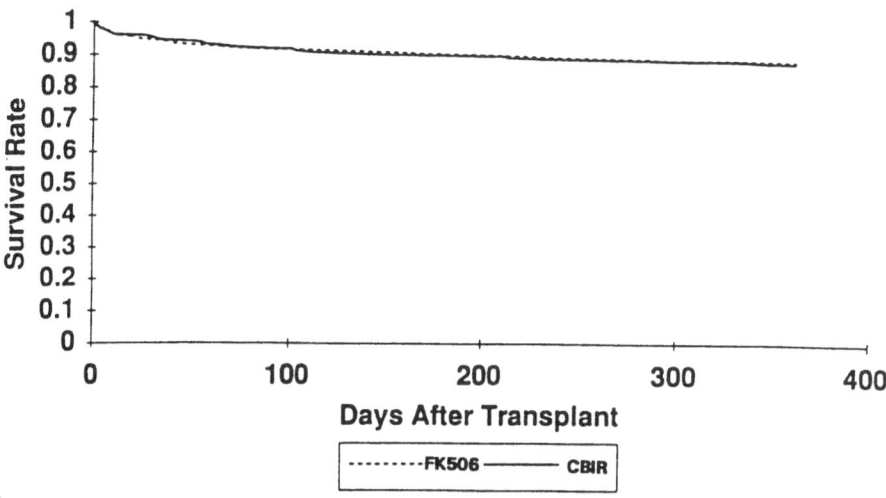

Figure 3-10. Kaplan-Meier estimate of patient survival for FK506 and CBI in US trial.

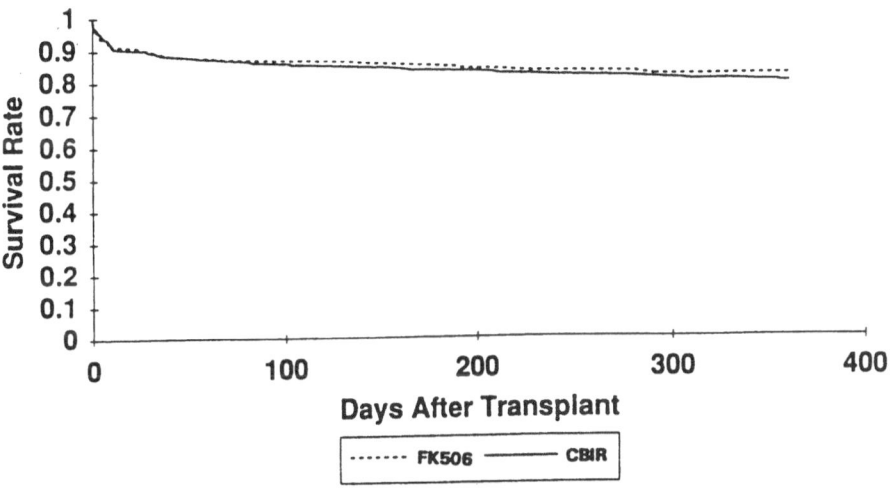

Figure 3-11. Kaplan-Meier estimate of graft survival for FK506 and CBIR in US trial.

In the US study, muromonab-CD3 (OKT3) was administered for steroid-resistant rejection and usage was significantly lower in the FK506 group than in the CBIR group (Figure 3-14).

The main adverse effects in both studies were renal dysfunction, abnormal glucose metabolism, neurological symptoms, and gastrointestinal

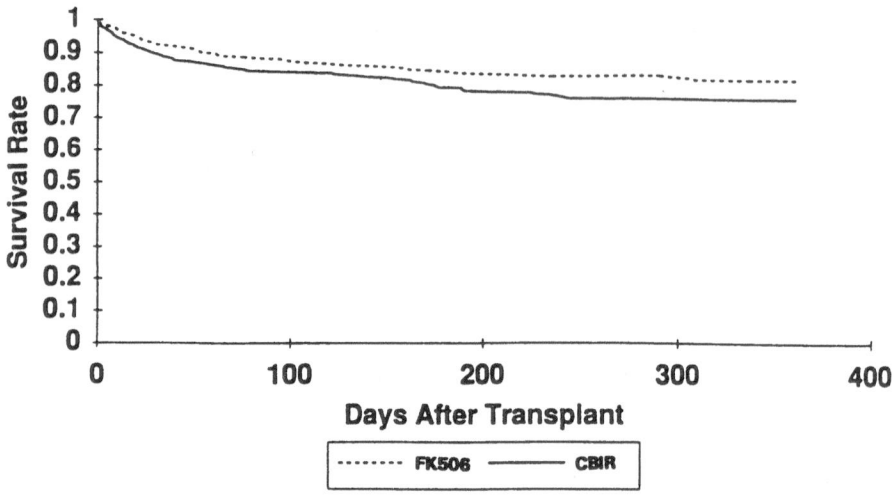

Figure 3-12. Kaplan-Meier estimate of patient survival for FK506 and CBIR in European trial.

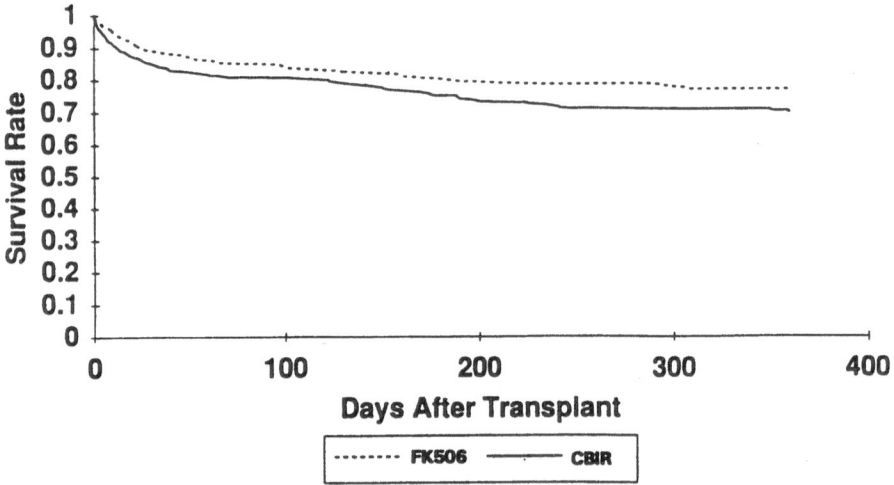

Figure 3-13. Kaplan-Meier estimate of graft survival for FK506 and CBIR in European trial.

symptoms (European FK506 Multicenter Liver Study Group, 1994; US Multicenter FK506 Liver Study Group, 1994).

Other Organ Transplantation Studies

Fujisawa expanded its clinical trials of FK506 to include patients undergo-

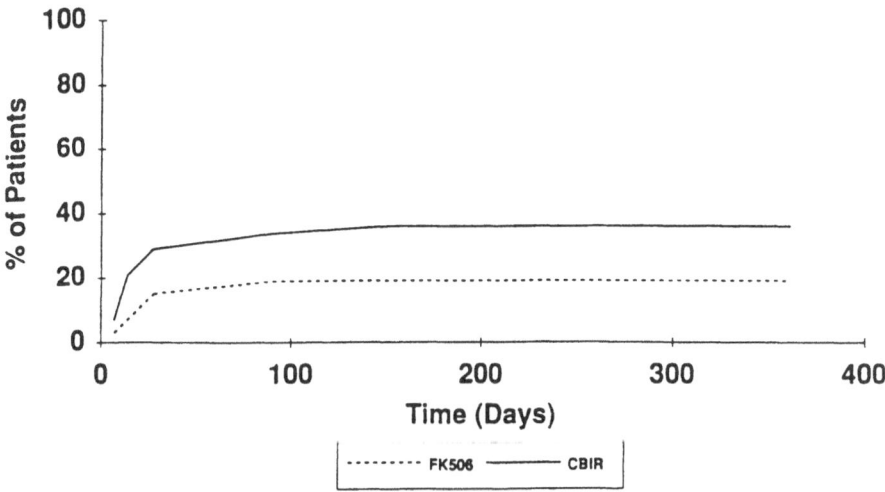

Figure 3-14. Kaplan-Meier estimate of patients (%) requiring treatment with OKT3 for steroid resistant rejection in US trial (P = 0.0001 for difference between group).

ing transplants of other organs, such as kidney and bone marrow, while the University of Pittsburgh studied FK506 immunosuppression after multi-organ transplantations and xenotransplantations. These trials explored the efficacy and safety of FK506 in the prophylaxis of rejection in primary transplant patients and rescue of allografts undergoing rejection were conducted. Some exploratory studies of FK506 in the treatment of autoimmune diseases were also initiated, e.g., nephritis syndrome, primary biliary cirrhosis (PBC), multiple sclerosis, and multiple scleroderma.

Closing Comments

In this chapter, we described the studies on FK506 from the initial screening through clinical development. Ever since the discovery of penicillin a half century ago, microbial products have been agents with potential therapeutic benefits. In the immunology field, several useful immunosuppressants of microbial origin have been discovered to date: CsA, RAP, and superguarin. However, these compounds were detected first as antibiotics or anti-cancer agents, and later their immunosuppressive properties were discovered. We succeeded in discovering the immunosuppressive compound FK506 by using the MLR assay method, which is considered to be the in vitro model of graft rejection. This fact highlights the necessity for a rational approach in the search for novel immunological agents, and also demonstrates the versatility of microorganisms.

The discovery of FK506 provided a useful drug for the treatment of graft rejection and autoimmune disease, as well as an effective probe for immunological investigations. As such, FK506 has given us new insights into the mechanism of T-cell activation, which is one of the most mysterious and fascinating topics in modern immunology.

Acknowledgment. Appreciation is expressed to Mary Beth Holms, M.S., for editorial assistance.

REFERENCES

Ben-Nun A, Wekerle H, Cohern IR (1981): The rapid isolation of clonable antigen-specific T lymphocyte lines capable of mediating autoimmune encephalomyelitis. *Eur J Immunol* 11:195–199

Bierer BE, Mattila PS, Standaert RF, Hezenberg LA, Burakoff SJ, Crabtree G, Schreiber SL (1990a): Two distinct signal transmission pathways in T lymphocytes are inhibited by complexes formed between an immunophilin and either FK506 or rapamycin. *Proc Natl Acad Sci USA* 87:9231–9235

Bierer BE, Schreiber SL, Burakoff SJ (1991): The effect of the immunosuppressant FK506 on alternative pathways of T-cell activation. *Eur J Immunol* 21:439–445

Bierer BE, Somers PK, Wandless TJ, Burakoff SJ, Schreiber SL (1990b): Probing immunosuppressant action with a nonnatural immunophilin ligand. *Science* 250:556–559

Bijsterbosch MK, Klaus GGB (1985): Cyclosporine does not inhibit mitogen-induced inositol phospholipid degradation in mouse lymphocytes. *Immunology* 56:435–440

Bram RJ, Hung DT, Martin PK, Schreiber SL, Crabtree GR (1993): Identification of the immunophiles capable of mediating inhibition of signal transduction by cyclosporin A and FK506: Roles of calcineurin binding and cellular location. *Mol Cell Biol* 13:4760–4769

Carrieri G, Murase N, Woo J, Nalesnik M, Azzarone A, Funakoshi Y, Thomson AW, Todo S, Starzl T (1991): Effect of FK506 in the prophylaxis of autoimmune glomerulonephritis in NZB/WF1 mice. *Transplant Proc* 23:3357–3359

Carroll PB, Strasser S, Alejandro R (1991): The effect of FK506 on cyclophosphamide-induced diabetes in the NOD mouse model. *Transplant Proc* 23:3348–3350

Clipstone NA, Crabtree GR (1992): Identification of calcineurin as a key signalling enzyme in T-lymphocyte activation. *Nature* 357:695–697

Crabtree GR (1989): Contingent genetic regulatory events in T lymphocyte activation. *Science* 243:355–361

Deguchi K, Takeuchi H, Miki H, Yamada A, Touge T, Terada S, Nishioka M (1991): Effects of FK506 on acute experimental allergic encephalomyelitis. *Transplant Proc* 23:3360–3362

De Paulis A, Stellato C, Cirillo R, Ciccarelli A, Oriente A, Marone G (1992): Anti-inflammatory effect of FK506 on human skin mast cells. *J Invest Dermatol* 99:723–728

Dumont FJ, Melino MR, Staruch MJ, Koprak SL, Fischer PA, Sigal NH (1990): The immunosuppressive macrolides FK506 and rapamycin act as reciprocal antagonists in murine T cells. *J Immunol* 144:1418–1424

Durand DB, Shaw JP, Bush MR, Replogle RE, Belagaje R, Crabtree GR (1988): Characterization of antigen receptor response elements within the interleukin-2 enhancer. *Mol Cell Biol* 8:1715–1724

Dutz JP, Fruman DA, Burakoff SJ, Bierer BE (1993): A role for calcineurin in degranulation of murine cytotoxic T lymphocytes. *J Immunol* 150:2591–2598

Emmel EA, Verweij CL, Durand DB, Higgins KM, Lacy E, Crabtree GR (1989): Cyclosporin A specifically inhibits function of nuclear proteins involved in T-cell activation. *Science* 246:1617–1620

Engers HD, Glasebrook AL, Sorenson GD (1982): Allogeneic tumor rejection induced by the intravenous injection of Lyt-2$^+$ cytolytic T lymphocyte clones. *J Exp Med* 156:1280–1285

Ericzon B-G, Wijnen RMH, Kubota K, Kootstra G, Groth CG (1991): Effect of FK506 on glucose metabolism in the cynomolgus monkey: Studies in pancreatic transplant recipients and nontransplanted animals. *Transplant Proc* 23:511

European FK506 Multicenter Liver Study Group (1993): A European, multicenter randomized study to compare the efficacy and safety of FK506 with that of cyclosporine in patients undergoing prmary liver tranplantation: Six month results. *Amer Soc Transplant Surgeons Abstract*: 63

European FK506 Multicenter Liver Study Group (1994): Randomized trial comparing tacrolimus (FK506) and cyclosporin in prevention of liver allograft rejection. *Lancet* 344:423–428

Fischer G, Wittmann-Liebold B, Lang K, Kiefhaber T, Schmid FX (1989): Cyclophilin and peptidyl-prolyl cis-trans isomerase are probably identical protein. *Nature* 337:476–478

Flanagan WM, Corthesy B, Bram RJ, Crabtree GR (1991): Nuclear association of a T cell transcription factor blocked by FK506 and cyclosporin A. *Nature* 352:803–807

Flavin T, Ivens K, Wang J, Gutierrez J, Hoyt EG, Billingham M, Morris RE (1991): Initial experience with FK506 as an immunosuppressant for nonhuman primate recipients of cardiac allografts. *Transplant Proc* 23:531–532

Forrest MJ, Jewell ME, Koo GC, Sigal NH (1991): FK506 and cyclosporin A: Selective inhibition of calcium ionophore-induced polymorphonuclear leukocyte degranulation. *Biochem Pharmacol* 42:1221–1228

Fraser JD, Irving BA, Crabtree GR, Weiss A (1991): Regulation of interleukin-2 gene enhancer activity by the T cell accessory molecule CD28. *Science* 251: 313–316

Fretz H, Albers MW, Galat A, Standaert RF, Lane WS, Burakoff SJ, Bierer BE, Schreiber SL (1991): Rapamycin and FK506 binding proteins (immunophilins). *J Am Chem Soc* 113:1409–1411

Fruman DA, Klee CB, Bierer BE, Burakoff SJ (1992): Calcineurin phosphatase activity in T lymphocytes is inhibited by FK506 and cyclosporin A. *Proc Natl Acad Sci USA* 89:3686–3690

Fujita T, Shibuya H, Ohashi T, Yamanishi K, Taniguchi K (1986): Regulation of human interleukin-2 gene: Functional DNA sequences in the 5' flanking region for the gene expression in activated T lymphocytes. *Cell* 46:401–407

Fung JJ, Todo S, Jain A, McCauley J, Alessiani M, Scotti C, Starzl TE (1990): Conversion from cyclosporine to FK506 in liver allograft recipients with cyclosporine-related complications. *Tranplant Proc* 22(S1):6–12

Gillis S, Smith KA (1977): Long term culture of tumor-specific cytotoxic T cells. *Nature* 268:154–156

Goldfeld AE, Flemington EK, Boussiotis VA, Theodos CM, Titus RG, Strominger JL, Speck SH (1992): Transcription of the tumor necrosis factor α gene is rapidly induced by anti-immunoglobulin and blocked by cyclosporin A and FK506 in human B cells. *Proc Natl Acad Sci USA* 89:12198–12201

Grenier FC, Luczkiw J, Bergmann M, Lunetta S, Morrison M, Blonski D, Shoemaker K, Kobayashi M (1991): A whole blood FK506 assay for the IMx analyzer. *Transplant Proc* 23:2748–2749

Gudas VM, Carmichael PG, Morris RE (1989): Comparison of the immunosuppressive and toxic effects of FK506 and cyclosporine in xenograft recipients. *Transplant Proc* 1072–1073

Handschumacher RE, Harding MW, Rice J, Drugge RJ (1984): Cyclophilin: A specific cytosolic binding protein for cyclosporin A. *Science* 226:544–546

Harding MW, Galat A, Uehling DE, Schreiber SL (1989): A receptor for the immunosuppressant FK506 is a cis-trans peptidyl-prolyl isomerase. *Nature* 341:758–760

Harrison RK, Stein RL (1990): Substrate specificities of the peptidyl prolyl cis-trans isomerase activities of cyclosporin and FK506 binding protein: Evidence for the existence of a family of distinct enzymes. *Biochemistry* 29:3813–3816

Hatanaka H, Iwami M, Kino T, Goto T, Okuhara M (1988a): FR900520 and FR900523, novel immunosuppressants isolated from a Streptomyces I. *J Antibiot* 41:1586–1591

Hatanaka H, Kino T, Hashimoto M, Tsurumi Y, Kuroda A, Tanaka H, Goto T, Okuhara M (1988b): FR65814, a novel immunosuppressant isolated from a *Penicillium* strain. *J Antibiot* 41:999–1008

Hatanaka H, Kino T, Miyata S, Inamura N, Kuroda A, Goto T, Tanaka H, Okuhara M (1988c): FR900520 and FR900523, novel immunosuppressants isolated from a Streptomyces II. *J Antibiot* 41:1592–1601

Hatanaka H, Kino T, Asano M, Goto T, Tanaka H, Okuhara M (1989): FK506 related compounds produced by Streptomyces tsukubaensis No. 9993. *J Antibiot* 42:620–622

Heitman J, Movva NR, Hall MN (1991): Targets for cell cycle arrest by the immunosuppressant rapamycin in yeast. *Science* 253:905–909

Hildebrandt A, Meiser B, Human P, Reichenspurner H, Rose A, Odell J, Reichart B (1991): FK506: Short- and long-term treatment after cardiac transplantation in nonhuman primates. *Transplant Proc* 23:509–510

Hirai O, Miyamae Y, Izumi H, Miyamoto A, Takashima M, Zaizen K, Ohara K, Noguchi H, Bakke JP, Rudd CJ, Mirsalis JC (1992): Mutagenicity test of tacrolimus (FK506). *The Clinical Report* 26:179–191

Hirano Y, Fujihira S, Ohara K, Katsuki S, Noguchi H (1992a): Morphological and functional changes of islets of Langerhans in FK506-treated rats. *Transplantation* 53:889–894

Hirano Y, Ishida H, Maeda H, Adachi A, Kimura M, Yamada A, Ohara K, Noguchi H (1992b): Antigenicity and local irritation studies of tacrolimus (FK506). *The Clinical Report* 26:173–177

Hultsch T, Albers MW, Schreiber SL, Hohman RJ (1991): Immunophilin ligands demonstrate common features of signal transduction leading to exocytosis or transcription. *Proc Natl Acad Sci USA* 88:6229–6233

Iga C, Okajima K, Takeda Y, Tezuka K (1990): Prolonged survival of small intestinal allograft in the rat with cyclosporine A, FK506, and 15-deoxyspergualin. *Transplant Proc* 22:1658

Imai K, Sato K, Nakayama Y, Takishima T, Osakabe T, Yokota K, Yamagishi K, Uchida H, Hiki Y, Kakita A (1991): Histopathological study of canine pancreaticoduodenal allotransplantation with FK506, cyclosporine, and triple regimen immunosuppression. *Transplant Proc* 23:1589–1592

Imventarza O, Todo S, Eiras G, Ueda Y, Furukawa H, Wu YM, Zhu Y, Oks A, Demetris J, Starzl TE (1990): Renal transplantation in baboons under FK506. *Transplant Proc* 22(S1):64–65

Inamura N, Nakahara K, Kino T, Goto T, Aoki H, Yamaguchi I, Kohsaka M, Ochiai T (1988): Prolongation of skin allograft survival in rats by a novel immunosuppressive agent, FK506. *Transplantation* 22:2125–2126

Isai H, Painter DM, Sheil AGR (1990): FK506 and orthotopic liver transplantation in the rat. *Transplantation* 45:206–209

Iwasaki K, Shiraga T, Nagase K, Hirano K, Nozaki K, Noda K (1991): Pharmacokinetic study of FK506 in the rat. *Transplant Proc* 23:2757–2759

Iwasaki K, Shiraga T, Nagase K, Tozuka Z, Noda K, Sakuma S, Fujitsu T, Shimatani K, Sato A, Fujioka M (1993): Isolation, identification and biological activities of oxidative metabolites of FK506, a potent immunosuppressive macrolide lactone. *Drug Metab Dispo* 21:971–977

Jain AB, Venkataramanan R, Cadoff E, Warty V, Iwasaki K, Nagase K, Krajack A, Imventarza O, Todo S, Fung JJ, Starzl TE (1990): Effect of hepatic dysfunction and T-tube clamping on FK506 pharmacokinetics and trough concentrations. *Transplant Proc* 22(S1):57–59

Jain J, McCaffrey PG, Valge-Archer VE, Rao A (1992): Nuclear factor of activated T cells contains Fos and Jun. *Nature* 356:801–804

Jain J, McCaffrey PG, Miner Z, Kerppola TK, Lambert JN, Verdine GL, Curran T, Rao A (1993): The T-cell transcription factor $NFAT_p$ is a substrate for calcineurin and interacts with Fos and Jun. *Nature* 365:352–355

Kagayama A, Tanimoto S, Fujisaki J, Kaibara A, Kaname O, Iwasaka K, Hirano Y, Hata T (1993): Oral absorption of FK506 in rats. *Pharm Res* 6:40–43

Karsh J, Klippel JH, Plotz PH, Decker JL, Wright DG, Flye MW (1981): Lymphapheresis in rheumatoid arthritis. *Arthritis Rheum* 24:867–873

Katayama Y, Takao M, Onoda K, Hiraiwa T, Yada I, Namikawa S, Kusagawa M (1991): Immunosuppressive effects of FK506 and 15-deoxyspergualin in rat lung transplantation. *Transplant Proc* 23:349–353

Kawashima H, Fujino Y, Mochizuki M (1990): Antigen-specific suppressor cells induced by FK506 in experimental autoimmune uveoretinitis in the rat. *Invest Ophthal Vis Sci* 31:2500–2507

Kawauchi M, Gundry SR, Alonso de Begona J, Beierle F, Feikes R, Bailey LL (1991): Xenotransplantation in newborn goats with FK506. *Transplant Proc* 23:3293–3295

Kay JE, Doe SEA, Benzie R (1989): The mechanism of action of the immunosuppressive drug FK506. *Cell Immunol* 124:175–181

Kenmochi T, Asano T, Enomoto K, Goto T, Nakagori T, Sakamoto K, Horie H, Ochiai T, Isono K (1988): The effect of FK506 on segmental pancras allografts in mongrel dogs. *Transplant Proc* 20(S1):223–225

Kino T, Hatanaka H, Hashimoto M, Nishiyama M, Goto T, Okuhara M, Kohsaka M, Aoki H, Imanaka H (1987a): FK506, a novel immunosuppressants isolated from a Streptomyces. I. Fermentation, isolation, and physico-chemical and biological characteristics. *J Antibiot* 40:1249–1255

Kino T, Hatanaka H, Miyata S, Inamura N, Nishiyama M, Yajima T, Goto T, Okuhara M, Kohsaka M, Aoki H, Ochiai T (1987b): FK506, a novel immunosuppressants isolated from a Streptomyces. II. Immunosuppressive effect of FK506 in vitro. *J Antibiot* 40:1256–1265

Klee CB, Dreatta GF, Hubbard MJ (1989): Calcineurin. *Adv Enzymol* 61:149–200

Klintmalm GBG (1993): U.S. multicenter prospective randomized trial comparing FK506 to cyclosporine after liver transplantation: Primary outcome analysis. *Amer Society Transplant Surgeons Abstract*:33

Klintmalm GBG, Nert JR, Husberg BS, Gonwa TA, Tillery GW (1989): Rejection in liver transplantation. *Hepatology* 10:978–985

Kobayashi C, Kanai A, Nakajima A, Okumura K (1989): Suppression of corneal graft rejection in rabbits by a new immunosuppressive agent, FK506. *Transplant Proc* 21:3156–3158

Kobayashi M, Tamura K, Katayama N, Nakamura K, Nagase K, Hane K, Tutumi T, Niwa M, Tanaka H, Iwasaki K, Kohsaka M (1991): FK506 assay past and present—Characteristic of FK506 ELISA. *Transplant Proc* 23:2725–2729

Kuroki H, Ikuta Y, Akiyama M (1989): Experimental studies of vascularized allogeneic limb transplantation in the rat using a new immunosuppressive agent, FK506: Morphological and immunological analysis. *Transplant Proc* 21:3187–3190

Lagodzinski Z, Górski A, Wasik M (1990): Effect of FK506 and cyclosporine on primary and secondary skin allograft survival in mice. *Immunology* 71:148–150

Lautenschlager I, Höckerstedt K, Mäkisalo H, Orko R, Taskinen E (1991): Efficiency of FK506 and CyA to prevent acute cellular rejection of pig liver allografts. *Transplant Proc* 23:2233–2235

Lewis D, Jenkins R, Burke P, Winn KM, Shaffer D, Lopez R, Monaco AP (1991): FK506 rescue therapy in liver transplant recipients with drug-resistant rejection. *Transplant Proc* 23(S6):2989–2991

Liu J, Farmer JD, Lane WS, Friedman J, Weissman I, Schreiber SL (1991): Calcineurin is a common target of cyclophilin-cyclosporin A and FKBP-FK506 complexes. *Cell* 66:807–815

Mattila PS, Ullman KS, Fiering S, Emmel EA, McCutcheon M, Crabtree GR, Herzenberg LA (1990): The actions of cyclosporin A and FK506 suggest a novel step in the activation of T lymphocytes. *EMBO J* 9:4425–4433

McDiarmid S, Klintmalm G, Busuttil R (1991): FK506 rescue therapy in liver transplantation: Outcome and complications. *Transplant Proc* 23(S6):2996–2999

Miller JFAP (1962): Effect of neonatal thymectomy on the immunological responsiveness of the mouse. *Proc R Soc Lond B* 156:415–428

Miyahara H, Hotokebuchi T, Arita C, Arai K, Sugioka Y, Takagishi K, Kaibara N (1991): Comparative studies of the effects of FK506 and cyclosporin A on passively transferred collagen-induced arthritis in rats. *Clin Immunol Immunopathol* 60:278–288

Monden M, Gotoh M, Kanai T, Valdiva LA, Umeshita K, Endh W, Nakano Y, Kawai M, Ohzato H, Ukei T, Dono K, Tono T, Murata M, Wang KS, Okamura J, Tanimoto Y, Hashimoto M, Mori T (1990): A potent immunosuppressive effect of FK506 in orthotopic liver transplantation in primates. *Transplant Proc* 22(S1):66–71

Morgan DA, Ruscetti FW, Gallo R (1976): Selective in vitro growth of T lymphocytes from normal human bone marrows. *Science* 193:1007–1008

Morikawa K, Oseko F, Morikawa S (1992): The distinct effects of FK506 on the activation, proliferation, and differentiation of human B lymphocytes. *Transplantation* 54:1025–1030

Moriyama S, Shimizu N, Teramoto S (1989): Experimental tracheal allotransplantation using omentopexy. *Transplant Proc* 21:2595–2600

Morris RE (1991): Rapamycin: FK506's fraternal twin or distant cousin? *Immunol Today* 12:137–140

Morris RE, Wu J, Shorthouse R (1990): A study of the contrasting effects of cyclosporine, FK506, and rapamycin on the suppression of allograft rejection. *Transplant Proc* 22:1638–1641

Murase N, Kim D-G, Todo S, Cramer DV, Fung JJ, Starzl T (1990a): Suppression of allograft rejection with FK506. I. Prolonged cardiac and liver survival in rats following short-course therapy. *Transplantation* 50:186–189

Murase N, Lieberman I, Nalesnik M, Mintz D, Todo S, Drash AL, Starzl TE (1990b): Prevention of spontaneous diabetes in BB rats with FK506. *Lancet* 336:373–374

Murase N, Todo S, Lee P-H, Lai H-S, Chapman F, Nalesnik MA, Makowka L, Starzl TE (1987): Heterotopic heart transplantation in the rat receiving FK506 alone or with cyclosporine. *Transplant Proc* 19(S6):71–75

Nakajima K, Sakamoto K, Ochiai T, Asano T, Isono K (1989): Effects of 15-deoxy-spergualin and FK506 on the histology and survival of hamster-to-rat cardiac xenotransplantation. *Transplant Proc* 21:546–548

Ochiai T, Nagata M, Nakajima K, Sakamoto K, Asano T, Isono K (1987a): Prolongation of canine renal allograft survival by treatment with FK506. *Transplant Proc* 19(S6):53–56

Ochiai T, Nakajima K, Nagata M, Hori S, Asano T, Isono K (1987b): Studies of the induction and maintenance of long-term graft acceptance by treatment with FK506 in heterotopic cardiac allotransplantation in rats. *Transplantation* 44:734–738

Ochiai T, Nakajima K, Sakamoto K, Nagata M, Gunji Y, Asano T, Isono K, Sakamaki T, Hamaguchi K (1989): Comparative studies on immunosuppressive activity of FK506, 15-deoxyspergualin and cyclosporine. *Transplant Proc* 21:829–832

Ohara K, Billington R, James RW, Dean GA, Nishiyama M, Noguchi H (1990): Toxicologic evaluation of FK506. *Transplant Proc* 22:83–86

Ohara K, Iwanami K, Noguchi H, Hooks WN, Peters DH, Morrow J, Gopinath C (1992a): Fifty-two-week oral toxicity study of tacrolimus (FK506) in rats. *The Clinical Report* 26:143–157

Ohara K, Iwanami K, Noguchi H, Billington R, Horner SA, Smith T, Buist DP, Crook D, Gopinath C, Anderson A, Dawe S (1992b): 52-week oral toxicity study of tacrolimus (FK506) in baboons. *The Clinical Report* 26:399–407

Okuhara M, Goto T, Hatanaka H, Yuasa S, Nakatani I (1990): FR901154 and FR901155 substances, a process for their production and pharmaceutical composition containing the same European Patent No. 353678

Okuhara M, Kino T (1994): Immunomodulators. In: *The Discovery of Natural Products with Therapeutic Potential*, Gullo VP, ed. London: Butterworth-Heinemann

O'Keefe SJ, Tamura J, Kincaid RL, Tocci MJ, O'Neill EA (1992): FK506- and CsA-sensitive activation of the interleukin-2 promotor by calcineurin. *Nature* 357:692–694

Paulus HE, Machleder HI, Levine S, Yu DTY, MacDonald NS (1977): Lymphocyte involvement in rheumatoid arthritis. *Arthritis Rheum* 20:1249–1262

Saegusa T, Ohara K, Noguchi H, York RG, Weisenberger WP, Sahardein JL (1992): Reproductive and developmental studies of tacrolimus (FK506) in rats and rabbits. *The Clinical Report* 26:159–171

Saitoh Y, Fujisawa T, Ogawa T, Urabe N, Yamaguchi Y, Kimizuka G (1990): Morphological rejection phases and cytotoxic activity in peripheral blood lymphocytes in canine lung allo-transplantation. *Japn J Surgery* 20:205–211

Schreiber SL, Crabtree GR (1992): The mechanism of the action of cyclosporin A and FK506. *Immunol Today* 13:136–142

Sehgal SN, Baker H, Vezina C (1975): Rapamycin (AY-22,989), a new antifungal antibiotic. II. Fermentation, isolation and characterization. *J Antibiot* 28:727–732

Shaw JP, Utz PJ, Durand DB, Toole JJ, Emmel EA, Crabtree GR (1988): Identification of a putative regulator of early T cell activation genes. *Science* 241:202–205

Shaw B, Markin R, Stratta R, Langnas A, Donovan J, Sorrell M (1991): FK506 for rescue treatment of acute and chronic rejection in liver allograft recipients. *Transplant Proc* 23(S6):2994–2995

Sher LS, Pan S-H, Hoffman AL, Villanil FG, Howard TK, Podesta LG, Makowka L (1991): Liver transplantation. In: *Handbook of Transplantation Management*, Pan S, Hoffman A, Makowka L, eds. Austin: RG Landes Co

Siekerka JJ, Hung SHY, Poe M, Lin CS, Sigal NH (1989): A cytosolic binding protein for the immunosuppressant FK506 has peptidyl-prolyl isomerase activity but is distinct from cyclophilin. *Nature* 341:755–757

Somers PK, Wandless TJ, Schreiber SL (1991): Synthesis and analysis of 506BD, a high-affinity ligand for the immunophilin FKBP. *J Am Chem Soc* 113:8045–8057

Starzl T, Todo S, Fung J, Demetris AJ, Venkataramanan R, Jain A (1989): FK506 for liver, kidney, and pancreas transplantation. *Lancet* II(8670):1000–1004

Suzuki N, Sakane T, Tsunematsu T (1990): Effects of a novel immunosuppressive agent, FK506, on human B cell activation. *Clin Exp Immunol* 79:240–245

Takahashi N, Hayano T, Suzuki M (1989): Peptidyl-prolyl cis-trans isomerase is the cyclosporin A-binding protein cyclophilin. *Nature* 337:473–475

Tamura K, Kobayashi M, Hashimoto K, Kojima K, Nagase K, Iwasaki K, Kaizu T, Tanaka H, Niwa M (1987): A highly sensitive method to assay FK506 levels in plasma. *Transplant Proc* 19:23–29

Tanaka H, Kuroda A, Marusawa H, Hatanaka H, Kino T, Goto T, Hashimoto M (1987): Structure of FK506: A novel immunosuppressant isolated from Streptomyces. *J Am Chem Soc* 109:5031–5033

Thompson CB, Lindsten T, Ledbetter JA, Kunkel SL, Young HA, Emerson SG, Leiden JM, June CH (1989): CD28 activation pathway regulates the production of multiple T-cell-derived lymphokines/cytokines. *Proc Natl Acad Sci USA* 86:1333–1337

Tocci MJ, Matkovich DA, Collier KA, Kwok P, Dumont F, Lin S, Degudicibus S, Siekierka JJ, Chin J, Hutchinson NI (1989): The immunosuppressant FK506 selectively inhibits expression of early T cell activation genes. *J Immunol* 143:718–726

Todo S, Fung JJ, Demetris AJ, Jain A, Venkataramanan R, Starzl TE (1990): Early trials with FK506 as primary treatment in liver transplantation. *Transplant Proc* 22(S1):13–16

Todo S, Podesta L, ChapChap P, Kahn D, Pan C-E, Ueda Y, Okuda K, Imventarza O, Casavilla A, Demetris AJ, Makowska L, Starzl TE (1987): Orthotopic liver transplantation in dogs receiving FK506. *Transplant Proc* 19(S6):64–67

Tropschung M, Barthelmess IB, Neupert W (1989): Sensitivity to cyclosporin A is mediated by cyclophilin in Neurospora crassa and Saccharomyces cervisiae. *Nature* 342:953–955

Tsuchimoto S, Kusumoto K, Nakajima Y, Kakita A, Uchino J, Natori T, Aizawa M (1989): Orthotopic liver transplantation in rats receiving FK506. *Transplant Proc* 21:1064–1065

Ueda Y, Todo S, Eiras G, Furukawa H, Imventarza O, Wu YM, Oks A, Zeevi A, Oguma S, Starzl TE (1990): Induction of graft acceptance after dog kidney or liver transplantation. *Transplant Proc* 22(S1):80–82

U.S. Multicenter FK506 Liver Study Group (1993): Use of FK506 for the prevention of recurrent allograft rejection after successful conversion from cyclosporine for refractory rejection. *Transplant Proc* 25(S1):635–637

U.S. Multicenter FK506 Liver Study Group (1994): A comparison of tacrolimus (FK506) and cyclosporine for immunosuppression in liver transplantation. *N Engl J Med*: 331:1110–1115

Venkataramanan R, Jain A, Cadoff E, Warty V, Iwasaki K, Nagase K, Krajack A, Imventarza O, Todo S, Fung JJ, Starzl TE (1990): Pharmacokinetics of FK506: Preclinical and clinical studies. *Transplant Proc* 22(S1):52–56

Vezina C, Kudelski A, Sehgal SN (1975): Rapamycin (AY-22,989), a new antifungal antibiotic. I. Taxonomy of producing streptomycete and isolation of the active principle. *J Antibiot* 28:721–726

Walliser P, Benzie CR, Kay JE (1989): Inhibition of murine B-lymphocyte proliferation by the novel immunosuppressive drug FK506. *Immunology* 68:434–435

Weaver CT, Unanue ER (1990): The costimulatory function of antigen-presenting cells. *Immunol Today* 11:40–55

Wicker LS, Boltz RC Jr, Matt V, Nichols EA, Peterson LB, Sigal NH (1990): Suppression of B cell activation by cyclosporin A, FK506 and rapamycin. *Eur J Immunol* 20:2277–2283

Yamashita T, Maeda Y, Ishikawa T, Ohira M, Nakagawa H, Nagai Y, Park I, Yamamoto Y, Yoshikawa K, Cai R, Umeyama K (1991): Prolongation of pancreaticoduodenal allograft survival in rats by treatment with FK506. *Transplant Proc* 23:3219–3220

Yasunami Y, Ryu S, Kamei T, Konomi K (1989): Effects of a novel immunosuppressive agent, FK506, on islet allograft survival in the rat. *Transplant Proc* 21:2720

Yasunami Y, Ryu S, Kamei T (1990): FK506 as the sole immunosuppressive agent for prolongation of islet allograft survival in the rat. *Transplantation* 49:682–686

Yokota K, Takishima T, Sato K, Osakabe T, Nakayama Y, Uchida H, Aso K, Masaki Y, Ohbu M, Okudaira M (1989): Comparative studies of FK506 and cyclosporine in canine orthotopic hepatic allograft survival. *Transplant Proc* 21:1066–1068

4

Discovery of Sulindac, a Reversible Prodrug, as a Second-Generation Indomethacin

Tsung Ying Shen

Introduction

Indomethacin, a forerunner of a proliferative family of nonsteroidal anti-inflammatory drugs (NSAIDs), was first marketed three decades ago in Europe. The discovery of sulindac some ten years later (Shen et al., 1972) illustrated the strategy and process involved in the continuing search for a second product after an initial breakthrough. A new indomethacin congener with some unique properties was certainly needed to meet the growing demand for better NSAIDs in the 1970s by rheumatologists worldwide. Our objective then was to fulfill the usual expectations of a second-generation new drug: improvements in efficacy, safety, and pharmacodynamics.

Sulindac, as shown in Figure 4-1, is the last of a series of six clinical candidates derived from the original indole lead (Shen and Winter, 1977). The discovery of sulindac also marked a transition in the research and development of NSAIDs. Advances in pharmacological models of polyarthritis, better understanding of the biochemical mechanism of action of aspirin and indomethacin and the increasing sophistication of clinical pharmacology provided new rationale and experimental systems for medicinal chemists to investigate novel chemical structures as superior NSAIDs. To improve the traditional, and mostly empirical, structure-activity relationship (SAR) studies, researchers also began to explore quantitative approaches trying to correlate physicochemical constants, x-ray structures, and simple computer models with the complex in vivo biological data. It was still an era before

The Search for Anti-Inflammatory Drugs
Vincent J. Merluzzi and Julian Adams, Editors
© Birkhäuser Boston 1995

the vogue of *de novo* drug design, but a very fruitful golden period of new drug discovery through a semi-empirical approach with a multidisciplinary team effort.

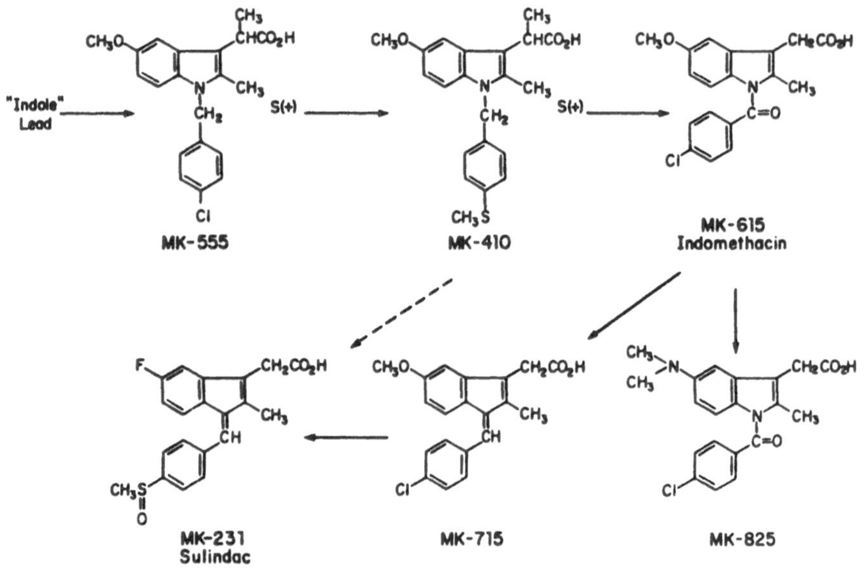

Figure 4-1. Clinical precursors of sulindac derived from the original indole lead.

As current anti-inflammatory research is progressing along several new directions, including those described in other chapters in this volume, to develop more effective anti-arthritic agents, some critical issues and problem-solving strategies involved in the discovery of sulindac are briefly analyzed below to provide some historical perspectives. The implications of some more recent biological and clinical observations with sulindac, including two serendipitous discoveries which are unrelated to its anti-arthritic applications, are also discussed.

In the past two decades, more than 500 scientific and clinical papers on sulindac have been published. For this brief review, only a few representative publications particularly germane to the original discovery and development are listed in the reference section. Several more recent papers are also cited to update the information and as a source of newer references on some specific aspects of sulindac.

Biological Assays

The discovery and development of indomethacin was initially guided by a seven-day cotton-pellet assay (Shen and Winter, 1977), which measured the inhibition of the growth of granuloma tissue in rats. An acute carrageenan-induced paw edema assay in rats was established by my pharmacological colleagues Drs. Charles Winter and Gordon Van Arman in the early 1960s and was then used to characterize indomethacin and congeners, including the indene isosteres. This three-hour assay required less time and testing sample, was more sensitive to NSAID inhibition, and gave more quantitative results than the granuloma assay. (Later it was widely adopted by most laboratories as a standard protocol in their search for newer NSAIDs.) Soon afterwards, the Freund adjuvant-induced polyarthritis rat model was refined by Drs. Carl Pearson, Brian Newbould, and others for compound evaluation (Billingham and Davies, 1979). Its semi-chronic immunological characteristics and bony lesions in afflicted animals provided us with an arthritic assay, although the very high potency of indomethacin in this model tempered our expectation that the assay would detect the superior and the much sought after disease-modifying anti-arthritic activities. Later, the urate crystal-induced canine synovitis and topical mouse ear inflammation assays were also used in the secondary evaluation of promising compounds.

NSAIDs, like aspirin, are generally considered as agents with triple anti-inflammatory, analgesic, and antipyretic activities for the relief of the four cardinal symptoms of inflammation: *rubor*, *tumor*, *calor*, *et dolor*, defined by Galen (AD 129–199) in ancient times. The analgesic effect of indomethacin was first noticed in patients. This observation prompted Dr. Winter to adapt a modified Randall-Sellito procedure to measure the peripheral analgesia of NSAIDs (Shen and Winter, 1977). The question of separating the anti-inflammatory effect from the peripheral analgesic activity in yeast-injected rat paw was resolved by their time-course difference and protocol adjustments. The antipyretic potency was determined by a standard rabbit model. The in vitro cyclooxygenase inhibition assay using sheep seminal vesicle preparations, though not available for compound evaluation until 1972, was set up in time to provide a clear evidence regarding the prodrug nature of sulindac as described below.

Another important development in the biological evaluation of indomethacin congeners was the introduction of gastrointestinal (GI) irritation assays (Shen and Winter, 1977). The previous emphasis on potency enhancement was broadened to optimize the activity profile or therapeutic index for better clinical efficacy. Following the realization that the efficacy of indomethacin in humans was mainly limited by its gastrointestinal irritations, an acute gastric hemorrhage assay and a three-day intestinal perforation assay in the rat were developed. In spite of some clear species differences observed with laboratory animals, these assays provided an indication of the GI tolerance and therapeutic index of active compounds from the carrageenan edema assay. On the other hand, the acute LD_{50} determination, widely reported by other investigators, was not deemed to be very informative in our study. The potential nephrotoxicity of NSAIDs, especially papillary edema and necrosis, had been observed in the preclinical safety assessment of several indomethacin congeners and an aryl propionic acid analog (MK-830), but no short-term assay was available. Various speculations on the nature of indomethacin-induced effects on the central nervous system were considered, but no experimental systems were developed.

The pharmacologic profile of indomethacin, sulindac and its sulfide metabolites are shown in Table 4-1 as examples.

Table 4-1. Pharmacologic profiles of indomethacin, sulindac, and sulindac sulfide.*

Test	Indomethacin	Sulindac	Sulindac Sulfide
Cyclo-oxygenase IC_{50} μg/ml	0.2	Inactive	0.5
Carrageenan-induced paw edema (rat) ED_{50}	2.3	5.5	2.3
Adjuvant arthritis ED_{50}	0.27	0.55	0.36
Analgesia to yeast inflammation (rat)	2.2	1.7	1.0
Antipyresis (rat)	1.8	2.9	0.5
Gastric hemorrhage (rat)	5.4	27.4	6.8
Intestinal perforation (rat)	5	69.2	35.0

* In vivo values in mg kg^{-1} po.

Chemical Studies

Initial Investigations—MK-825 and MK-715

From the very beginning, in the absence of any biological understanding of the nature of the idiosyncratic side effects on the central nervous system (CNS) of indomethacin, a chemical strategy of altering the metabolic properties of indomethacin was adopted in our attempts to circumvent this clinical enigma.

The first clinical candidate thus selected was the 5-dimethylamino analog of indomethacin, MK-825 in Figure 4-1. Intuitively, it was surmised that replacing the 5-methoxy group in indomethacin by a more basic and metabolically labile dimethylamino substituent might alter the distribution and metabolic characteristics of the drug and thus reduce the CNS effects. Unfortunately, MK-825 turned out to be nearly identical with indomethacin in arthritic patients both in potency and in side effects. No further pursuit of close indomethacin analogs was considered.

Many indole derivatives have shown pronounced CNS effects. In fact the lead for indomethacin congeners came from a group of compounds originally synthesized in our laboratory as potential serotonin antagonists. It would seem logical then to replace the indole ring in indomethacin by an indole-equivalent in order to minimize the CNS side effects. Around that time, stimulated in part by the serendipitous discovery in Professor Cope's laboratory at MIT of a synthetic route to the dibenzocycloheptatriene ring system, medicinal chemists at the Merck, Sharp, and Dohme Research Laboratories (MSDRL) in West Point, PA had just succeeded in developing amitriptyline as an isostere of the tricyclic antidepressant imipramine. The heterocyclic ring nitrogen atom in imipramine was successfully replaced by an electronically equivalent methylene linkage in amitriptyline. By analogy, my West Point pharmacological colleague Dr. Winter became interested in the indene ring system. A chemical synthesis of the bright yellow indene isostere of indomethacin was readily accomplished by a simple procedure (Shen et al., 1972). It was indeed an active anti-inflammatory agent and was designated as MK-715 (Figure 4-1) for further study.

In animal assays as well as later in humans, MK-715 was found to be approximately $\frac{1}{4}$ to $\frac{1}{2}$ as potent as indomethacin, but was significantly less ulcerogenic and, interestingly, was free from any significant CNS side effects in patients. In other words, the poorly understood and troublesome headache and muzziness produced by indomethacin in some patients were finally avoided by this medicinal chemical approach, albeit through a biological mechanism that remains elusive.

However, the encouraging clinical findings of MK-715 were soon followed by the unsettling observation of crystalluria in Dr. Delbarre's clinic in Paris. Having a rigid and conjugated carbocyclic aromatic ring system, MK-715, as well as its acyl D-glucuronide metabolite, was extremely insoluble in water and produced crystalluria in patients taking more than 200 mg/day of MK-715. In short, while the exact indene isostere of indomethacin offered us a possible resolution of the GI and CNS problems, it now presented a distinct side effect of its own.

Aryl Propionic Acids—M-830

Chemically, MK-715, a 3-indenyl acetic acid, may also be considered as a carbocyclic aryl acetic acid derivative. Its structure is similar in a generic

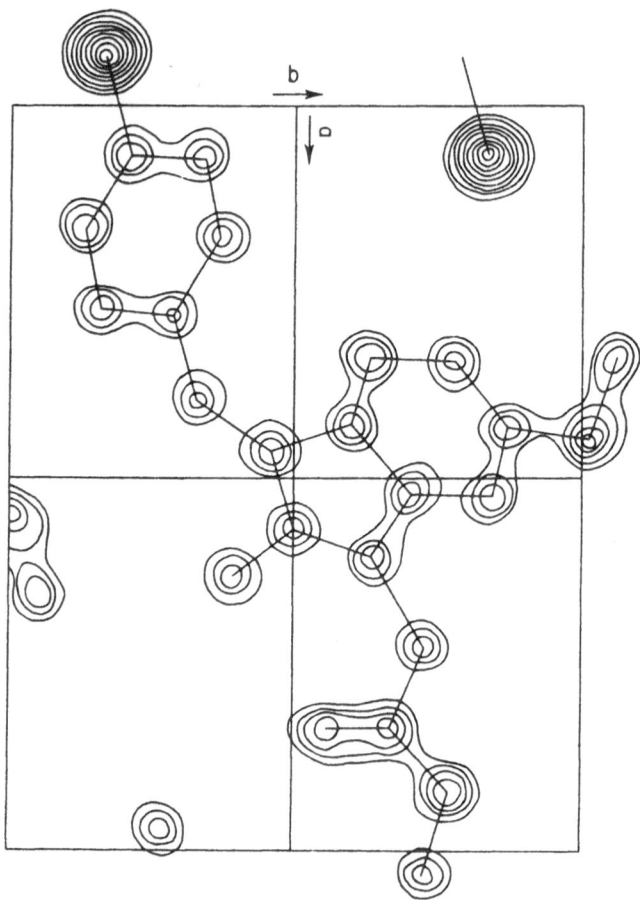

Figure 4-2. X-ray structure of MK-715.

sense to the big family of substituted phenyl propionic acids, e.g., ibuprofen
and naproxen. The pharmacological activities of MK-715 first showed us
that the indole ring in indomethacin is not essential for anti-inflammatory
activities. The x-ray structure of the rigid MK-715 (Figure 4-2) showed
that the benzylidenyl group is twisted out of the plane of the indene ring
(Hoogsteen and Trenner, 1970). The overall stereochemistry is very similar
to that of indomethacin (Shen and Winter, 1977). Without any knowledge
about the nature of the biological target(s) of indomethacin at that time,
we hypothesized that a nonplanar configuration is probably preferred for
indomethacin and other aryl acetic acids to interact with the active site(s).
The proliferation of a variety of aryl acids as anti-inflammatory agents
from many laboratories later clearly showed that, as we know today, the

structure-activity relationship of NSAIDs can indeed accommodate a broad range of aryl groups. The principal task for medicinal chemists then was to optimize the activity-enhancing substituents, which differed somewhat in each aryl series, and more importantly to improve the therapeutic index or safety margin of active compounds, under the usual constraints of having limited biological understanding and relying on animal assays of uncertain correlation with clinical results.

Soon after the selection of MK-715 for clinical trials, our preclinical laboratory effort was then diverted to the investigation of novel aryl acetic acids, again using the carrageenan edema assay as our primary biological screen. This effort led us to study biphenyl and the nonplanar 4-cyclohexylphenyl acetic acid derivatives, independent of the ibuprofen research at the Boots Laboratories in England, which culminated in the discovery of MK-830, S(+) 3-chloro-4-cyclohexylphenyl-a-propionic acid (Shen, 1984). MK-830 is one of the most potent aryl propionic acids known, being ten to twenty times more active, with regard to dosage, than indomethacin in our animal models. Its chiral specificity—the R(−) optical isomer is inactive both in vivo and in vitro—later lent some support to the proposal regarding prostaglandin synthesis inhibition as a mechanism of action for NSAIDs. Unfortunately, MK-830 was found to be highly ulcerogenic in the preclinical safety assessment and was abandoned. A wide variety of substituted aryl moieties were also examined in our laboratory and patented abroad, including some closely related to several newer NSAIDs successfully developed by other laboratories later. However, with limited resources, it was impractical for us to investigate each lead fully by the semi-empirical chemical and pharmacological approach. Of course, this is not an uncommon dilemma, encountered by many researchers in developing broad chemical leads like H_2-receptor antagonists, ACE inhibitors, β-lactam antibiotics, angiotensin II receptor antagonists, etc.

The recurring activity enhancing effect of a (halo)phenyl substituent in our anti-inflammatory agents, as in the structure of indomethacin, MK-825, 715, and 830, was also demonstrated in two other intervening synthetic projects assigned to us by the research management, namely, a more effective analog of dimethylsulfoxide (DMSO) and a super aspirin. The former project was a marketing-defense necessity. It was terminated after the observation of ocular toxicity of DMSO in canines, much to the relief of NSAID researchers and marketing people alike. The super aspirin project was prompted by an earlier suggestion from the eminent Harvard rheumatologist Dr. Walter Bower to George Merck following Merck's successful development of hydrocortisone, which became feasible in the late 1960s by the availability of more sensitive pharmacological assays. This medicinal chemical venture, with the persistent encouragement of Dr. Lewis Sarett, advanced progressively through the addition of halophenyl groups to the salicylate molecule and moved from 5-phenyl to 5-4'-fluorophenyl (flufenisal) and ultimately to 5-2'4'-diflurophenyl salicylic acid (diflunisal, MK-647) (Hannah et al., 1977; Winter et al., 1981). Diflunisal has since been marketed successfully as Dolobid (see Structure 4-A).

S (+) MK-830   Diflunisal

Structure 4-A. S(+) MK-830 and diflunisal.

More Soluble MK-715 Analogs

The disappointing case of MK-830 and a prolonged development of flufenisal and diflunisal renewed our interest in seeking more soluble MK-715 analogs for indomethacin-intolerant patients. In view of increasing competition, an intensive chemical and biological crash program was launched. New compounds were speedily synthesized and first tested orally in the rat carrageenan foot edema assay, followed by the gastric hemorrhage and intestinal perforation assays to assess GI tolerance. The three-week rat adjuvant arthritis assays, with both preventive and therapeutic protocols, were used to characterize the more active compounds. The aqueous solubility of newly designed analogs were readily determined by their UV absorption spectra.

The first approach was to increase the hydrophilicity or basicity of the substituents, e.g., hydroxyl, amino, pyridyl, etc., around the 1-benzylidenyl-indene ring and to reduce the rigidity of the highly conjugated system. It met with little success. Most of the structural changes reduced the anti-inflammatory potency significantly; the modest gains in solubility were more than offset by higher doses required for maintaining efficacy.

The Pharmacodynamic and Prodrug Approach

The successful solution eventually came through a fortuitous combination of several factors, including some prior SAR data of indomethacin analogs, our laboratory experience with the DMSO project, increasing emphasis of pharmacodynamics in drug development, and the timely discovery of the biochemical mechanism of action of NSAIDs (Shen and Winter, 1977; Shen, 1985).

First, it seemed that if one could use a metabolically-labile substituent to generate several drug metabolites, one might be able to avoid the accumulation of any metabolite beyond its solubility in the urine. The sequential metabolic transformation of the sulfide–sulfoxide–sulfone system came to mind:

$$\text{HS-R} \longleftarrow \text{CH}_3\text{S-R} \rightleftharpoons \text{CH}_3\text{SO-R} \longrightarrow \text{CH}_3\text{SO}_2\text{-R} \dashrightarrow \text{HOSO}_2\text{-R}$$

Sulfhydryl   **Sulfide**   **Sulfoxide**   **Sulfone**   **Sulfonic acid**

Our prior experience indicated that such a system would not only be compatible with the established SAR for indomethacin and indene analogs, but also provide a product with much improved solubility and distribution properties.

In the original study of indolylacetic acids, two prototypes of indomethacin, MK-555 and MK-410, were successively tested in patients (Figure 4-1). MK-555 was the first compound which demonstrated a correlation of the anti-inflammatory activity in the rat cotton-pellet granuloma model with clinical efficacy, thus validating our experimental approach to a new NSAID. Replacement of the p-chloro substituent in MK-555 with a methylthio group (a chlorine equivalent commonly used in the study of phenothiazine and tricyclic analogs at that time) increased the potency four times in rats and twice in humans. Of its two in vivo metabolites, the methysulfinyl derivative of MK-410 was equally potent as MK-410, whereas the methylsulfonyl metabolite was inactive in the rat assay. As the SAR of indole and indene series are generally parallel, the para-methylthio and methylsulfinyl (methylsulfoxide) analogs of MK-715 were indeed found to be fully active.

During the earlier MK-410 study, in an attempt to correlate some physicochemical properties of compounds with their biological activities, long before the age of QSAR, the pKa value and lipid-water distribution coefficient were determined. Although no simple correlation with anti-inflammatory effect was found in this QSAR exercise, it was noted that replacement of the para-chloro or methylsulfide substituent by a sulfoxide group markedly increased both the aqueous solubility (tenfold) and the water/lipid distribution coefficient (almost 100-fold) of MK-555 and 410 (Shen, 1965). The hydrophilicity of sulfoxides was later reaffirmed in our brief study of dimethyl-sulfoxide (DMSO) derivatives.

These considerations led us to propose that sulfoxide analogs of MK-715 would likely be active, more soluble, and transformed to multiple metabolites in vivo.

Selection of MK-231 (Sulindac)

Having formulated a strategy to improve the pharmacodynamics of MK-715 with a sulfoxide group, further optimization of the ring substituents was carried out to increase the potency of MK-715 (Shen and Winter, 1977). Among the most active analogs synthesized were the 5,6-difluoro, 5-methoxy-6-fluoro, and 5-fluoro analogs. All three compounds were comparable to indomethacin in potency with significantly less GI irritation in the animal models. Since their pharmacological profiles differed only slightly, instead of selecting one with the highest potency or slightly higher GI tolerance in the usual manner, a decision was made to put all three candidates into a one-month anticipatory safety assessment. This strategy was chosen for two practical reasons: First, it was difficult to extrapolate short-term toxicity of potential NSAIDs to long-term safety. A month-long semi-chronic evaluation of all three compounds, in spite of added expenses, would, it was hoped, yield more reliable data for selecting the best clinical candidate. Second, competitive pressure would not allow the time-consuming process of sequential safety evaluation of one candidate at a time, which might cost from nine to twelve months delay if the first and second compound should falter as many NSAIDs did. At the end of the one-month comparison, it turned out that the 5-fluoro analog, the least potent among the three candidates, was the only one with acceptable long-term safety in the gastrointestinal, renal, and hepatic systems (Van Arman et al., 1972). It was then designated as MK-231 (sulindac) for clinical trials. In retrospect, had we followed the usual practice by putting the two slightly more potent analogs into the regular safety assessment successively and discovered their toxicities after a year or two, it would be very unlikely for any research management to continue with the third analog (MK-231), which has similar chemical and biological properties. In other words, luck still played a key role in the empirical search for clinical candidates with acceptable long-term safety; the gamble of using an anticipatory safety assessment strategy was crucial in the successful selection of a satisfactory second-generation indomethacin.

Sulindac as a Reversible Prodrug

Mechanism of Action

The facile interconversion of sulindac and its equally active sulfide metabolite in vivo would not allow us to distinguish whether both sulindac and its sulfide are active or only one of the two is active per se. The potency of sulindac, relative to the sulfide, did appear to increase with the experimental duration of several animal models of inflammation (Table 4-2), suggesting that sulindac might be converted to the more active sulfide in vivo. Fortunately, the discovery by John Vane et al. in 1971 that aspirin and indomethacin inhibit the biosynthesis of prostaglandins in vitro provided a timely and critical test to clarify this question. In vitro, MK-231 does not inhibit the conversion of arachidonic acid to prostaglandins by the sheep seminal vesicle enzyme, whereas the sulfide is as active as indomethacin. From the data summarized in Table 4-2, sulindac is clearly a prodrug acting through its sulfide metabolite (Duggan et al., 1977b). The sulfone metabolite is inactive both in vitro and in vivo (Glavin and Sitar, 1986).

Table 4-2. Pharmacologic activities of sulindac and its sulfide metabolite.

Test	Sulindac (prodrug)	Sulfide metabolite (active species)
In vitro assays		
PG synthetase inhibition IC_{50} (μM)	inactive	2
Platelet aggregation versus ADP (μg ml^{-1})	inactive	0.5
Topical administration		
Intrasynovial (canine synovitis) ED_{50} (mg kg^{-1})	6.4	0.1
Ocular inflammation (rabbits)	weakly active	active
Relative potency in in vivo models ED_{50} (mg kg^{-1} po)		
Urate synovitis (canine), 2 h	45	5.3
Antipyresis (rat), 2 h	2.9	0.5
Carrageenan paw edema (rat), 3 h	5.5	2.3
Cotton pellet granuloma (rat), 7 days	5.4	> 3.4
Adjuvant arthritis (rat), 14 days	0.55	0.36
Correlation of drug concentration versus response		
Plasma level in carrageenan edema	poor	good

Like indomethacin, the sulindac sulfide is a substrate competitive inhibitor of the cyclooxygenase. In addition, the redox potential of the sulfide group renders it a scavenger of oxygen radicals, including those produced in the arachidonic acid pathway. In an elegant experiment using an ^{18}O-labeled intermediate, it was shown by Drs. Fred Kuehl and Robert Egan that the sulfide reacted with an ^{18}O-radical generated by the PGH$_2$ synthase to form ^{18}O-labeled sulindac stoichiometrically (Egan et al., 1980). This conversion demonstrated an interesting example of site-specific trapping of a reactive species and reverse prodrug formation at the active site of the target enzyme. Later, the sulfide was found to inhibit the 5-lipoxygenase at $> 5 \ \mu$M levels and several cellular systems of inflammation, but the in vivo significance of these modest inhibitory activities is not certain (Shen, 1985).

Pharmacodynamics of Sulindac

The methylsulfoxide group in sulindac is chiral. The two optical enantiomers are separable and equally active in vivo. The reversibility of the sulfoxide reduction justified the choice of using the racemic mixture, instead of one enantiomer, of sulindac in humans. Biochemically, the stereospecific oxidation of the sulfide metabolite by FAD-containing monooxygenase and cytochrome P$_{450}$ enzymes has been shown to produce the R(+) and S(−) enantiomers of sulindac, respectively (Light et al., 1982). Both enantiomers are readily reduced in many tissues back to the nonchiral sulfide. That means, regardless of which enantiomer of sulindac is given, after a short period the amount of two sulindac enantiomers and the sulfide metabolite in the body will not change significantly. The oxidation of sulindac to the stable sulfone metabolite is irreversible (Figure 4-3).

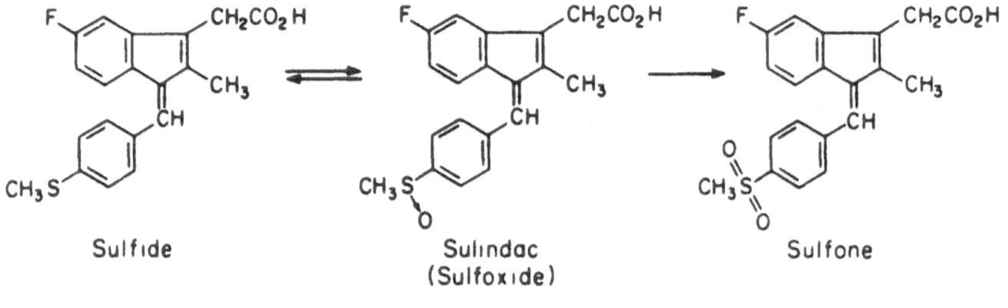

Figure 4-3. Metabolic transformations of sulindac.

The marked difference in aqueous solubility and partition coefficient between a hydrophilic sulfoxide molecule and its hydrophobic sulfide derivative mentioned earlier affects the tissue distribution of sulindac and its sulfide metabolite significantly. The insertion of the lipophilic sulindac sulfide, but not sulindac, into an artificial bilayer membrane system was clearly shown by differential scanning calorimetry and NMR studies. It implied that in vivo, sulindac tends to remain in the extracellular compartment whereas the active sulfide, like many lipophilic NSAIDs, diffuses through the lipid membrane and accumulates inside the inflammatory cells. This was indeed confirmed later with ^{3}H-labeled sulindac and ^{14}C-labeled sulfide in a mouse leukocyte system. The major urinary metabolites are the more polar sulindac, sulfone, and their glucuronides. Only a very small amount ($< 1\%$) of the sulfide metabolite is excreted through the kidney. The possible beneficial effect of such differential tissue distribution pattern on the level of renal prostaglandins will be discussed below.

Clinical Pharmacology Experiments

Sulindac is completely absorbed from the gastrointestinal tract. The facile interconversion of sulindac and its sulfide in vivo makes sulindac a reversible prodrug with some interesting pharmacodynamic properties in humans (Duggan et al., 1977a). As shown in a simplified diagram (Figure 4-4), sulindac, its sulfone metabolite, and a small amount of the sulfide metabo-

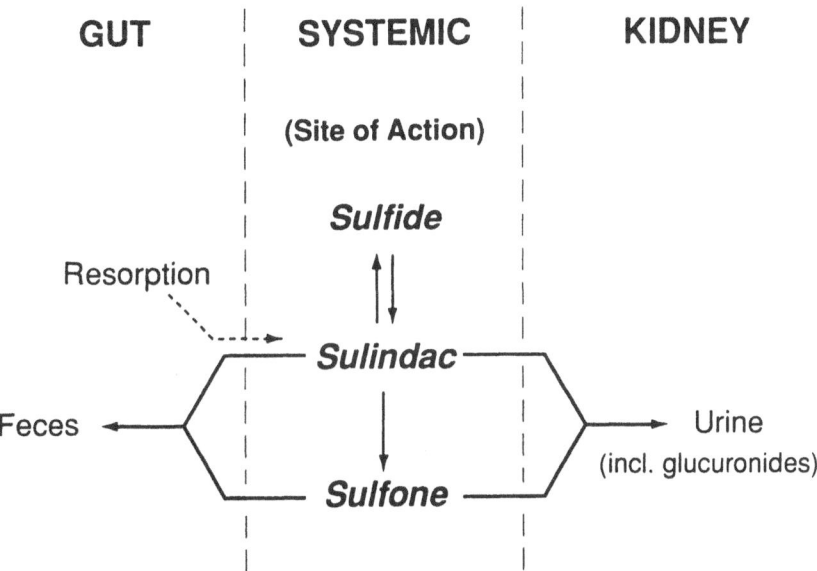

Figure 4-4. The pharmacodynamics of sulfide in humans.

lite are secreted in the bile and undergo extensive enterohepatic recycling. In the urinary excretion, approximately 20% of the dose appears as sulindac and its glucuronide and 15%–30% as sulfone and conjugates. Only negligible amounts of the sulfide metabolite are excreted in urine. The plasma concentrations of the sulfide reaches 7–8 μg/ml after a 300 mg dose and gradually declines with an effective half-life of > 12 hours (Figure 4-5). These data confirmed the preclinical laboratory findings and justified the desired bid (twice a day) dosing schedule, thus fulfilling one of our original objectives. More recently, important to any prodrug development, possible variations of the pharmacodynamics of sulindac in aged or renal failure patients have been investigated.

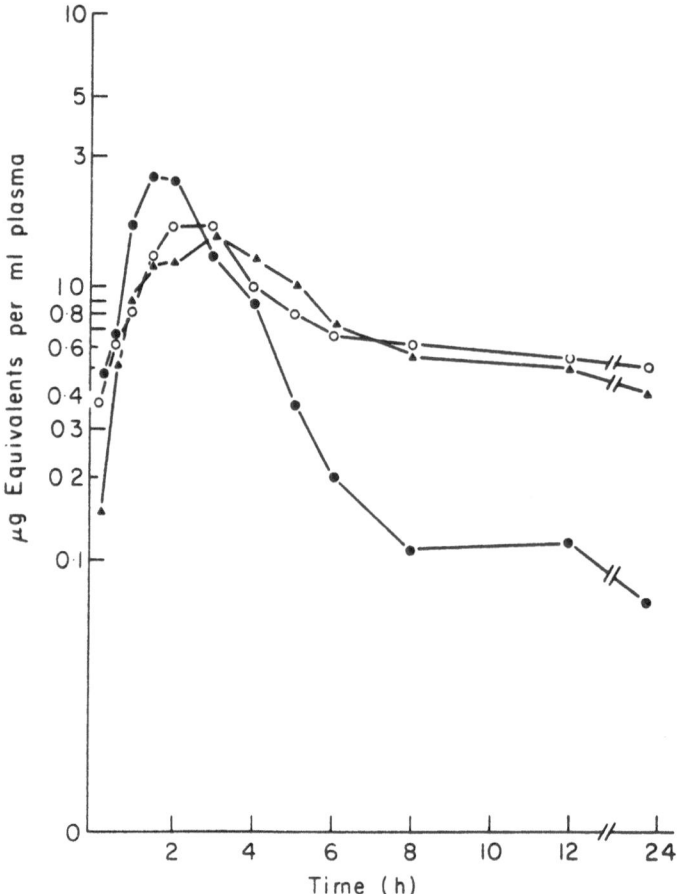

Figure 4-5. The mean plasma levels of sulindac (•), sulfide (○), and sulfone (▲).

The nonspecific inhibition of local prostaglandin synthesis by NSAIDs is a major contributor to their well-demonstrated gastrointestinal and renal toxicities. In theory, prodrug of NSAIDs which are irreversibly metabolized to the active species in vivo may reduce the initial exposure of topical GI tissues to active drugs but not the prolonged systemic drug effect, and therefore may be only partially effective in reducing GI irritations. With sulindac, several attempts have been made to correlate a differential distribution of sulindac and the active sulfide metabolite in the GI and renal tissues with local prostaglandin concentrations and side effects. So far, no definitive data have been published to show a clear difference in GI irritation after oral administration of sulindac or the active sulfide directly.

Premarketing Research Activities

Continuing their effort to determine the potential benefits of sulindac, research scientists at Merck laboratories also participated actively in the market introduction of this new drug. A well-coordinated effort supported by the Marketing Division, including collaborative research and clinical seminars, helped to launch this new drug in the mid 1970s worldwide. The reversible prodrug nature of sulindac kindled considerable interest in both rheumatologists and academic researchers, whose clinical pharmacological investigations generated much new data to confirm and extend our laboratory findings. The scientific rationale was also instrumental in facilitating new drug approval and favorable pricing from regulatory agencies in many countries including France, Italy, and Japan.

Clinical Evaluation

Five Claims and Overall Tolerance

At the time the clinical program of sulindac was planned, prior experience with the development of various NSAIDs and their clinical applications encouraged our medical colleagues to launch an ambitious full scale clinical trial which resulted the simultaneous allowance by the FDA of five claims, namely osteoarthritis, rheumatoid arthritis, ankylosing spondylitis, gouty arthritis, and acute painful shoulder. At 200–300 mg/day on a bid dosing schedule, sulindac (as Clinoril) is comparable to 50–150 mg/day of indomethacin on a three times a day schedule in efficacy but with reduced side effects, notably less gastrointestinal irritation and insignificant CNS disturbances. The GI side effects of NSAIDs, ranging from nausea and dyspepsia to ulceration and hemorrhage and their complex contributing factors, have been extensively investigated (Griffin et al., 1991; Henry et al., 1993; Savage et al., 1993). Overall, sulindac has been generally considered to be slightly better than several widely used NSAIDs, but not as safe as ibuprofen. The crystalluria propensity of its precursor, MK-715, was apparently circumvented. On the whole, the expectations from preclinical laboratory studies for this second-generation indomethacin were realized satisfactorily.

Long-Term Observations

While the medical acceptance of sulindac has been very gratifying, better understandings of its two potential side effects, nephro- and hepatotoxicity, which were even more difficult than gastrointestinal irritations to predict and modulate in laboratory experiments, were only clarified gradually after the long-term clinical data were accumulated and analyzed in the light of current knowledge.

HEPATOTOXICITY.   As summarized in a recent review by FDA investigators (Tarazi et al., 1993), the potential of producing hepatic injury is a class characteristic of NSAIDs, which are mostly hydrophobic aryl derivatives. Sulindac was selected for a comprehensive analysis first because several hundred cases of suspected hepatic injury in sulindac-treated patients had been reported to the FDA between 1978 and 1986, representing some 25% of the reported cases for all NSAIDs. Closer examination showed that three-quarters of the reports were considered inadequate or unconvincing for sulindac toxicity. The remaining 91 cases indicate that sulindac can lead to cholestatic or hepatocellular injury, most often because of immunological idiosyncrasy. The systemic hallmarks of hypersensitivity thus provide useful guidance for monitoring for potential hepatic injury during the first 4 to 8 weeks.

NEPHROTOXICITY.   Although the frequency, morbidity, and mortality of NSAID-induced nephrotoxicity are less serious than the gastropathy observed with many arthritic patients, it remains a clinical problem of some concern, especially in aged patients on chronic therapy (Stillman, 1989). The clinical manifestations of NSAID nephrotoxicity are complex, ranging from transient hemodynamic effects, acute interstitial nephritis to prolonged ischemic necrosis. The blockade of renal prostaglandin synthesis by NSAIDs contributes directly, or indirectly via immunological or hormonal disturbances, to these clinical syndromes. During our investigation of sulindac and congeners, several attempts to develop short-term animal models of papillary necrosis for compound selection were hampered by the complexity of the pathogenic mechanisms.

Renal prostaglandin synthesis occurs in both the cortex (producing mainly PGI_2) and the medulla (PGE_2). PGI_2 exerts a positive influence on glomerular filtration rate (GFR). PGE_2 inhibits the response to vasopressin and tubular reabsorption of sodium and chloride. It also increases renin release and potassium secretion. To minimize the multiple renal effects of various NSAIDs, in theory at least, some renal-sparing effects might be achieved by selective inhibition of the compartmentalized prostaglandin synthesis. In view of the pharmacodynamics of sulindac, considerable attention has been given to its effect on renal prostaglandin levels and functions by Drs. Carlo Patrono, M.J. Dunn (1990), and other investigators.

Less than 1% of the sulfide metabolite of sulindac is excreted in the urine. The conversion of sulfide to the inactive sulfoxide and sulfone by oxidative enzymes may occur prior to or in the kidney. In normal individuals, sulindac has less propensity to inhibit the production of cortical PGI_2 when compared with other NSAIDs. In a careful analysis by G. Cibattoni et al. (1987) of the urinary metabolites after chronic administration of sulindac in healthy subjects, the dose-dependent reduction (45%–85%) of extrarenal eicosanoids metabolites in the urine was not accompanied by a similar decrease of the renal metabolites, 6-keto-$PGF_{1\alpha}$ and TXB_2. It was suggested that the apparent differential inhibition, or renal-sparing effect, occurs only below a threshold plasma sulfide concentration, possibly between 4 and 9 $\mu g/ml$. Some discrepancies in observing this renal-sparing effect in different studies may be attributable to variations in the delicate analytical procedures and clinical protocols. In the past decade, some renal-sparing effect of sulindac in patients has been indicated. However, a clear correlation of the encouraging biochemical data with significant improvement of clinical safety remains a challenging task. Some individual variability in concentrations of urinary sulindac sulfide have been noted in 70 arthritic patients, possibly due to genetic and environmental factors (Brandli et al., 1991). Changing pharmacokinetics of sulindac and metabolites in patients with end-stage renal disease (Ravis et al., 1993), resulting in decreased plasma concentration of the active sulfide, may require higher doses for clinical efficacy (Gibson et al., 1987). Such findings further complicate the extrapolation from a laboratory concept of a reversible prodrug to significant clinical advantages in the complex nephrotoxicity area.

As a corollary, sulindac, distinct from many NSAIDs, was reported to exert little or no interference with the efficacy of hypertension treatment with diuretics, β-blockers, and angiotensin-converting enzyme inhibitors, whose antihypertensive actions involve renal prostaglandins (Oates, 1990).

New Developments

Cyclooxygenase-2 (COX 2)

The development of sulindac, as described above, was facilitated by the timely recognition of the arachidonic acid cyclooxygenase (COX) as a rational biochemical target for NSAIDs (Vane, 1971). However, without purified enzyme and its structural information, we could only postulate a hypothetical binding domain based on the experimental SAR in accordance with the x-ray structure of the rigid MK-715 (Figure 4-2), and a simple computer model of the substrate arachidonic acid (Figure 4-6) (Gund and Shen, 1977). These pictures provided a working model to assist the search for other cyclooxygenase inhibitors. To improve the selective toxicity of inhibitors of such an important and ubiquitous enzyme, our reversible prodrug strategy has produced some differential drug actions based on the pharmacodynamics of sulindac and its metabolites. The limiting factor of this approach is the uncertain pharmacodynamics due to species differences and heterogeneous patient populations.

In the early 1980s, following the elucidation of the 5-lipoxygenase pathway (Figure 4-7) and the roles of leukotrienes in inflammation and NSAID-induced gastrointestinal irritation, a new strategy was proposed to develop dual COX and 5-LO inhibitors as effective and less irritating NSAIDs (see *b* in Figure 4-7). As mentioned above, sulindac sulfide is a modest 5-LO inhibitor only. The clinical trials of several novel dual inhibitors from other laboratories and verification of the merits of this approach are currently in progress.

Finally, in this age of mechanism-based drug design, the long suspected, and hoped for, subtypes of cyclooxygenase have finally been characterized in the past three years (Figure 4-7). Cox-1 (prostaglandin H_2 synthase-1) is a constitutive enzyme which upon activation produces the antithrombogenic and cytoprotective PGI_2 by the endothelium and gastric mucosa. Inhibition of COX-1 by aspirin, indomethacin, and ibuprofen is associated with their well-known side effects. A second isozyme of cyclooxygenase, COX-2, is induced in many cell types and the synovium by stimuli to produce the pro-inflammatory mediators prostaglandins, thromboxane, etc. (Meade et al., 1993). Indomethacin and sulindac sulfide inhibit both COX-1 and -2. The inhibitory concentrations of different NSAIDs against each isozyme have been found to vary significantly with the source and preparation of the COX isozymes. Nevertheless, knowing the differences in the peptide sequence and x-ray structure of human COX-1 and -2 (Picot et al., 1994), and encouraged by the partial selectivity of some old NSAIDs, the stage is now set for the development of specific COX-2 inhibitors that would

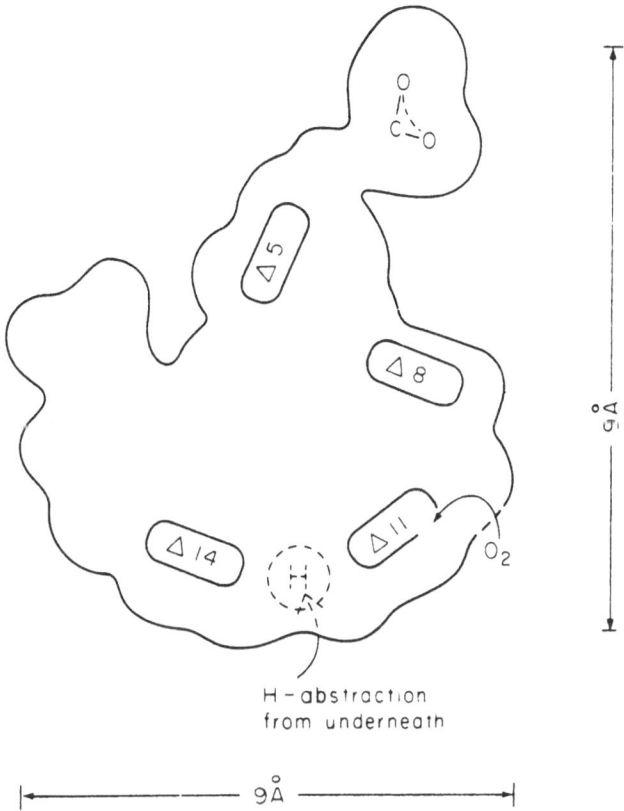

Figure 4-6. An early computer model of the hypothetical binding domain of arachidonic acid and inhibitors of cyclooxegenase.

be expected to be better tolerated by arthritic patients (Vane, 1994). Much progress has been made in many laboratories toward this goal (e.g., Masferrer et al., 1994). Not surprisingly, structures related to old cyclooxegenase inhibitors, including indomethacin analogs (Prasit, 1994), have provided some promising chemical leads.

Inhibition of Aldose Reductase

Diabetic neuropathy and retinopathy are two common complications of diabetes. Aldose reductase, which converts glucose to sorbitol in the polyol pathway, may be involved in the pathogenesis of both symptoms. Serendipitously, sulindac and its sulfide metabolite have been found by Abbott scientists and others to be effective inhibitors of the human lens aldose reductase in vitro at 0.4 and 0.8 μM, respectively (Chaudhry et al., 1983). The

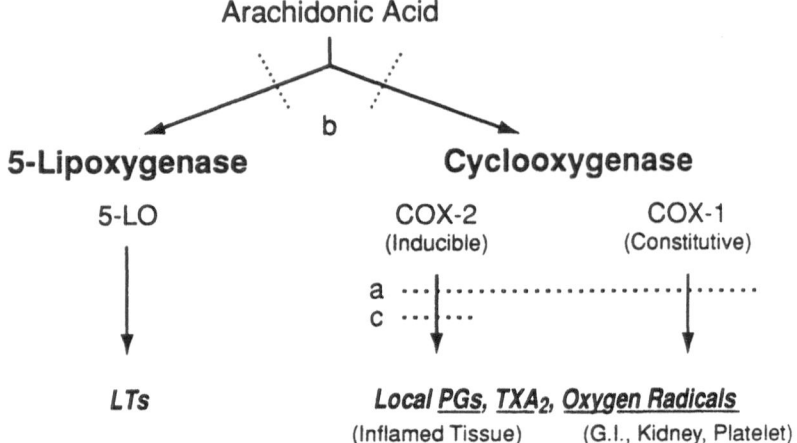

a: Nonspecific COX inhibitors (Aspirin, indomethacin)

b: Dual COX and 5-LO inhibitors

c: Selective COX-2 inhibitors

Figure 4-7. Different types of cyclooxegenase inhibitors.

potency of sulindac is comparable to those of two typical aldose reductase inhibitors, sorbinil and abrestatin. Indomethacin and aspirin showed very weak inhibitory activities at 10 μM, indicating a divergence of the structural requirements for cyclooxygenase and aldose reductase. In this case, sulindac is active per se; it forms two less-active sulfide and sulfone metabolites.

As an aldose reductase inhibitor with established anti-inflammatory and analgesic activities, sulindac was shown to be efficacious and safe in the therapy of diabetic neuropathy patients in preliminary clinical trials (Cohen and Harris, 1987). In a more recent investigation, the thickening of the retinal capillary basement membrane in diabetic cats was significantly decreased after treatment with sulindac for three months or longer.

Chemoprevention of Colorectal Cancer

Two recent population studies, involving 635,000 American and 11,000 Swedish rheumatoid arthritis patients, respectively, showed that frequent aspirin users had 40%–50% lower risk of incidence or death from cancers of the digestive tract, specifically of the stomach, colon, and rectum, than those who did not take aspirin or other NSAIDs. Although the possible contributions of other confounding factors as smoking, alcohol, diet, etc.

remain to be clarified, the question of potential cancer prevention with low dose of NSAIDs has stimulated much discussion recently (Boone et al., 1992).

In a more specific investigation, the beneficial treatment of colonic and rectal adenomas with sulindac in familial adenomatous polyposis has been firmly demonstrated (Giardiello et al., 1993). For example, in one study, after nine months of sulindac treatment at 300 mg/day the number and diameter of polyps decreased to 44% and 35% of baseline values, although the complete resolution observed in several previous studies was not confirmable (Spagnesi et al., 1994). The mechanism of adenoma regression is unknown (Parker et al., 1993). Inhibition of cell cycle progression from the G1 to the S phase and ornithine decarboxylase by sulindac or its metabolite(s) have been considered. Sulindac appeared to be more effective in reducing colonic polyps than indomethacin and other NSAIDs, again showing a dissociation of biological activities between indomethacin and its indene isostere in this unexpected and intriguing clinical effect. The direct effect of the sulfide and sulfone metabolites on polyps has not yet been clarified.

Conclusion

Drug discovery in the 1960s and 1970s was essentially a semi-empirical process. Incremental progress was generally made through a concerted effort of chemical synthesis and in vivo pharmacological assays, but novel ideas and serendipity also contributed to the success of any lead development. An element of luck was particularly important for the successful passage through the preclinical toxicological assessment and the satisfactory demonstration of efficacy and safety in the clinic, as the outcome of both are largely beyond the control of laboratory researchers. The discovery of sulindac as a new indomethacin congener with an improved safety profile illustrates such an interdisciplinary endeavor. Starting with a medicinal chemical approach based on some prior pharmacological findings, the evolution of a reversible prodrug concept was facilitated by some timely biochemical and pharmacodynamic advances. Broad clinical claims have been obtained with well-conducted clinical programs. However, the verification of potential clinical advantages of a reversible prodrug in heterogeneous patient populations proved to be a major challenge for many clinicians and researchers. Finally, the unexpected discovery of the aldose reductase inhibition and potential chemoprevention of colorectal adenoma by sulindac, distinct from other NSAIDs, shows that sulindac possesses a potential pharmacophore not recognized previously. Its perceived biochemical inactivity was defined only by the anti-inflammatory assays used. Some structures related to indomethacin have provided promising indole leads in the later search for novel lipoxygenase and COX-2 inhibitors. Many old drugs never die, but remain as intriguing and valuable probes in biomedical research.

Acknowledgments. Some 25 years after sulindac was first synthesized, the author again wishes to express his gratitude and appreciation of the important contributions, barely reflected in this brief review, made by his long-term chemical, biochemical, biological, pharmaceutical, medical, and marketing colleagues at Merck & Co., which together made sulindac a widely useful NSAID. Members of the former MSDRL management, particularly Drs. Max Tishler, L.H. Sarett, K.H. Beyer, and C.A. Stone, not only provided encouragement and wise counsel but also fostered an atmosphere of creative tension which made the final resolution of this crash program all the more gratifying. I also wish to acknowledge the subsequent moral support of our NSAID Task Force and the informative and stimulating interactions with prominent rheumatologists worldwide, whose enthusiasm in the application of a reversible prodrug to arthritic patients gave a real meaning to the laboratory exercise of drug discovery.

REFERENCES

Billingham MEJ, Davies GE (1979): Experimental models of arthritis in animals as screening tests for drugs to treat arthritis in man. In: *Anti-Inflammatory Drugs*, Vane JR, Ferreira SH, eds. New York: Springer-Verlag

Boone CW, Kelloff GJ, Steele VE (1992): Natural history of intraepithelial neoplasia in humans with implications for cancer chemoprevention strategy. *Cancer Research* 52:1651–1659

Brandli DW, Sarkissian E, Ng S-C, Paulus HE (1991): Individual variablity in concentrations of urinary sulindac sulfide. *Clin Pharmacol Ther* 50:650–655

Chaudhry PS, Cabrera J, Juliani HR, Varma SD (1983): Inhibition of human lens aldose reductase by flavonoids, sulindac and indomethacin. *Biochem Pharmacol* 32:1995–1998

Cibattoni G, Boss AH, Patrignani P, Catella F, Simonetti BM, Pierucci A, Pugliese F, Filabozzi P, Patrono C (1987): Effects of sulindac on renal and extrarenal eicosanoid synthesis. *Clin Pharmacol Ther* 41:380–383

Cohen KL, Harris S (1987) Efficacy and safety of nonsteroidal anti-inflammatory drugs in the therapy of diabetic neuropathy. *Arch Intern Med* 147:1442–1444

Duggan DE, Hare LE, Ditzler CA, Lei BW, Kwan KC (1977a): The disposition of sulindac. *Clin Pharmacol Ther* 21:326–335

Duggan DE, Hooke KF, Risley ER, Shen TY, Van Arman CG (1977b): Indentification of the biologically active form of sulindac. *J Pharmacol Exp Ther* 201:8–13

Egan RW, Gale PH, Vanden Heuvel WJA, Baptista EM, Kuehl FA Jr (1980): Mechanism of oxygen transfer by prostaglandin hydroperoxidase. *J Biol Chem* 255:323–326

Giardiello FM, Hamilton SR, Krush AJ, Piantadosi S, Hylind LM, Celano P, Booker SV, Robinson CR, Offerhaus GJ (1993): Treatment of colonic and rectal adenomas with sulindac in familial adenomatous polyposis. *New England J Med* 328:1313–1316

Gibson TP, Dobrinska MR, Lin JH, Entwistle LA, Davies RO (1987): Biotransformation of sulindac in end-stage renal disease. *Clin Pharmacol Ther* 42:82–88

Glavin GB, Sitar DS (1986): The effects of sulindac and its metabolites on acute stress-induced gastric ulcers in rats. *Toxicol Appl Pharmacol* 83:386–389

Gund P, Shen TY (1977): A model for the prostaglandin synthetase cyclooxygenas site and its inhibition by antiinflammatory arylacetic acids. *J Med Chem* 20:1146–1152

Griffin MR, Piper JM, Daugherty JR, Snowden M, Ray WA (1991): Nonsteroidal anti-inflammatory drug use and increased risk for peptic ulcer disease in elderly persons. *Ann Intern Med* 114:257–263

Hannah J, Ruyle WV, Jones H, Matzuk AR, Kelly KW, Witzel BE, Holtz WJ, Houser RW, Shen TY, Sarett LH (1977): The discovery of diflunisal. *Brit J Clin Pharmacol* 4:7s–13s

Henry D, Dobson A, Turner C (1993): Variablity in the risk of major gastrointestinal complications from nonaspirin nonsteroidal anti-inflammatory drugs. *Gastroenterology* 105:1078–1088

Hoogsteen K, Trenner NR (1970): Structure and conformation of cis and trans isomers of 1(p-chlorobenzylidene)-2-methyl-5-methoxyl-3-indenylacetic acid. *J Org Chem* 35:521–523

Light DR, Waxman DJ, Walsh C (1982): Studies on the chirality of sulfoxidation catalyzed by bacterial flavoenzyme cyclohexane monooxygenase and hog liver flavin adenine dinucleotide containing monooxygenase. *Biochem* 21:2490–2498

Mansour SZ, Hatchell DL, Chandler D, Saloupis P, Hatchell MC (1990): Reduction of basement membrane thickening in diabetic cat retina by sulindac. *Investigative Ophthalmol* 31:457–463

Masferrer JL, Zweifel BS, Manning PT, Hauser SD, Leahy KM, Smith WG, Isakson PC, Seibert K (1994): Selective inhibition of inducible cyclooxygenase 2 in vivo is antiinflammatory and nonulcerogenic. *Proc Natl Acad Sci USA* 91:3228–3232

Meade LA, Smith WL, DeWitt DL (1993): Differential inhibition of prostaglandin endoperoxide synthase (cyclooxygenase) isozymes by aspirin and other nonsteroidal antiinflammatiry drugs. *J Biol Chem* 268:6610–6614

Oates JA (1990): Cyclo-oxygenase inhibitors and blood pressure. Interaction with antihypertensive drugs. In: *Hypertension: Pathophysiology, Diagnosis, and Management*, Laragh LH, Brenner BM, eds. New York: Raven Press

Parker AL, Kadakia SC, Maccini DM, Cassaday MA, Angueria CF (1993): Disappearance of duodenal polyps in Gardner's Syndrome with sulindac therapy. *Amer J Gastroenterol* 88:93–94

Patrono C, Dunn MJ (1990): The clinical significance of renal prostaglandin synthesis. *Kidney Int* 32:1–12

Picot D, Loll PJ, Garavito RM (1994): The X-ray crystal structure of the membrane protein prostaglandin H2 synthase-1. *Nature* 367:243–249

Prasit P (1994): L-745,337, a selective COX-2 inhibitor. *Abstr. 208th National Meeting of Amer Chem Soc*, Washington, DC, August 21–25, 1994, MEDI 272

Ravis WR, Diskin CJ, Campagna KD, Clark CR, McMillian CL (1993): Pharmacokinetics and dialyzability of sulindac and metabolites in patients with end-stage renal failure. *J Clin Pharmacol* 33:527–534

Savage RL, Moller PW, Ballantyne CL, Wells JE (1993): Variation in the risk of peptic ulcer complications with nonsteroidal antiinflammatory drug therapy. *Arthritis Rheumatism* 36:84–90

Shen TY (1965): Synthesis and biological activity of some indomethacin analogs. In: *Non-Steroidal Antiinflamatory Drugs*, Garattini S, Dukes MNG, eds. New York: Excerpta Medica Foundation

Shen TY (1984): The proliferation of non-steroidal antiinflamatory drugs (NSAIDs). In: *Discoveries in Pharmacology*, Vol. 2, Parnham MJ, Bruinvels J, eds. Amsterdam: Elsevier/North Holland Biomedical Press

Shen TY (1985): Indomethacin, sulindac and their analogs. In: *Anti-Inflammatory and Anti-Rheumatic Drugs*, Rainford K, ed. Boca Raton, Florida: CRC Press

Shen TY, Winter, CA (1977): Chemical and biological studies on indomethacin, sulindac and their analogs. In: *Advances in Drug Research*, Harper NJ, Simmonds AB, eds. New York: Academic Press

Shen TY, Witzel BE, Jones H, Linn BO, McPherson J, Greenwald R, Fordice M, Jacobs A (1972): Synthesis of a new antiinflammatory agent, cis-5-fluoro-2-methyl-1-[p-(methylsulfinyl)-benzylidenyl]indene-3-acetic acid. *Fed Proc* 31:577

Spagnesi MT, Tonelli F, Dolara P, Caderni G, Valanzano R, Anastasi A, Bianchini F (1994): Rectal proliferation and polyp occurrence in patients with familial adenomatous polyposis after sulindac treatment. *Gastroenterology* 106:362–366

Stillman MT (1989): Interaction and selection of therapeutic agents in the elderly: NSAIDs and the aging kidney. *Scand J Rheumatology Suppl* 82:33–38

Tarazi EM, Harter JG, Zimmerman HJ, Ishak KG, Eaton RA (1993): Sulindac-associated hepatic injury: Analysis of 91 cases reported to the Food and Drug Administration. *Gastroenterology* 104:569–574

Van Arman CG, Risley EA, Nuss GW (1972): Pharmacologic properties of an anti-inflammatory agent 5-fluoro-2-methyl-1-(p-methyl sulfinyl benzylidene)-inden-3-yl acetic acid. *Fed Proc* 31:577

Vane JR (1971): Inhibition of prostaglandin synthesis as a mechanism of action for aspirin-like drugs. *Nature (New Biol)* 231:232

Vane J (1994): Towards a better aspirin. *Nature* 367:215–217

Winter CA, Shen TY, Tocco DJ, Robertson RT, Shackleford RW (1981): Diflunisal. In: *Pharmacological and Biochemical Properties of Drug Substances*, Vol. 3, Goldberg ME, ed. Washington, DC: American Pharmaceutical Association

5

The Discovery of Zileuton (Leutrol®)

Dee W. Brooks and George W. Carter

Introduction

Abbott Laboratories made a strong commitment and significant investment in expanding pharmaceutical discovery research in the early 1980s. Research programs were initiated with the goal to develop pharmaceuticals that would provide treatment for diseases with currently unmet needs. The generation of research conceptions and scientific discussions of potential strategies were outwardly invisible contrasted with the large capital investments in building and equipping laboratories and the aggressive recruitment of talented scientific staff to augment discovery research. These ideas and strategies when transformed by skillful experimentation would, it was hoped, lead to the discovery of novel pharmaceuticals. This report describes the research program initiated at Abbott in 1982 which led to the discovery of zileuton (Leutrol®), the first selective orally active 5-lipoxygenase inhibitor to demonstrate efficacy in humans.

Evolution of the Intervention Hypothesis

The metabolism of arachidonic acid results in the formation of a variety of molecules with potent biological actions. The oxidation of arachidonic acid by the enzyme cyclooxygenase initiates the biosynthetic pathway to the prostaglandins, prostacyclin, and thromboxanes. The large class of therapeutic agents referred to as nonsteroidal anti-inflammatory drugs (NSAIDs) effectively inhibit cyclooxygenase. The discovery of an alternative pathway of arachidonic acid oxidation (Murphy et al., 1979; Samuelsson, 1983) and the structural elucidation of these metabolites (Corey et al., 1980a, 1980b)

The Search for Anti-Inflammatory Drugs
Vincent J. Merluzzi and Julian Adams, Editors
© Birkhäuser Boston 1995

established the leukotriene biosynthetic pathway (Figure 5-1) and the chemical composition of slow-reacting substance of anaphylaxis (SRS-A), first described in 1938 from cobra venom-treated guinea pig lung tissue (Feldberg and Kellaway, 1938) and later from antigen challenge of sensitized guinea pig lung tissue (Kellaway and Trethewie, 1940). Numerous studies of SRS-A had occurred in the forty years leading up to these landmark delineations of the biosynthesis and chemical structure of the leukotrienes (Orange and Austen, 1969).

Arachidonic acid is normally esterified in membrane phospholipids and is released in response to some extracellular stimuli through the actions of phospholipases. 5-Lipoxygenase catalyzes the stereoselective addition of molecular oxygen to arachidonic acid to form 5S-hydroperoxy-6,8-*trans*-11, 14-*cis*-eicosatetraenoic acid (5-HPETE) (Samuelsson et al., 1987). A second catalytic step dehydrates 5-HPETE to form the reactive epoxide, leukotriene LTA_4 (Rouzer et al., 1986). Further enzymatic metabolism of this reactive epoxide involves stereospecific hydration by LTA_4 hydrolase to provide LTB_4 or conjugation with glutathione via a specific glutathione-S-transferase to produce LTC_4. Successive proteolytic cleavage steps convert LTC_4 to LTD_4 and LTE_4.

The formation of leukotrienes by leukocytes involved in inflammatory conditions and the activity of leukotrienes, which induce disease-like symptoms and pathophysiology, supports the premise that they are important mediators of inflammatory and allergic diseases (Lewis et al., 1990). One important factor was the different activity profile for the peptidyl leukotrienes, LTC_4, LTD_4, LTE_4, and the dihydroxy leukotriene, LTB_4 (Lewis and Austen, 1984). The peptidyl LTs have potent bronchoconstricting action, being 100–1000-fold more potent than histamine on human airways (Dahlen et al., 1983). The leukotrienes have been shown to increase mucus production (Marom et al., 1982). Peptidyl leukotrienes act as mediators of airway edema and affect vascular permeability (Lewis et al., 1990). LTB_4 is a potent inflammatory mediator via chemokinesis, chemotaxis, aggregation and degranulation effects on various leukocytes (Ford-Hutchinson et al., 1980).

Supporting evidence for the involvement of leukotrienes in inflammatory and allergic conditions has been discussed in numerous publications as the field has evolved. The list of potential indications with leukotriene involvement covers most inflammatory and allergic disorders. Considerable evidence suggests that leukotriene-induced responses contribute to the pathophysiology in asthma. Elevated levels of leukotrienes have been identified in biological fluids from patients with asthma (Drazen and Austen, 1987). Inhaled LTC_4 and LTD_4 in normal and asthmatic subjects were

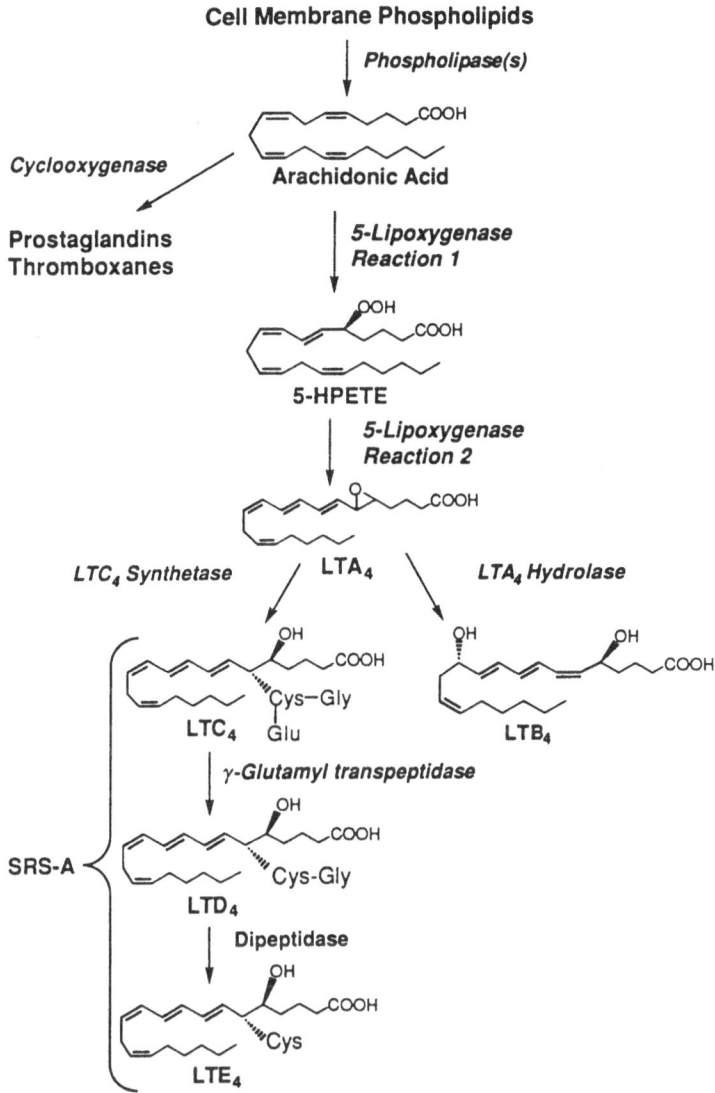

Figure 5-1. Leukotriene biosynthesis.

found to induce bronchoconstriction similar to that observed in asthmatic patients following antigen inhalation (Lewis et al., 1990). Leukotrienes have also been implicated in a variety of other inflammatory disorders, such as inflammatory bowel disease, rheumatoid arthritis, and psoriasis (Lewis et al., 1990). Despite all of this circumstantial evidence, confirmation of the pathophysiological role of these mediators requires selective blockage of leukotriene biosynthesis or receptor antagonism.

The discovery of clinically useful agents able to abrogate the effects of leukotrienes offered a new therapeutic intervention. Several strategies to specifically test this hypothesis were apparent. Inhibition of critical enzymes in the biosynthetic cascade was one strategy. Antagonizing the actions of leukotrienes at the receptor level provided another approach. At the beginning of this endeavor it was not clear which of the leukotrienes or their subsequent metabolites were important mediators of a particular pathological condition. Therefore, we favored an initial strategy which would prevent the effects of all the leukotrienes and their metabolites. Since the enzyme 5-lipoxygenase catalyzes the first step in the leukotriene biosynthetic pathway, selective inhibition of this enzyme provided a definitive means to limit the effects of all leukotrienes and their metabolites. We focused our investigation on the discovery of an orally active, selective 5-lipoxygenase inhibitor suitable for clinical study to evaluate the role of leukotrienes in human diseases.

5-Lipoxygenase

The biochemical characterization of 5-LO (arachidonate: oxygen 5-oxido-reductase EC1.13.11.34) proved to be challenging due to the unstable nature of the enzyme. 5-Lipoxygenase was first purified to homogeneity from rat basophilic leukemia cells (Goetz et al., 1985). It was subsequently purified, sequenced, cloned, and expressed from other sources including human polymorphonuclear leukocytes (Matsumoto et al., 1988; Rouzer et al., 1988). The amino acid sequence of both human and rat 5-LO have been determined, and there is 93% homology between these two enzymes, which consist of 674 amino acids with a molecular weight of 78 K (Dixon et al., 1988).

The enzyme was believed to reside in the cytosol until cell activation induces a calcium-dependent translocation to a putative cell membrane receptor (Rouzer and Kargman, 1988). Support for this translocation phenomena was demonstrated with the discovery of a novel 18 K membrane protein named 5-lipoxygenase activating protein (FLAP) (Miller et al., 1990). The presence of both 5-LO and FLAP was required for LT biosynthesis in intact cells (Dixon et al., 1990). This activation process provided another target amenable to drug intervention (Rouzer et al., 1990).

A hypothetical description of the catalytic reactions of 5-LO are shown in Figure 5-2. The first step, the oxidative catalysis, is derived by analogy from that proposed for the soybean 15-LO (Gibian and Gallaway, 1977). Both the 5- and 15-LO are known to contain a nonheme iron atom (Percival, 1991). No structure determination has been made for 5-lipoxygenase.

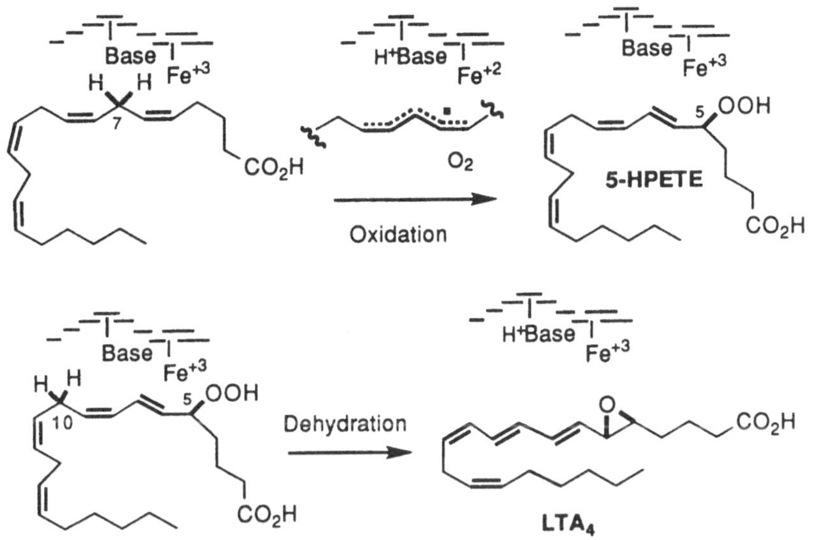

Figure 5-2. Catalytic reactions of 5-lipoxygenase.

Discovery Strategy

Since no structural information existed for 5-LO, the inhibitor design strategy involved intuitive medicinal chemistry guided by biological evaluation in leukotriene inhibition assays. The conceptual design of selective inhibitors of 5-LO was based on creating molecular entities with structural components complementary to one or more of the following hypothetical active-site entities: (1) an iron atom and accompanied ligands involved in oxygen transfer to the 5-alkene; (2) a basic group which removes the 7-pro-S hydrogen; (3) a lipophilic binding region which interacts with the polyunsaturated fatty acid chain of arachidonate; and (4) a coordinating binding site for a carboxylate.

Numerous inhibitors of 5-LO have been reported in the scientific and patent literature since the discovery of this enzyme (Batt, 1992; Musser and Kreft, 1992). Very few of the early inhibitors progressed to clinical trials due to problems with toxicity or lack of oral bioavailability. Another problem was that many of the inhibitors identified by random screening of compound libraries were nonselective antioxidants. The general wish list of desirable features sought in an inhibitor included the following: (1) selective and potent inhibition of 5-lipoxygenase; (2) reversible, dose-related inhibition; (3) oral bioavailability; (4) satisfactory duration of action; (5) no significant toxicity; (6) minimal metabolism; and (7) simple structure with a cost-effective synthesis.

Several different structural classes of inhibitors were investigated during the course of our hunt for a clinical candidate. Direct evaluation of 5-lipoxygenase catalysis was conducted using the supernatant from sonicated rat basophilic leukemia (RBL) cells (Jakschik et al., 1980) and measuring 5-HPETE formation via the reduced form 5-HETE (Carter et al., 1991). Results from this *in vitro* RBL assay provided leads that we elaborated into several inhibitor series over the period of 1984–1987: 2-aryl-1,3-diones (**1**) (Brooks et al., 1992c), pyrimido-pyrimidines (**2**) (Brooks and Basha, 1990), tetrahydro-2H-pyridazinones (**3**) (Brooks et al., 1992a, 1992b), triazinones (**4**) (Brooks et al., 1990), and 4-hydroxythiazoles (**5**) (Kerdesky et al., 1991). These avenues of investigation did not culminate in the identification of a clinical candidate.

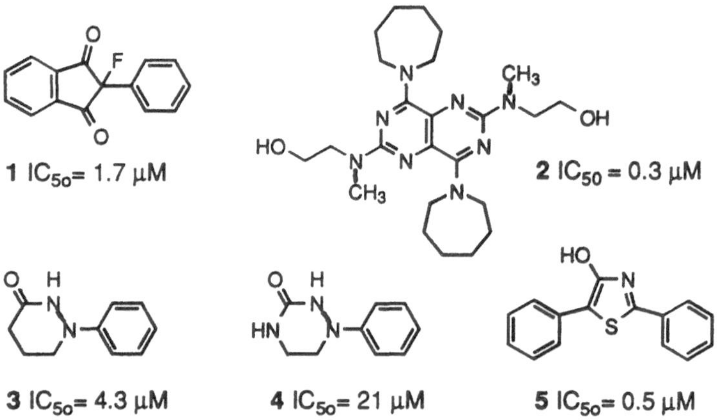

Figure 5-3. Early Abbott 5-lipoxygenase inhibitor leads.

Initial rational inhibitor design at Abbott (1982–1984) centered on using eicosanoid templates such as 15-HETE (Haviv et al., 1987). Others used eicosanoids as a starting point for the design of leukotriene receptor antagonists (Shaw and Krell, 1991). An alternative strategy specific for 5-LO took the first product, 5-hydroperoxy-6,8,11,14-eicosatetraenoic acid (5-HPETE), as a scaffold for inhibitor conception. Since the enzyme catalyzes the dehydration of 5-HPETE, this scaffold was both a product and a substrate of the same enzyme. Analogs were proposed as potential inhibitors by replacing the hydroperoxy group with alternative functional groups that might interfere with catalysis. This offered an intuitive strategy to design catalytic site inhibitors. Direct 5-LO inhibition was evaluated using the RBL assay. The data for representative examples are given in Figure 5-4. These results clearly demonstrated the preferred inhibitory properties of the hydroxamic acid function and further showed that inhibitory potency could be substantially enhanced by positioning the hydroxamate group on a 5-HETE template (Kerdesky et al., 1985, 1987). Optimal potency was achieved with a methylene spacer at the 5-position. This result was initially viewed as proprietary information within Abbott but was a very short-lived secret, as others had also independently discovered the inhibitory properties of hydroxamates.

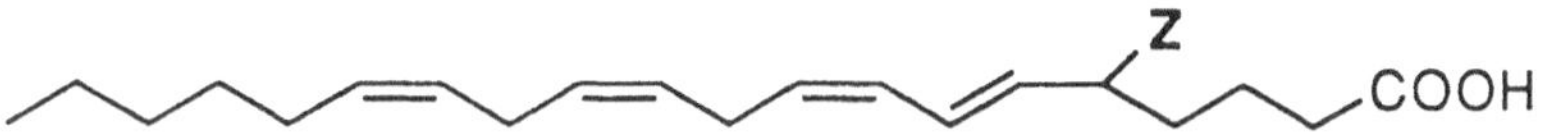

Compound	Z	IC_{50} μM[a]
6 5-HETE	OH	77
7	CH_2OH	29
8	$OCOCH_3$	68
9	SH	7.6
10	CH_2SH	11
11	NH_2	25
12	NHOH	21
13	=NOH	11
14	$=NNHCONH_2$	7
15	=O	18
16	CONHOH	1.4
17	$CH_2CONHOH$	0.19
18	$CH_2CH_2CONHOH$	2.8
19	CH_2COOH	96
20 5-HETE hydroxamic acid		2.1
21 arachidonohydroxamic acid		2.2
22 15-HETE		7.3
23 phenidone		2.1
24 BW-755c		1.3

[a] RBL 20,000 x g supernatant 5-lipoxygenase inhibition

Figure 5-4. 5-Lipoxygenase inhibition by 5-HPETE analogs

It was not surprising that several researchers derived the same logical starting point in the design of 5-LO inhibitors, which was the definition of molecular entities that would interact with a purported iron moiety in the active site. Compounds that contained functional groups that would bind to Fe^{3+} offered a source of potential inhibitors. Corey and co-workers provided the first published report outlining this strategy with the synthesis of arachidonohydroxamic acid and the discovery of its potent *in vitro* inhibition of 5-LO and further demonstrated the discovery of low molecular weight, nonlipid hydroxamates that were potent *in vitro* inhibitors (Corey et al., 1984). With the hydroxamate inhibitor concept generally divulged, the challenge remained to create a hydroxamate analog with properties sufficient to warrant clinical evaluation.

Pharmacological Optimization of Hydroxamate Inhibitors

The optimization challenge appeared to be fairly accessible because aryl-hydroxamates (Figure 5-5) were readily synthesized and found to be potent inhibitors of 5-LO activity *in vitro* in our broken cell, RBL assay (Summers et al., 1987b). The hydroxamate function was shown to be essential for inhibitory activity using 2-naphthylcarbohydroxamic acid **25** as a representative example with an IC_{50} of 14 μM. The corresponding functions of COOH, $CONH_2$, $CONHNH_2$, and $COCH_2OH$ led to inactive compounds. Replacing the hydroxyl of the hydroxamate with a methoxy group also led to a loss of inhibitory activity. Adding the methyl substitutent on the nitrogen atom of the hydroxamate led to an unexpected tenfold increase in inhibitory potency for hydroxamate **26**, $IC_{50} = 1.3$ μM. Adding an olefin to extend the planar aryl lipophilic array led to another order of magnitude improvement in potency providing the naphthylpropenehydroxamate **27**, $IC_{50} = 0.10$ μM. The biphenylhydroxamic acid **28**, $IC_{50} = 4.1$ μM was also an interesting starting point. By adding methyl substituents to force a perpendicular conformational alignment of the phenyl rings, significant enhancement in potency was observed as in the mesityl analog **29**, $IC_{50} = 0.29$ μM. Adding the conjugated olefin again improved potency providing **30** (A-61442), $IC_{50} = 0.022$ μM. At this time (April 1985), to our knowledge, A-61442 was the most potent 5-lipoxygenase inhibitor.

Evaluation of *In Vivo* Inhibition by Oral Administration

With novel, readily synthesized, small molecule inhibitors identified with potent *in vitro* inhibitory activity, the next stage of the drug discovery process was to evaluate *in vivo* activity. Few animal models existed at

Increasing Inhibitory Potency

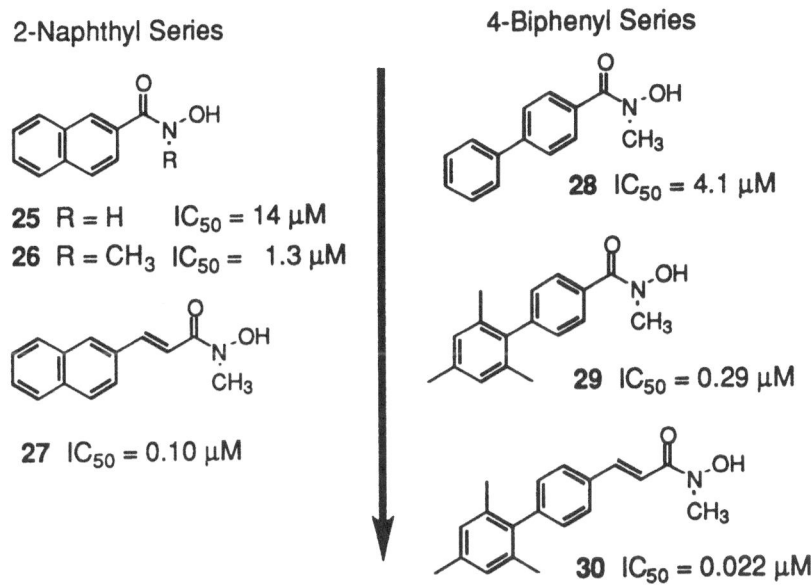

2-Naphthyl Series

4-Biphenyl Series

25 R = H IC_{50} = 14 µM
26 R = CH_3 IC_{50} = 1.3 µM

27 IC_{50} = 0.10 µM

28 IC_{50} = 4.1 µM

29 IC_{50} = 0.29 µM

30 IC_{50} = 0.022 µM

Figure 5-5. Enhancing inhibitor potency in simple arylhydroxamates.

the onset of this research program that specifically involved leukotriene mediated effects and at the same time offered adequate throughput for efficient evaluation of numerous synthetic compounds. The rat anaphylaxis model was developed for this purpose by adapting a method of inducing SRS-A formation in the peritoneal cavity of rats (Orange et al., 1968). Leukotrienes were quantitated from the peritoneal fluids collected in response to an induced antibody-antigen reaction (Young et al., 1991). Up to 1985, we found that very few reportedly potent *in vitro* inhibitors of 5-LO demonstrated oral activity in this model. Two widely used reference inhibitors were found to be orally effective in this model: **23** (phenidone, Blackwell et al., 1978) had an ED_{50} of 17 mg/kg, and **24** (BW 755c, Higgs et al., 1979) had an ED_{50} of 36 mg/kg. Inhibitors based on phenidone analogs were also studied at Abbott (Brooks et al., 1992a, 1992b) and Sterling (Hlasta et al., 1991), and the general observation of hematologic toxicity precluded clinical development of lead compounds.

Activity in this rat model allowed direct evaluation of leukotriene inhibition, which was dependent on a combination of parameters including: inhibitory potency, bioavailability, distribution to the site of measurement, and rate of clearance or metabolism to inactive compounds. It was very disappointing when A-61442 and other potent *in vitro* hydroxamate inhibitors did not exhibit satisfactory inhibition of leukotrienes in this model (Summers et al., 1987a). A-61442 (**30**) showed no significant *in vivo* inhibition of LTC_4 when orally dosed at 100 mg/kg. The less potent naphthylpropene-hydroxamate **27** did provide 66% LTC_4 inhibition at this high dose. Intravenous administration of 20 mg/kg of **27** to rats revealed a rapid hydrolysis of the hydroxamate function to the corresponding carboxylate with an estimated half-life of about five minutes. The inactive carboxylate had a half-life of about two hours. Similar results were found in dogs and cynomolgus monkeys. Pharmacokinetic analyses conducted in rats for several potent hydroxamate inhibitors demonstrated good absorption but rapid metabolic hydrolysis of the hydroxamate function to the corresponding inactive carboxylate. The explanation of poor *in vivo* leukotriene inhibition upon oral administration of these inhibitors was clearly due to rapid metabolic hydrolysis of the essential hydroxamate function.

The strategic decision was implemented to directly examine the time course of plasma levels of orally and intravenously administered inhibitors in rats. The goal was to generate data that might provide a better understanding of structural parameters influencing both the magnitude and time course of plasma inhibitor concentrations. Inhibitors were orally administered usually at 100 mg/kg, and small blood samples were taken from the tail vein at set time points. The concentration of parent drug and metabolites were determined by high-performance liquid chromatography (HPLC).

An interesting observation emerged from the analysis of numerous hydroxamate inhibitors (Figure 5-6). Structure-activity relationships revealed a general trend of slower rates of metabolic hydrolysis for 2-arylpropionic hydroxamates. One example was the nonsteroidal anti-inflammatory drug, ibuproxam **31**, a weak 5-LO inhibitor, $IC_{50} = 4.8 \mu M$, which appeared to be relatively resistant to metabolic hydrolysis when orally administered to the rat. This represented the birth of our realization of the importance of a methyl-branch-linking group connecting the lipophilic aryl template to the hydroxamate pharmacophore. In order to achieve this current level of oral bioavailability we had to sacrifice inhibitor potency elements such as the conjugated olefin link which suffered rapid metabolism. We chose to enhance the lipophilicity of the aryl template as a means to regain inhibitor potency for the orally bioavailable 2-arylpropionate series. Replacing the isobutyl group of ibuproxam with a phenylmethoxy substituent provided the inhibitor **32** (A-63788) with an IC_{50} of 0.28 μM. Oral administration of A-63788 at 100 mg/kg to rats gave mean plasma levels of 8 μM (1 h), 6 μM (3 h) and 4 μM (6 h). In the rat anaphylaxis model, A-63788 inhibited leukotriene biosynthesis with an ED_{50} of 40 mg/kg, which was comparable to the reference inhibitor BW 755c (24) but still well short of the potency one would desire for a clinical candidate. The hydrolysis of the hydroxamate appeared to be a severely limiting hurdle for this series of inhibitors. What at the beginning looked like a quick entry to potent inhibitors was now proving to be a more formidable challenge.

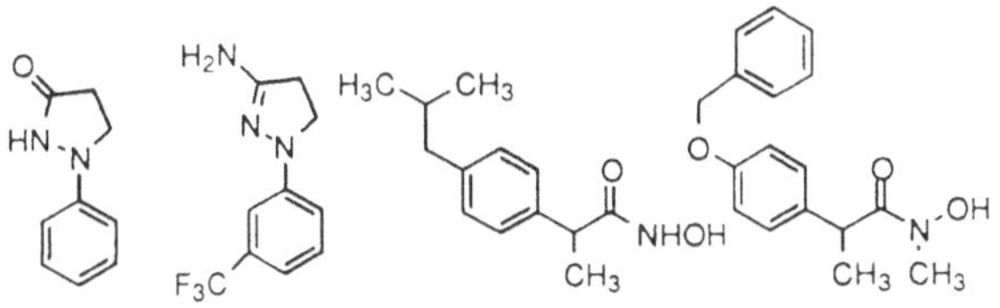

Assay	23 phenidone	24 BW 755C	31 ibuproxam	32 (A-63788)
RBL 5-LO IC_{50}	1.9 µM	2.0 µM	4.8 µM	0.28 µM
Rat anaphylaxis ED_{50}	17 mg/kg	36 mg/kg	0% at 100 mg/kg	40 mg/kg
Plasma levels of hydroxamate[a]			27 µM	8 µM
Plasma levels of carboxylate metabolite[a]			146 µM	51 µM

[a] measured at 1 hour after a 100 mg/kg oral dose in rats

Figure 5-6. Comparative evaluation of *in vitro* and *in vivo* activity for early 5-lipoxygenase inhibitors.

Optimization of *In Vivo* Inhibition of Hydroxamates

The propensity for metabolic hydrolysis of these hydroxamates would likely remain a limiting factor for any further development of this series since the carboxylates were inactive against 5-LO. A new approach was launched by suggesting a structural variation involving repositioning the lipophilic template-link group at nitrogen instead of at the carbonyl, as shown in Figure 5-7. The terms normal or type A hydroxamic acids and retro or type B hydroxamic acids were used to distinguish these two classes of hydroxamate inhibitors.

Most retro acetohydroxamates (type B) arising from this modification exhibited superior *in vivo* potency over their normal hydroxamate analogs (type A), which correlated with improved plasma concentration time courses after oral administration in the rat. A representative comparison between the normal analog, 32 (A-63788) and the corresponding retro analog, 33 (A-63162) is shown in Figure 5-7. Both compounds had similar *in vitro* inhibitory activity but the retro hydroxamate, A-63162 had greater

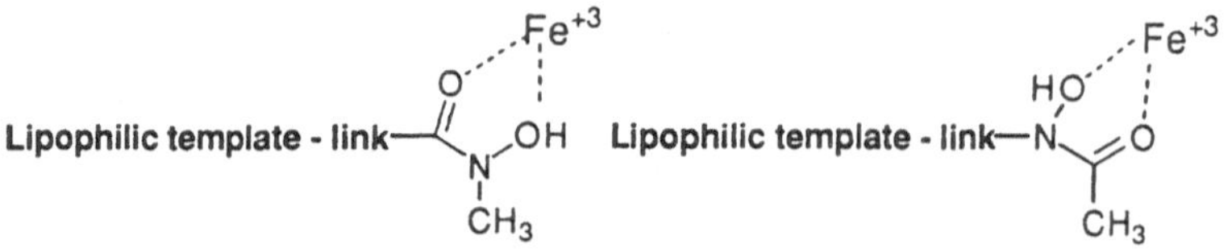

"Normal" Hydroxamate or Type A   "Retro" Hydroxamate or Type B

	32 (A-63788)	33 (A-63162)
Assay		
RBL 5-LO IC_{50}	0.28 μM	0.37 μM
Rat anaphylaxis ED_{50}	40 mg/kg	8 mg/kg
Plasma levels after oral dose in rats [a]	8 μM	110 μM
Plasma half life, 20 mg/kg iv		
rat	0.4 h	0.8 h
dog	0.5 h	0.9 h
monkey	0.3 h	0.7 h

[a] measured at 1 hour after a 100 mg/kg oral dose in rats

Figure 5-7. Orientation of groups on the hydroxamate function and a representative comparison of two examples, A-63788 and A-63162.

than tenfold improved plasma concentrations (110 μM, 1 h; 140 μM, 3 h; 95 μM, 6 h) following the same oral dose of 100 mg/kg in rats. More important to us was the fivefold increase in *in vivo* potency in the rat anaphylaxis model providing an ED_{50} of 8 mg/kg, the best we had seen to date for any inhibitor class. Thus, we had discovered that the structural orientation on the hydroxamate dramatically influenced the rate of metabolic hydrolysis (Summers et al., 1988b). However this represented only an incremental improvement in activity as we were somewhat disappointed by the relatively short plasma half-life found upon intravenous administration (Figure 5-7), about a twofold improvement over the normal hydroxamate congener. Further optimization of the series would require identification of other metabolic routes and structural modifications to prevent this metabolism.

A-63162 was synthesized with a ^{14}C at the methylene carbon of the benzyloxy substituent and evaluated in the rat. Radiolabeled hippuric acid (benzoyl glycine) accounted for almost 90% of the administered dose. This observation validated a suspected oxidative metabolism at the benzyloxy methylene leading to an unidentified intermediate which was subsequently cleaved to benzoic acid, then conjugated to glycine and excreted. Conjugation at the hydroxy group of the hydroxamate function with glucuronic acid was another important route of metabolism. In rats the glucuronide of A-63162 was excreted via the bile and cleaved back to the hydroxamate in the gastrointestinal tract, which was then reabsorbed. Preventing this enterohepatic recirculation in rats by bile duct cannulation resulted in a significantly shorter observed plasma duration for A-63162.

Addition of a methyl group at the benzyloxy site of oxidation provided A-63954, an inseparable mixture of diastereoisomers. As expected, this mixture exhibited four to fivefold longer plasma duration than A-63162 upon intravenous administration in rat, 4.4 h, and monkcy, 2.8 h. The *in vitro* inhibitory potency of the A-63954 mixture against the broken cell 5-LO was 0.5 μM and in the rat anaphylaxis model an ED_{50} of 27 mg/kg was observed. Extended duration of action was achieved but the potency of inhibition *in vivo* was not adequate.

The retro congener offered a renewed opportunity to further optimize the pharmacological properties of hydroxamate 5-LO inhibitors. A vigorous structure-activity analysis examined the template, link, and acyl group with respect to both *in vitro* and *in vivo* inhibition and pharmacokinetics (Summers et al., 1988a). From this investigation of hydroxamates, A-63162 remained the most potent *in vivo* inhibitor in the rat anaphylaxis model.

Discovery of 5-lipoxygenase inhibitors was a competitive arena with teams entered from numerous pharmaceutical companies. Independently and almost simultaneously, researchers at the Wellcome Research Laboratories developed a similar optimization strategy for hydroxamates (Garland and Salmon, 1991; Salmon and Garland, 1991). Their first compounds **34** (BW A137C) and **35** (BW A4C) (Figure 5-8) were published (Jackson et al., 1988) shortly after our report on **33** (A-63162) (Summers et al., 1988b). These investigators had also discovered the improved oral activity resulting from the retro hydroxamate orientation. One point of departure was in the selection of compounds to highlight. The Wellcome group selected BW A4C as a preferred inhibitor on the basis of the enhanced *in vitro* potency provided by the conjugated olefin link. We had previously relinquished the olefin link group for the improved metabolic stability provided by the methyl branch. The results with BW A4C in our inhibition assays (broken cell 5-LO, $IC_{50} = 0.14\,\mu$M and rat anaphylaxis, $ED_{50} = 11$ mg/kg) were very comparable to A-63162. The plasma half-life for BW A4C was approximately half that observed for A-63162 in dog and monkey but similar in the rat. Both research teams were intent on the advancement of a

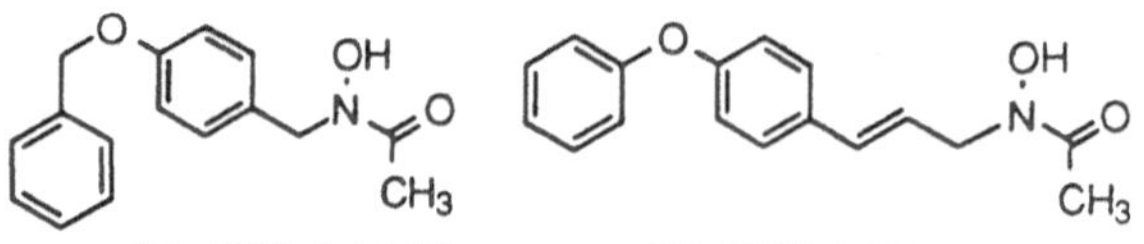

34 BW A137C   **35** BW A4C

Figure 5-8. Hydroxamate inhibitors discovered at Wellcome Research Laboratories.

hydroxamate clinical candidate, but the optimization process was far from complete.

N-Hydroxyurea Series of 5-LO Inhibitors

Expanding the structural diversity of inhibitors beyond the hydroxamates was an important aspect of the optimization process in view of the difficulties previously experienced. A logical strategy was to investigate bioisosteres of the hydroxamate pharmacophore that might provide potent inhibitors and offer additional resistance against hydrolytic, oxidative, and conjugative metabolism. A comparison of structural analogs of hydroxamates and the corresponding N-hydroxyureas revealed some interesting differences in biological properties. A representative comparison of simple naphthalene analogs was conducted as shown in Figure 5-9. The normal hydroxamate **36** and the retro hydroxamate **37** had similar *in vitro* potency against the broken cell 5-LO, IC_{50}'s of 0.59 and 0.54 μM, respectively. Also, as previously explained, the retro hydroxamate **37** had better *in vivo* activity in the rat anaphylaxis model than the normal congener **36**. The two regioisomers of the corresponding N-hydroxyureas were also compared. The normal or type A, N-hydroxyurea **38**, ($IC_{50} = 4.6$) was a significantly less active inhibitor, almost tenfold less potent than the hydroxamates. Making matters worse, no detectable plasma levels of **38** were found one hour after oral dosing in rats, and the observation of no activity *in vivo* was understandable. These disappointing observations might have led to a strategic decision to abandon N-hydroxyureas as potential inhibitors had this study come earlier in our program. However, armed with the knowledge of the retro hydroxamate modification, our curiosity guided us to examine the retro or type B, N-hydroxyurea **39**. The *in vitro* potency of the N-hydroxyurea **39** was slightly less than the hydroxamate **37**; however it was about twofold more effective *in vivo* with an ED_{50} of 10 mg/kg, which was comparable to A-63162, our best hydroxamate inhibitor. Thus the retro N-hydroxyurea became an exciting new pharmacophore for further optimization.

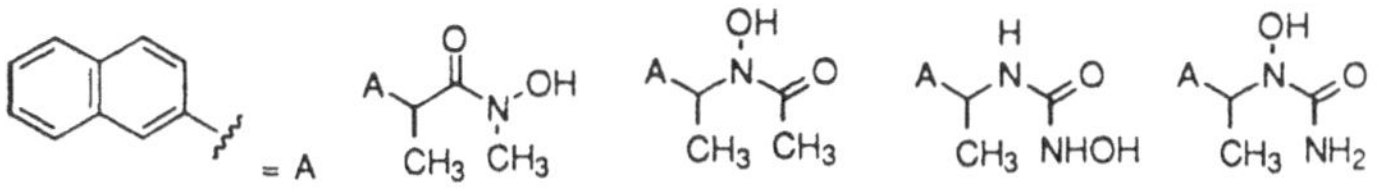

Assay	36	37	38	39
RBL 5-LO IC$_{50}$	0.59 μM	0.54 μM	4.60 μM	0.87 μM
Rat anaphylaxis ED$_{50}$	30 mg/kg	19 mg/kg	inactive	10 mg/kg
Plasma levels at 1 hour	59 μM[a]	65 μM[a]	not detected[b]	89 μM[b]

[a] measured after an oral dose of 437 μmol/kg in rats
[b] measured after an oral dose of 200 μmol/kg in rats

Figure 5-9. Comparison of naphthalene hydroxamate and N-hydroxyureas.

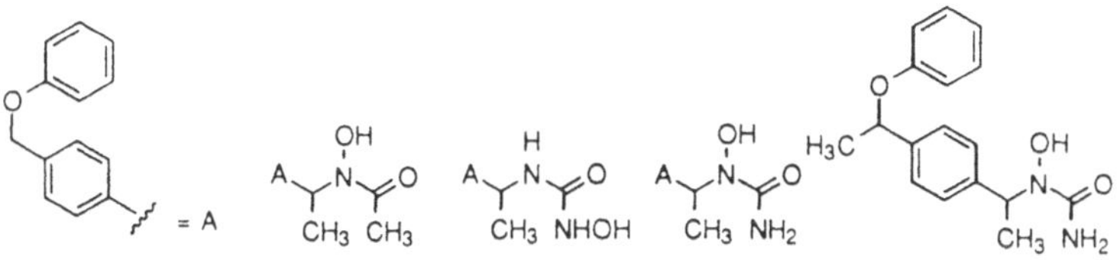

Assay	33(A-63162)	40	41	42
RBL 5-LO IC$_{50}$	0.37 μM	1.20 μM	0.42 μM	0.62 μM
Rat anaphylaxis ED$_{50}$	8 mg/kg	>60 mg/kg	7 mg/kg	9 mg/kg
Plasma levels[a]	67 μM	11 μM	46 μM	224 μM

[a] measured at 1 hour after a 200 μmol/kg oral dose in rats

Figure 5-10. Comparison of benzyloxyphenyl hydroxamates and N-hydroxyureas.

The N-hydroxyurea modification was applied to the benzyloxyphenyl template of our best hydroxamate, A-63162. A comparison of analogs is shown in Figure 5-10. Confirming the previous lesson, the normal N-hydroxyurea **40** was about threefold less potent *in vitro* than A-63162, had poor oral bioavailability in the rat and was inactive *in vivo* at a screening dose. The retro N-hydroxyurea **41** showed comparable *in vitro* and *in vivo* activity to A-63162, and the plasma half-life in the rat was about twofold longer for the N-hydroxyurea **41**. Being aware of the metabolism of the benzyloxy group in A-63162, we added the methyl substituent to block this, providing the N-hydroxyurea **42** (A-64666) as a mixture of diastereoisomers. This mixture showed comparable *in vitro* and *in vivo* inhibition potency to A-63162. One significant advantage of **42** was improved duration of plasma levels upon intravenous administration in several species:

rat, 5.6 h; dog, 3.3 h; and cynomolgus monkey, 2.0 h. A-64666 was the longest acting orally active 5-lipoxygenase inhibitor we had discovered at this time.

Various other combinations of hydroxylamine derivatives were examined as illustrated in Figure 5-11, but the simple *N*-hydroxyurea pharmacophore with X = O and Y = NH_2 offered the most promising inhibitors.

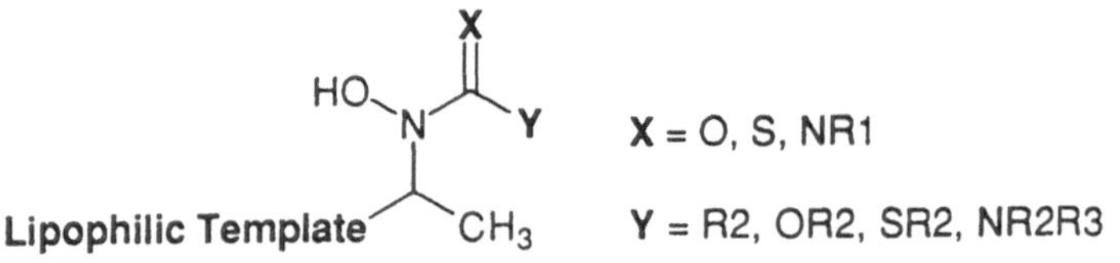

$$X = O, S, NR1$$

$$Y = R2, OR2, SR2, NR2R3$$

Figure 5-11. Hydroxylamine derivatives evaluated for 5-lipoxygenase activity.

Having defined the preferred pharmacophore, *N*-hydroxyurea, numerous combinations were envisioned for the template and link units in our simple descriptor model, Template–Link–Pharmacophore. Optimization of the pharmacological properties of this series was facilitated by a dedicated team effort providing high chemical throughput and rapid biological feedback. Figure 5-12 depicts an overview of the structure-activity plan. One part examined lipophilic aryl and heteroaryl templates building from our previous experience to maximize inhibitory potency. A second part examined the link group connecting the template and *N*-hydroxyurea. A third part examined substitution on the urea. Hundreds of *N*-hydroxyureas were evaluated.

The following general observations resulted from our experience with hundreds of hydroxamate and *N*-hydroxyurea inhibitors. The template served to provide the hydrophobic component for enzyme interaction as visualized by replacing that of the natural substrate arachidonate. As found in the hydroxamate series (Summers et al., 1990), increased hydrophobicity was closely correlated to increased inhibitory potency *in vitro*. However, specific structural parameters beyond mere lipophilicity were apparent which provided specific potency improvements. Aryl templates were often superior to saturated ring systems. Heterocyclic templates consisting of groups such as furan, thiophene, and fused systems such as benzofuran and benzothiophene provided more potent templates than nitrogen containing heterocycles. The linking group functioned to insulate the hydroxylamine derived pharmacophore from the aryl template to prevent putative arylhydroxylamine metabolites. In actuality the template and link group could not be optimized in isolation as they mutually defined the pharmacophore

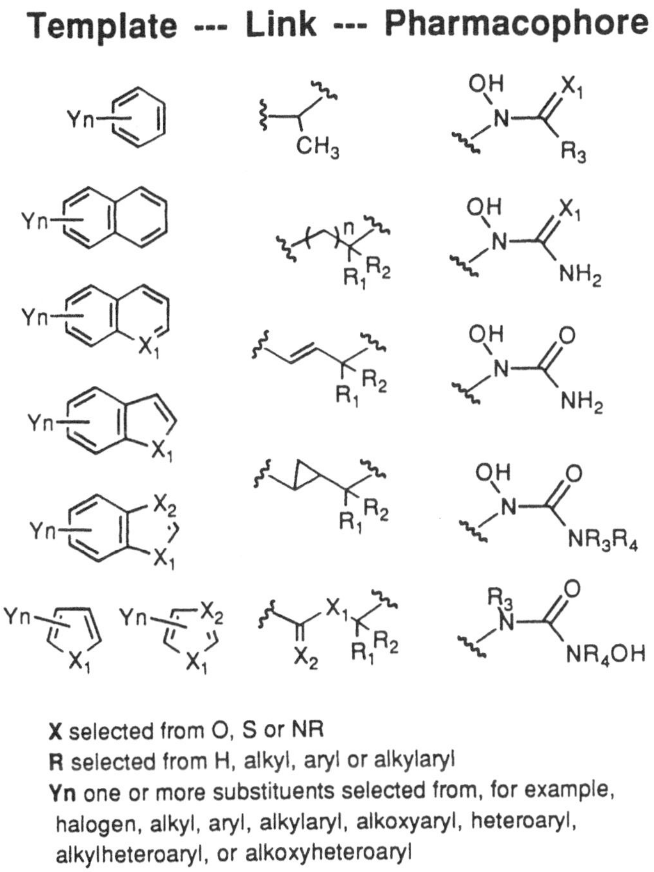
Template --- Link --- Pharmacophore
Yn
CH3
OH
N
X1
R3
NH2
R1
R2
n
O
X2
NR3R4
NR4OH
X selected from O, S or NR
R selected from H, alkyl, aryl or alkylaryl
Yn one or more substituents selected from, for example, halogen, alkyl, aryl, alkylaryl, alkoxyaryl, heteroaryl, alkylheteroaryl, or alkoxyheteroaryl

Figure 5-12. The structure-activity plan for optimization of the *N*-hydroxyurea series.

orientation. Optimization of potency involved matching both components to maximize van der Waals interactions in the inhibitor-enzyme complex and position the potential iron-binding pharmacophore for maximum interaction.

Chemistry

The structure-activity evaluation of *N*-hydroxyureas required the efficient synthesis of hundreds of analogs. The synthetic methods we found useful for the preparation of inhibitors are summarized in Figure 5-13. One useful synthetic route used the known method of reduction of oximes by pyridine:borane to the corresponding hydroxylamine, which was then transformed to the *N*-hydroxyurea by common procedures. In some cases where the oximes were not available or the reducing conditions were incompatible with substrate functionalities we had to innovate alternative synthetic methods. We studied the addition of aryl and heteroaryl organometallic groups to oxime ether intermediates to provide hydroxylamine precursors, but these methods were not broadly applicable and often provided poor yields of product (Basha and Brooks, 1987; Rodriques et al., 1988). Another interesting addition reaction involved O-tetrahydropyranylacetylnitrone, a reagent we invented to serve as an electrophilic hydroxylamine equivalent (Basha et al., 1991). Several unique heteroaryl templates could be incorporated into N-hydroxyurea inhibitors rapidly on a laboratory scale by this method. The most useful general method we developed for the synthesis of N-hydroxyureas converted alcohol substrates via a modified Mitsunobu reaction. The reagent we invented to accomplish this was N-O-bis(phenoxycarbonyl) hydroxylamine (Stewart and Brooks, 1992). In a simple, two-step procedure, N-hydroxyureas could be prepared from alcohol precursors, thus facilitating rapid synthetic throughput for inhibitor optimization.

Selection Criteria for a Clinical Candidate

At the start of our program in 1982, we did not expect to be first in the clinic with a selective orally active 5-LO inhibitor because we had entered this area later than many others. However, as our program persevered through the discovery hurdles and witnessed the failure of several series, we arrived at the first half of 1987 realizing that no selective, orally active 5-LO inhibitor had yet successfully cleared Phase I clinical trials. We believed that the opportunity to select the first clinical candidate that might validate the hypothesized importance of leukotrienes in human disease might be met by one of our preferred N-hydroxyurea or hydroxamate inhibitors. The question we faced was, "What biochemical, pharmacological, and pharmacokinetic data would justify the selection of our first 5-LO clinical development candidate?"

Biochemically, we desired an inhibitor with sub-micromolar inhibition that was reversible and selective for 5-lipoxygenase. Many of the N-hydroxyurea and hydroxamate compounds met these criteria with *in vitro* IC$_{50}$'s in the range of 0.1 to 1.0 μM. The N-hydroxyureas were generally more selective than hydroxamates against 5-LO compared to related enzymes: cyclooxygenase, 12-LO, and 15-LO. The ability of inhibitors to block *in vivo* leukotriene formation in the rat anaphylaxis model provided a means to more rigorously select potential candidates. Very few inhibitors

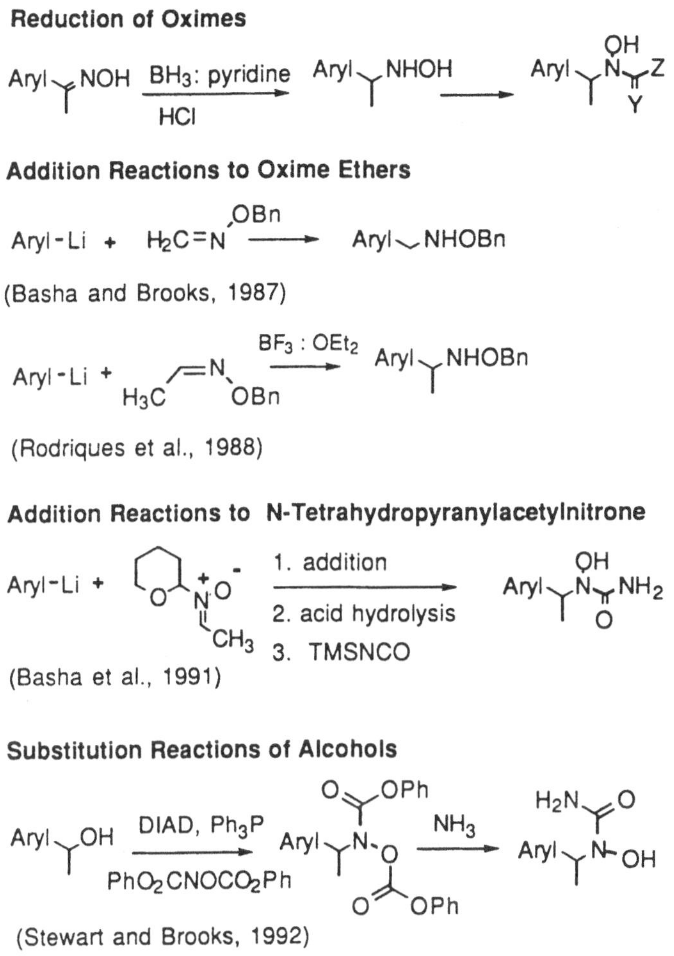

Reduction of Oximes

$$\text{Aryl} \underset{}{\overset{}{\diagdown}} \text{NOH} \quad \xrightarrow[\text{HCl}]{\text{BH}_3: \text{pyridine}} \quad \text{Aryl} \underset{}{\overset{}{\diagdown}} \text{NHOH} \quad \longrightarrow \quad \text{Aryl} \underset{}{\overset{\text{OH}}{\diagdown}} \text{N} \underset{Y}{\overset{}{\diagdown}} Z$$

Addition Reactions to Oxime Ethers

$$\text{Aryl-Li} + \text{H}_2\text{C}{=}\text{N} \diagup^{\text{OBn}} \longrightarrow \text{Aryl} \diagdown \text{NHOBn}$$

(Basha and Brooks, 1987)

$$\text{Aryl-Li} + \quad \overset{}{\underset{\text{H}_3\text{C}}{\diagup}}{=}\text{N} \diagdown_{\text{OBn}} \quad \xrightarrow{\text{BF}_3 : \text{OEt}_2} \quad \text{Aryl} \diagdown \text{NHOBn}$$

(Rodriques et al., 1988)

Addition Reactions to N-Tetrahydropyranylacetylnitrone

$$\text{Aryl-Li} + \quad \overset{}{\underset{\text{CH}_3}{\diagdown}} \quad \xrightarrow[\text{2. acid hydrolysis}]{\text{1. addition}} \quad \text{Aryl} \overset{\text{OH}}{\underset{\text{O}}{\diagdown}} \text{N} \diagdown \text{NH}_2$$

3. TMSNCO

(Basha et al., 1991)

Substitution Reactions of Alcohols

$$\text{Aryl} \diagdown \text{OH} \quad \xrightarrow[\text{PhO}_2\text{CNOCO}_2\text{Ph}]{\text{DIAD, Ph}_3\text{P}} \quad \text{Aryl} \underset{\underset{\text{OPh}}{\overset{\text{O}}{\diagdown}}}{\overset{\overset{\text{O} \diagdown \text{OPh}}{}}{\diagdown}} \text{N} \diagdown \text{O} \quad \xrightarrow{\text{NH}_3} \quad \text{Aryl} \overset{\text{H}_2\text{N} \diagdown \text{O}}{\diagdown} \text{N} \diagdown \text{OH}$$

(Stewart and Brooks, 1992)

Figure 5-13. Synthetic methods for *N*-hydroxyurea inhibitor synthesis.

provided oral ED_{50}'s better than 10 mg/kg in this model of antigen-antibody stimulated leukotriene formation.

Another useful method was *ex vivo* evaluation of leukotriene biosynthesis in blood samples over a time course after oral administration of inhibitors (Sweeney et al., 1987). For those few inhibitors that met the 10 mg/kg *in vivo* potency limit, we measured *ex vivo* leukotriene formation in blood samples stimulated by calcium ionophore A23187 taken over a time course after oral administration of inhibitor in several species, including rat, dog, and monkey. Plasma levels of inhibitor were also measured over the time course and compared to the *ex vivo* inhibition observed. Inhibition usually correlated closely with the plasma level of inhibitor. Both inactive and active metabolites could often be identified by HPLC analysis of these blood samples. The inhibitor with the best overall profile of potency in the rat anaphylaxis model and most effective *ex vivo* duration of inhibition in various species would likely be a potential candidate choice. Since, selective inhibition of 5-lipoxygenase was desired, cyclooxygenase products were also measured in these assays.

Leukotriene-dependent animal disease models were not well established and not yet validated as a test for a clinically effective leukotriene inhibitor. With this in mind, we did evaluate the most promising inhibitors in a few pharmacological paradigms. For anti-inflammatory properties, cell influx, edema formation, and leukotriene production were examined using a rat pleural Arthus reaction with oral administration of inhibitor. Also the arachidonic acid-induced ear edema mouse model was used to confirm anti-inflammatory activity. The *N*-hydroxyureas were among the first inhibitors to show effective oral activity in this model, which had normally been used with topical application of inhibitors. Two models of bronchospasm in the guinea pig were used to characterize orally administered inhibitors, one using intravenously administered arachidonic acid and the other using aerosolized antigen to elicit airway obstruction.

Another important aspect of the strategy for the selection of a clinical candidate was conducting pilot toxicity assessments. Previous lessons had taught about methemoglobinemia as the nemesis of other classes of inhibitors (Brooks et al., 1992a, 1992b). We also routinely requested evaluation in the Ames test. Promising inhibitors would be evaluated for acute toxicity in mice and rats. Further evaluation with large daily oral doses, often in the 400 mg/kg range, for a week or two in a small group of rats, was conducted to examine the toxic liabilities of various hydroxamate and *N*-hydroxyurea analogs and provide a means to crudely distinguish any structure-toxicity differences among inhibitors.

The value of the *N*-hydroxyurea pharmacophore was not realized until after *in vivo* evaluation and pharmacological comparisons of several inhibitor series were conducted. Exploration of heterocyclic templates (Figure 5-12) to replace the initially studied benzenoid groups proved to be important. The rationale was to provide flat, delocalized π-electron systems containing heteroatoms that might offer better bioavailability due to electron pair solvation or polarization effects while maintaining the overall lipophilic character required for effective inhibitory binding. The 2-benzo[b]thiophene was a good choice as a heteroaryl naphthalene template surrogate, and from our prior experiences this was matched to the methyl branch link. This prototype, *N*-1(1-benzo[b]thien-2-ylethyl)-*N*-

hydroxyurea, coded as A-64077, subsequently became Abbott's first 5-LO inhibitor clinical candidate, zileuton (Brooks et al., 1989; Carter et al., 1989). It is intriguing when looking retrospectively at all the medicinal chemistry and pharmacological evaluation conducted to support the selection of zileuton as a clinical candidate, that in fact it was one of the first N-hydroxyureas to be synthesized by our team.

Synthesis of Zileuton

The first synthesis of zileuton was accomplished by Bruce P. Gunn, in July 1986. An efficient synthesis of zileuton involved conventional synthetic procedures from 2-acetylbenzo[b]thiophene first by oxime formation followed by reduction to the hydroxylamine and finally urea formation using standard procedures, as outlined in Figure 5-14.

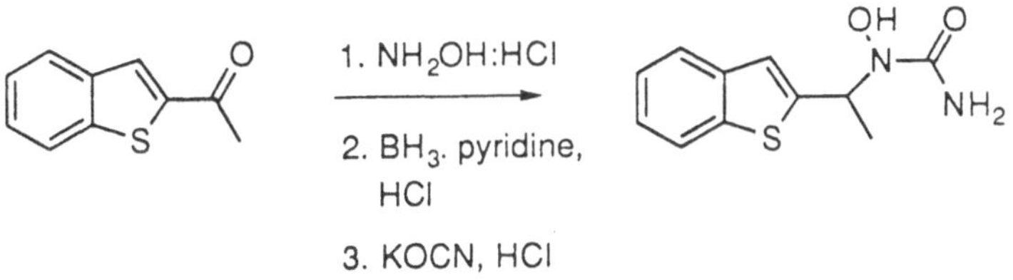

Figure 5-14. Synthesis of zileuton.

Characterization of Zileuton

The 5-lipoxygenase inhibitory activity of zileuton has been described in detail (Carter et al., 1991). Zileuton was effective in the desired submicromolar range in several biochemical assays of 5-LO inhibitory activity. In the cell lysate from rat basophilic leukemia cells, zileuton inhibited 5-HETE formation in a concentration-dependent manner. Zileuton's inhibitory potency was dependent on the arachidonic acid substrate concentration which would be expected of a competitive active-site inhibitor. In this assay, zileuton had IC_{50}'s of 92 nM, 400 nM, and 670 nM at 6 μM, 30 μM, and 65 μM arachidonate concentrations, respectively (Bell et al., 1993a). Zileuton inhibited calcium ionophore-stimulated LTB_4 formation in purified human neutrophils with an IC_{50} of 600 nM. This inhibition could be removed by washing cells previously treated with zileuton prior to ionophore stimulation. This served as additional evidence of the reversible nature of zileuton's inhibition. Zileuton inhibited leukotriene production by a variety of cell types such as rat peritoneal mast cells, guinea pig eosinophils, and guinea pig lung mast cells with IC_{50}'s from 0.5 to 1.0 μM.

The mechanism of 5-lipoxygenase inhibition by zileuton is not known and has been the subject of considerable speculation. A redox mechanism of inhibition is unlikely since no correlation was seen between the cyclic voltametric anodic peak potential and 5-lipoxygenase inhibitory potency for N-hydroxyureas. Current biochemical data would support the conclusion that zileuton is a direct, competitive, active-site inhibitor of 5-lipoxygenase.

Negligible inhibitory activity toward related enzymes such as human platelet 12-LO, soybean and reticulocyte 15-LO, and sheep seminal vesicle cyclooxygenase (CO) occurred with zileuton up to 100 μM. Arachidonate metabolism was evaluated in human whole blood stimulated with calcium ionophore. Zileuton selectively inhibited the biosynthesis of products from 5-lipoxygenase (LTB$_4$, IC$_{50}$= 0.7 μM) in comparison to 12-lipoxygenase (12-HETE, IC$_{50}$ = 30 μM), 15-lipoxygenase (15-HETE, IC$_{50}$ = 32 μM), and cyclooxygenase (TXB$_2$, IC$_{50}$ = 20 μM). Zileuton had no effect on myeloperoxidase activity, neutrophil degranulation, mast cell histamine release, and phospholipase A$_2$ activity (Bell et al., 1993a).

Zileuton showed preferred activity over hundreds of congeners in inhibiting leukotriene production in the rat peritoneal cavity after antigen challenge in passively sensitized animals. With a one-hour oral pretreatment time, zileuton had an ED$_{50}$ of 3 mg/kg. A single oral dose of 10 mg/kg achieved significant leukotriene inhibition for up to eight hours. Zileuton showed similar inhibitory potency for both LTE$_4$ and LTB$_4$ generation in the peritoneal cavity in this anaphylaxis model (Bell et al., 1993b). This would be expected for an inhibitor whose site of action is prior to the divergence of leukotriene classes, such as acting on 5-LO and not further downstream in the leukotriene biosynthesis pathway. In this model, a 70 mg/kg oral dose of zileuton did not inhibit the formation of the cyclooxygenase product TXB$_2$, further confirming the selective nature of the inhibition.

Zileuton inhibited cellular influx and edema formation in rats undergoing a pleural Arthus reaction. The compound exhibited anti-inflammatory activity in the mouse inhibiting arachidonic acid-induced ear edema with an oral ED$_{50}$ of 31 mg/kg. These pharmacological effects were accompanied by significant leukotriene inhibition without inhibition of cyclooxygenase products further supporting the selectivity of action.

Zileuton inhibited antigen-induced contraction of the guinea pig trachea but not contractions induced by other stimuli such as histamine or acetylcholine. The oral ED$_{50}$ for zileuton to inhibit arachidonic acid-induced bronchospasm in the guinea pig was 18 mg/kg (Malo et al., 1989). Zileuton was also effective against aerosolized antigen-induced bronchospasm with an ED$_{50}$ of 12 mg/kg (Bell et al., 1993a). Late phase bronchoconstriction and airway hyperresponsiveness in allergic sheep was significantly inhibited by pretreatment with zileuton (Abraham et al., 1992). A single oral dose of 10 mg/kg also inhibited eosinophil influx into the lung at early times after challenge. These results support the utility of zileuton in inhibiting both contractile and inflammatory effects of leukotrienes in pulmonary tissues.

One of the factors favoring the selection of zileuton was the consistently good oral bioavailability in a number of animal species. In the rat the compound was well absorbed, with 40% bioavailability and an plasma elimination half-life of 2.3 hours. It had an even better pharmacokinetic profile in the dog, with 88% bioavailability and an elimination half-life of 7 hours. The elimination half-life of 0.3 hours in the monkey was quite short. Metabolism studies in monkey using $[^{14}C]$-labeled zileuton revealed that the compound was readily absorbed after oral administration, but suffered extensive first-pass metabolism to produce two diastereomeric glucuronic acid conjugates, which were excreted in the urine (Machinist et al., 1989, unpublished results).

Zileuton inhibited the *ex vivo* production of LTB_4 in dog, monkey, and sheep (Figure 5-15). In the dog, a single 5-mg/kg oral dose provided complete inhibition of *ex vivo* LTB_4 formation throughout a 24-hour period. A similar long duration of biochemical action was observed in sheep (Abraham et al., 1992). In the monkey, effective but short duration of *ex vivo* inhibition was observed as expected, given the rapid elimination of zileuton in this species.

Clinical Results

In September 1987, after successful completion of the required animal safety studies, Phase I clinical trials were initiated with zileuton. The safety and pharmacokinetics of zileuton were evaluated using single rising oral doses up to 800 mg (Rubin et al., 1991). An important design element of the trial was to measure the time course of *ex vivo* leukotriene inhibition and drug plasma levels. The *ex vivo* methods used in our animal studies were modified for clinical application. Blood samples from subjects were collected at various times and stimulated with calcium ionophore A23187, and the plasma was analyzed for LTB_4. Zileuton was the first selective 5-lipoxygenase inhibitor to demonstrate dose-dependent inhibition of LTB_4 formation in humans (Figure 5-16). A single 800-mg dose provided significant LTB_4 inhibition for greater than 6 hours. An associative relationship between *ex vivo* LTB_4 inhibition and plasma drug levels in blood samples was confirmed. No inhibition of the cyclooxygenase-derived TXB_2 was found at the 800-mg dose. The major route of metabolism for zileuton in

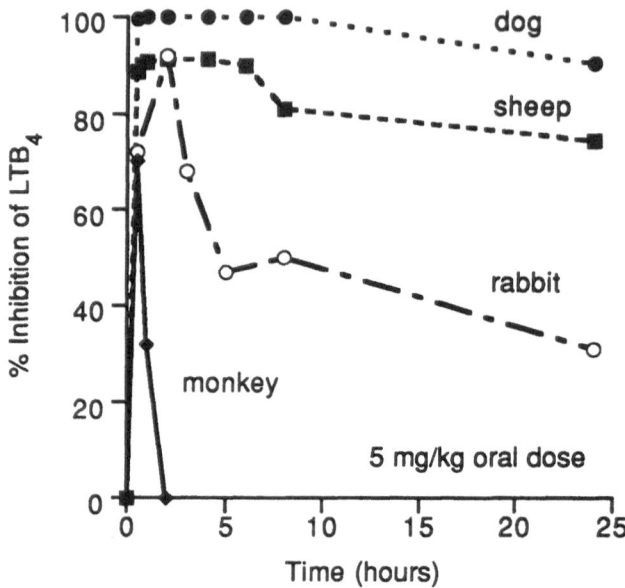

Figure 5-15. *Ex vivo* LTB$_4$ inhibition in calcium ionophore-stimulated blood samples from dog, sheep, rabbit, and monkey given 5-mg/kg oral doses of zileuton.

humans was found to be glucuronidation of the *N*-hydroxyurea function followed by elimination in the urine (Braeckman et al., 1989). The oral half-life for zileuton in humans was two to three hours. The metabolism and route of elimination of zileuton in humans appeared to be similar to that observed in cynomolgus monkeys; however, we were gratified to find that zileuton was longer lived in humans.

Zileuton cleared multiple-dose Phase I studies with no significant problems. In one study, zileuton was administered at 600 mg four times daily to healthy volunteers for fourteen days and blood samples were evaluated for *ex vivo* LTB$_4$ formation (Sirois et al., 1991). Zileuton inhibited *ex vivo* LTB$_4$ greater than 70% throughout the study. This inhibition was reversible after removal of the drug as *ex vivo* LTB$_4$ formation returned to control levels.

In early 1989, Phase II clinical trials were initiated to provide a preliminary evaluation of potential therapeutic applications for zileuton. In patients with ulcerative colitis, a single 800-mg dose provided significant reduction in LTB$_4$ levels in rectal dialysis fluid (Laursen et al., 1990). The first demonstration of efficacy for zileuton in ulcerative colitis was obtained with an oral dose of 800 mg twice daily for four weeks (Collawn et al., 1992). In allergic rhinitis patients, a single 800-mg oral dose pro-

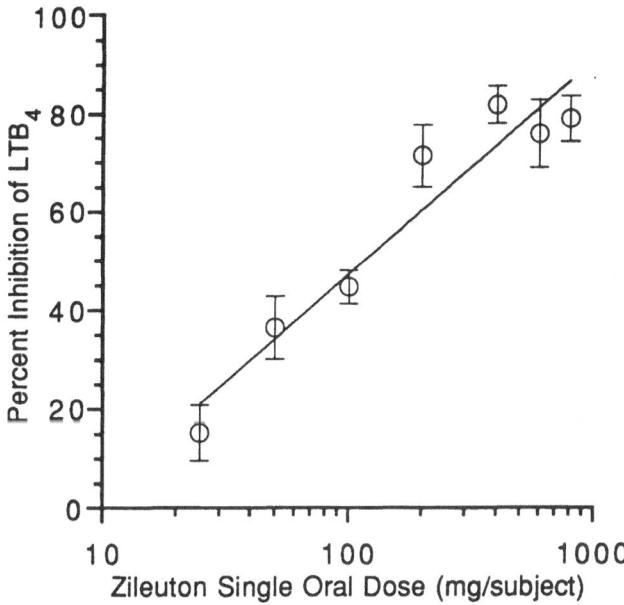

Figure 5-16. *Ex vivo* LTB_4 inhibition in calcium ionophore-stimulated blood samples from human volunteers after single oral doses of zileuton.

tected from a vigorous congestive response to intranasal antigen challenge (Knapp, 1990). Relief of symptoms was demonstrated in a small trial of patients with rheumatoid arthritis (Weinblatt et al., 1992). In an initial trial with asthmatics subjected to cold-air challenge, zileuton offered a significant protective effect against cold air induced bronchospasm (Israel et al., 1990).

The pathophysiological role of leukotrienes in asthma was clearly confirmed by the results from clinical studies with zileuton. A multicenter clinical trial with 129 mild to moderate asthmatics treated with zileuton administered 600 mg four times daily for four weeks provided significant improvement versus placebo in mean forced expiratory volume in a second (FEV_1) of 13.4%, improved morning peak expiratory flow rate by 10%, decreased daily β-agonist use by 24%, and improved symptom scores by 37% (Israel et al., 1993b). In addition, an acute improvement of airway obstruction was observed with the first 600-mg oral dose. The mean FEV_1 improved within 30 minutes and continued for the two-hour observation time with a maximum increase of 14.6% at one hour post dose. During the course of this study, excreted LTE_4 in urine was measured as an indication of whole body leukotriene inhibition. Zileuton given at 2.4 gram per day significantly decreased urinary LTE_4 levels by approximately 40% compared to the control lead-in values. Urinary LTE_4 levels remained unchanged in the placebo group. This method of evaluating leukotriene inhibition showed less blockade than the greater than 70% inhibition of *ex vivo* stimulated LTB_4 in plasma samples at the 2.4-gram daily dose. These results provided the first validation of efficacy for a 5-lipoxygenase inhibitor in asthma.

The potential pathological role of leukotrienes in aspirin- (ASA) sensitive asthmatics was investigated (Israel et al., 1993a). This subgroup of asthmatics upon ingestion of ASA or other cyclooxygenase inhibitors suffer bronchoconstriction and often additional symptoms involving naso-ocular, gastrointestinal, or dermal reactions. Previous studies (Christie et al., 1991) of ASA-sensitive asthmatics revealed sixfold higher basal levels of urinary LTE_4 compared to control asthmatics. After ASA ingestion, a fourfold increase in urinary LTE_4 levels occurred within three to six hours compared to placebo in ASA-sensitive asthmatics. Control non-ASA-sensitive asthmatic subjects' urinary LTE_4 levels were unaffected by ASA. A group of eight asthmatic patients were selected with known sensitivity to ASA and hyperexcretion of urinary LTE_4. In a double-blind crossover trial the effect of zileuton given orally 600 mg four times daily for six to eight days versus placebo was evaluated against ASA challenge. This group of asthmatic subjects had predominantly normal baseline pulmonary function (FEV_1 greater than 80%). Not surprising, ingestion of zileuton did not significantly alter baseline pulmonary function. Patients given placebo when challenged by ASA suffered a decrease in FEV_1 of 18.6% from pre-ASA measurements. Zileuton treatment blunted this ASA-induced decrease in FEV_1 to 4.4% and reduced the mean maximal urinary LTE_4 levels after ASA challenge by 68%. Zileuton treatment also prevented all nasal, gastrointestinal, and dermal responses to ASA challenge, whereas the placebo-treated subjects challenged with ASA had a 2.5-fold increase in nasal symptoms, five of eight patients showed gastrointestinal symptoms and three of eight patients showed angioedema. The results of this small trial clearly support the involvement of leukotrienes as major pathological mediators in ASA-induced asthma.

These clinical studies support the hypothesis of chronic overproduction of leukotrienes in asthmatics. Leukotriene biosynthesis inhibitors thus offer rapid improvement by blocking this chronic formation of pathological leukotriene mediators. From the promising results achieved thus far, it is likely that 5-lipoxygenase inhibitors will offer additional benefits in the treatment of asthma, since existing therapies do not block leukotriene formation.

Conclusion

Our multidisciplinary discovery research team contributed productive, intuitive, medicinal chemistry and focused pharmacological evaluation to the successful achievement of the discovery of zileuton. The Immunoscience Venture Team at Abbott created and guided the development activities of zileuton. The clinical studies with zileuton have provided the first evidence of efficacy for an orally active selective 5-LO inhibitor. The benefits of leukotriene blockade by zileuton in asthma have been confirmed in controlled Phase III clinical trials (Israel et al., 1994). The filing of the NDA for zileuton (Leutrol®) with the US Food and Drug Administration is scheduled for 1994. Future clinical studies with zileuton will likely define other therapeutic applications for this new class of drugs in treating inflammatory and allergic diseases. From this competitive area of drug discovery research by many pharmaceutical companies, several new chemical entities that inhibit leukotriene biosynthesis or block leukotriene receptors have progressed to the stage of clinical evaluation (Brooks, 1994). At this time, no entity operating by this specific modality has been approved yet for marketing as a treatment for human disease.

Acknowledgments. The collective contributions of the skillful, dedicated members of the 5-Lipoxygenase Project Team made the discovery of zileuton possible. Dr. Paul Rubin and Dr. Louis Dubé led the Immunoscience Venture Team responsible for the clinical development of zileuton. Recognition is extended to all the unnamed contributors that supported the discovery and development of zileuton from many functions within Abbott Laboratories. We thank the physicians who conducted the clinical research and recognize the many patients who volunteered to participate in the clinical trials. We are very appreciative of the inspiration and encouragement provided by Dr. K. Frank Austen over the course of this discovery adventure.

REFERENCES

Abraham W, Ahmed A, Cortes A, Sielcsak M, Hinz W, Bouska J, Lanni C, Bell R (1992): The 5-lipoxygenase inhibitor zileuton blocks antigen-induced late airway responses, inflammation and airway hyperresponsiveness in allergic sheep. *Eur J Pharmacol* 217:119–126

Basha A, Brooks DW (1987): Reaction of O-benzyl-N-methylenehydroxylamine with organolithium compounds, a CH_2^+-NH^+ equivalent. *J Chem Soc Chem Commun* 305–306

Basha A, Ratajczyk JD, Brooks DW (1991): Addition of organolithium compounds to N-THP protected nitrone. *Tetrahedron Letters* 32:3783–3786

Batt DG (1992): 5-Lipoxygenase inhibitors and their anti-inflammatory activities. In: *Progress in Medicinal Chemistry*, Ellis GP, Luscombe DK, eds. Amsterdam: Elsevier

Bell RL, Lanni C, Malo PE, Brooks DW, Stewart AO, Hansen R, Rubin P, Carter GW (1993a): Preclinical and clinical activity of zileuton and A-78773. *Ann N Y Acad Sci* 696:205–215

Bell RL, Young PR, Albert D, Lanni C, Summers JB, Brooks DW, Rubin P, Carter GW (1993b): The discovery and development of zileuton: An orally active 5-lipoxygenase inhibitor. *Int J Immunopharmac* 14:505–510

Blackwell GJ, Flower RJ (1978): 1-Phenyl-3-pyrazolidone: An inhibitor of cyclo-oxygenase and lipoxygenase pathways in lung and platelets. *Prostaglandins* 16:417–425

Braeckman RA, Granneman R, Rubin PR, Kesterson JW (1989): Pharmacokinetics and metabolism of the new 5-lipoxygenase inhibitor A-64077 after single oral administration in man. *J Clin Pharmacol* 29(9): Abst 22

Brooks DW (1994): Progress with investigational drugs for the treatment of pulmonary and inflammatory diseases. *Expert Opin Invest Drugs* 3:185–190

Brooks DW, Albert DH, Dyer RD, Bouska JB, Young P, Rotert, G, Machinist JM, Carter GW (1992a): 1-Phenyl-[2H]-tetrahydropyridazin-3-one, A-53612, a selective orally active 5-lipoxygenase inhibitor. *Bioorganic & Medicinal Chemistry Letters* 2:1353–1356

Brooks DW, Basha A (1990): Pyrimido-pyrimidine 5-lipoxygenase inhibitors. US Patent 4,963,541

Brooks DW, Basha A, Gunn, BP, Bhatia, PA (1990): Triazinone lipoxygenase compounds. US patent 4,970,210

Brooks DW, Basha A, Kerdesky FAJ, Holms JH, Ratajczyk, JD, Bhatia PA, Moore JL, Martin JG, Schmidt SP, Albert DH, Dyer RD, Young P, Carter GW (1992b): Structure-activity relationships of the pyridazinone series of 5-lipoxygenase inhibitors. *Bioorganic & Medicinal Chemistry Letters* 2:1357–1360

Brooks DW, Schmidt SP, Dyer RD, Young PR, Carter GW (1992c): Structural analysis of 2-aryl-1,3-dione compounds as inhibitors of 5-lipoxygenase. *Bioorganic & Medicinal Chemistry Letters* 2:1309–1314

Brooks DW, Summers JB, Gunn BP, Rodriques KE, Martin JG, Martin MB, Mazdiyasni H, Holms JH, Stewart AO, Moore JL, Young PR, Albert DH, Bouska JB, Malo PE, Dyer RD, Bell RL, Rubin P, Kesterson J, Carter GW (1989): The discovery of A-64077, a clinical candidate for treating diseases involving leukotriene mediators. *International Chemical Congress of Pacific Basin Societies*, Honolulu, Abstract BIOS 34

Carter GW, Young PR, Albert DH, Bouska J, Dyer R, Bell RL, Summers JB, Brooks DW (1991): 5-Lipoxygenase inhibitory activity of zileuton. *J Pharmacol Exp Ther* 256:929–937

Carter GW, Young PR, Albert DH, Bouska J, Dyer R, Bell RL, Summers JB, Brooks DW, Rubin P, Kesterson J (1989): A-64077, a new orally active 5-lipoxygenase inhibitor. In: *Leukotrienes and Prostanoids in Health and Disease, New Trends in Lipid Mediators Research*, Zor U, Naor Z, Danon A, eds. Basel: Karger

Christie PE, Tagari P, Ford-Hutchinson AW, Charlesson S, Chee P, Arm JP, Lee TH (1991): Urinary leukotriene E$_4$ concentrations increase after aspirin challenge in aspirin-sensitive asthmatic subjects. *Am Rev Respir Dis* 143:1025–1029

Collawn C, Rubin P, Perez H, Bobadilla J, Cabrera G, Reyes E, Borovoy J, Kershenobich D (1992): Phase II study of the safety and efficacy of a 5-lipoxygenase inhibitor in patients with ulcerative colitis. *Am J Gasteroenterol* 87:342–346

Corey EJ, Cashman JR, Kantner SS, Wright SW (1984): Rationally designed, potent competitive inhibitors of leukotriene biosynthesis. *J Amer Chem Soc* 106:1503–1504

Corey EJ, Clark D, Goto G, Marfat A, Mioskowski C, Samuelsson B, Hammarstrom S (1980a): Stereospecific total synthesis of a "slow reacting substance" of anaphylaxis, leukotriene C-1. *J Amer Chem Soc* 102:1436–1439

Corey EJ, Marfat A, Goto G, Brion F (1980b): Leukotriene B: Total synthesis and assignment of stereochemistry. *J Amer Chem Soc* 102:7984–7985

Dahlen S-E, Hansson G, Hedquist P, Bjorck T, Granstrom E, Dahlen B (1983): Allergen challenge of lung tissue from asthmatics elicits bronchial contraction that correlates with the release of leukotrienes C$_4$, D$_4$, and E$_4$. *Proc Natl Acad Sci USA* 80:1712–1716

Dixon RAF, Diehl RE, Opas E, Rands E, Vickers PJ, Evans JF, Gillard JW, Miller DK (1990): Requirement of a 5-lipoxygenase-activating protein for leukotriene synthesis. *Nature* 343:282–284

Dixon RAF, Jones RE, Diehl RD, Bennet CD, Kargman S, Rouzer CA (1988): Cloning of the cDNA for human 5-lipoxygenase. *Proc Natl Acad Sci USA* 85:416–420

Drazen JM, Austen KF (1987): Leukotrienes and airway responses. *Amer Rev Resp Dis* 136:985–98

Feldberg W, Kellaway CH (1938): Liberation of histamine and formation of a lecithin-like substance by cobra venom. *J Physiol* 94:187–226

Ford-Hutchinson AW, Bray MA, Dorg MV, Shipley MF, Smith MTH (1980): Leukotriene B$_4$: A potent chemokinetic and aggregation substance released from polymorphonuclear leukocytes. *Nature* 286:264–265

Garland LG, Salmon JA (1991): Hydroxamic acids and hydroxyureas as inhibitors of arachidonate 5-lipoxygenase. *Drugs of the Future* 16:547–558

Gibian MR, Galaway RA (1977): In: *Bio-organic Chemistry*, van Tamlen EE, ed. Vol. 1, pp. 117–136. New York: Academic Press

Goetz AM, Faver L, Bouska J, Bornemeier D, Carter GW (1985): Purification of a mammalian 5-lipoxygenase from rat basophilic leukemia cells. *Prostaglandins* 29:689–701

Haviv F, Ratajczyk JD, DeNet RW, Martin YC, Dyer RD, Carter GW (1987): Structural requirements for the inhibition of 5-lipoxygenase by 15-hydroxyeicosa-5,8,11,13-tetraenoic acid analogues. *J Med Chem* 30:254–263

Higgs GA, Flower RJ, Vane JR (1979): A new approach to anti-inflammatory drugs. *Biochem Pharmac* 28:1959–1961

Hlasta DL, Casey FB, Ferguson EW, Gangell SJ, Heimann MR, Jaeger EP, Kullnig RK, Gordon RJ (1991): 5-Lipoxygenase inhibitors: The synthesis and structure-activity relationships of a series of 1-phenyl-3-pyriazolidinones. *J Med Chem* 34:1560–1570

Israel E, Cohn J, Drazen J, Group ZS (1994): Effects of zileuton treatment on chronic stable asthma: Results of a 13-week double-blind placebo-controlled trial. *J Allergy Clin Immunol* 93:A793

Israel E, Dermarkarian R, Rosenberg M, Sperling R, Taylor G, Rubin P, Drazen JM (1990): The effects of a 5-lipoxygenase inhibitor on asthma induced by cold dry air. *New Eng J Med* 323:1740–1744

Israel E, Fischer AR, Rosenberg MW, Lilly CM, Callery JC, Shapiro J, Cohn, J, Rubin P, Drazen JM (1993a): The pivotal role of 5-lipoxygenase products in the reaction of aspirin-sensitive asthmatics to aspirin. *Am Rev Respir Dis* 148:1447–1451

Israel E, Rubin P, Kemo JP, Grossman J, Pierson W, Siegel SC, Tinkelman D, Murray JJ, Busse W, Segal AT, Fish J, Kaiser HB, Ledford D, Wenzel S, Rosenthal R, Cohn J, Lanni C, Pearlman H, Karahalios P, Drazen JM (1993b): The effect of inhibition of 5-lipoxygenase by zileuton in mild-to-moderate asthma. *Ann Intern Med* 119:1059–1066

Jackson WP, Islip PJ, Kneen G, Pugh A, Wates PJ (1988): Acetohydroxamic acids as potent, selective, orally active 5-lipoxygenase inhibitors. *J Med Chem* 31:499–500

Jakschik BA, Sun FF, Lee LH, Steinhoff MM (1980): Calcium stimulation of a novel lipoxygenase. *Biochem Biophy Res Commun* 95:103–110

Kellaway CH, Trethewie ER (1940): The liberation of a slow reacting smooth muscle-stimulating substance in anaphylaxis. *Q J Exp Physiol* 30:121–145

Kerdesky FAJ, Holms JH, Moore JL, Bell RL, Dyer RD, Carter GW, Brooks DW (1991): 4-Hydroxythiazole inhibitors of 5-lipoxygenase. *J Med Chem* 34:2158–2165

Kerdesky FAJ, Holms JH, Schmidt SP, Dyer RD, Carter GW (1985): Eicosatetraene-hydroxamates: Inhibitors of 5-lipoxygenase. *Tetrahedron Lett* 18:2143–2146

Kerdesky FAJ, Schmidt SP, Holms JH, Dyer RD, Carter GW, Brooks DW (1987): Synthesis and 5-lipoxygenase inhibitory activity of 5-hydroperoxy-6,8,11,14-eicosatetraenoic acid analogues. *J Med Chem* 30:1177–1186

Knapp HR (1990): Reduced allergen-induced nasal congestion and leukotriene synthesis with an orally active 5-lipoxygenase Inhibitor. *New Eng J Med* 323:1745–1748

Laursen LS, Naesdal J, Bukhave K, Lauritsen K, Rask-Madsen J (1990): Selective 5-lipoxygenase inhibiton in ulcerative colitis. *Lancet* 335:683–685

Lewis RA, Austen KF (1984): Molecular determinants for functional responses to the sulfido-peptide leukotrienes: Metabolism and receptor subclasses. *J Allerg Clin Immunol* 74:369–372

Lewis RA, Austen KF, Soberman RJ (1990): Leukotrienes and other products of the 5-lipoxygenase pathway, biochemistry and relation to pathobiology in human diseases. *New Eng J Med* 323:645–655

Malo PE, Shaughnessy TK, Bell RL, Bouska J, Hinz W, Majest S, Summers JB, Brooks DW, Carter GW (1989): The effect of 5-lipoxygenase inhibition on arachidonic acid induced (AAI) bronchoconstriction in the anesthetized guinea pig. *Pharmacologist* 31:172

Marom Z, Shelhamer JH, Bach MK (1982): Slow reacting substances, leukotrienes C_4 and D_4, increase the release of mucus from human airways *in vitro*. *Am Rev Respir Dis* 126:449–451

Matsumoto T, Funk CD, Radmark O, Hoog J, Jornvall H, Samuelsson B (1988): Molecular cloning and amino acid sequence of human 5-lipoxygenase. *Proc Natl Acad Sci USA* 85:26–30

Miller DK, Gillard JW, Vickers PJ, Sadowski S, Leveille C, Mancini JA, Charleson P, Dixon RAF, Ford-Hutchinson AW, Fortin R, Gauthier JY, Rodkey J, Rosen R, Rouzer C, Sigal IS, Strader CD, Evans JF (1990): Identification and isolation of a membrane protein necessary for leukotriene production. *Nature* 343:278–281

Murphy RC, Hammarstrom S, Samuelsson B (1979): Leukotriene C: A slow-reacting substance from murine mastocytoma cells. *Proc Natl Acad Sci USA* 76:4275–4279

Musser JH, Kreft AF (1992): 5-Lipoxygenase: Properties, pharmacology, and the quinolinyl(bridged)aryl class of inhibitors. *J Med Chem* 35:2501–2524

Orange RP, Austen KF (1969): Slow reacting substance of anaphylaxis. *Adv Immunol* 10:105–144

Orange RP, Valentine MD, Austen KF (1968): Antigen-induced release of slow reacting substance of anaphylaxis (SRS-Arat) in rats prepared with homologous antibody. *J Exp Med* 127:767–782

Percival MD (1991): Human 5-lipoxygenase contains an essential iron. *J Biol Chem* 266:10058–10061

Rodriques KE, Basha A, Summers JB, Brooks DW (1988): Addition of aryllithium compounds to oxime ethers. *Tetrahedron Letters* 29:3455–3458

Rouzer CA, Ford-Hutchinson AW, Morton HE, Gillard JW, Rycut H, Fortin R, Gauthier JY, Rodkey J, Rosen R, Sigal IS, Strader CD, Evans JF (1990): MK886, a potent and specific leukotriene biosynthesis inhibitor blocks and reverses the membrane association of 5-lipoxygenase in ionophore-challenged leukocytes. *J Biol Chem* 265:1436–1342

Rouzer CA, Kargman S (1988): Translocation of 5-lipoxygenase to the membrane in human leukocytes challenged with ionophore A23187. *J Biol Chem* 263:10980–10988

Rouzer CA, Matsumoto T, Samuelsson B (1986): Single protein from human leukocytes possesses 5-lipoxygenase and leukotriene A_4 synthase activities. *Proc Natl Acad Sci USA* 83:857–861

Rouzer CA, Rands E, Kargman S, Jones RE, Register RB, Dixon RAF (1988): Characterization of cloned human leukocyte 5-lipoxygenase expressed in mammalian cells. *J Biol Chem* 263:10135–10140

Rubin P, Dube L, Braeckman R, Swanson L, Hanson R, Albert D, Carter G (1991): Pharmacokinetics, safety and ability to diminish leukotriene synthesis by zileuton, an inhibitor of 5-lipoxygenase. In: *Progress in Inflammation Research and Therapy*, Ackerman NR, Bonney RJ, Welton AF, eds. Basel: Birkhäuser Verlag

Salmon JA, Garland LG (1991): Leukotriene antagonists and inhibitors of leukotriene biosynthesis as potential therapeutic agents. *Prog Drug Res* 37:9–90

Samuelsson B (1983): Leukotrienes: Mediators of immediate hypersensitivity reactions and inflammation. *Science* 220:568–575

Samuelsson B, Dahlen S-E, Lindgen JA, Rouzer CA, Serhan CN (1987): Leukotrienes and lipoxins: Structures, biosynthesis, and biological effects. *Science* 237:1171–1176

Shaw A, Krell RD (1991): Peptide leukotrienes: Current status of research. *J Med Chem* 34:1235–1242

Sirois P, Borgeat P, Lauziere M, Dube L, Rubin P, Kesterson J (1991): Effect of Zileuton (A-64077) on the 5-lipoxygenase activity of human whole blood *ex vivo*. *Agents and Actions* 34:117–120

Stewart AO, Brooks DW (1992): N,O-bis(phenoxycarbonyl)hydroxylamine: A new reagent for the direct synthesis of substitued *N*-hydroxyureas. *J Org Chem* 57:5020–5023

Summers JB, Gunn BP, Martin JG, Martin MB, Mazdiyasni H, Stewart AO, Young PR, Bouska JB, Goetze AM, Dyer RD, Brooks DW, Carter GW (1988a): Structure-activity analysis of a class of orally active hydroxamic acid inhibitors of leukotriene biosynthesis. *J Med Chem* 31:1960–1964

Summers JB, Gunn BP, Martin JG, Mazdiyasni H, Stewart AO, Young PR, Goetze AM, Bouska JB, Dyer RD, Brooks DW, Carter GW (1988b): Orally active hydroxamic acid inhibitors of leukotriene biosynthesis. *J Med Chem* 31:3–5

Summers JB, Gunn BP, Mazdiyasni H, Goetze AM, Young PR, Bouska JB, Dyer RD, Brooks DW, Carter, GW (1987a): *In vivo* characterization of hydroxamic acid inhibitors of 5-lipoxygenase. *J Med Chem* 30:2121–2126

Summers JB, Kim KH, Mazdiyasni H, Holms JH, Ratajczyk JD, Stewart AO, Dyer RD, Carter GW (1990): Hydroxamic acid inhibitors of 5-lipoxygenase: Quantitative structure-activity relationships. *J Med Chem* 33:992–998

Summers JB, Mazdiyasni H, Holms JH, Ratajczyk JD, Dyer RD, Carter GW (1987b): Hydroxamic acid inhibitors of 5-lipoxygenase. *J Med Chem* 30:574

Sweeney FJ, Eskra JD, Carty TJ (1987): Development of a system for evaluating 5-lipoxygenase inhibitors using human whole blood. *Prostaglandin Leukotriene Med* 28:73–93

Weinblatt ME, Kremer JM, Coblyn JS, Helfgott S, Maier AL, Petrillo G, Henson B, Rubin P, Sperling R (1992): Zileuton, a 5-lipoxygenase inhibitor in rheumatoid arthritis. *J Rheum* 19:1537–1541

Young PR, Bell RL, Lanni C, Summers JB, Brooks DW, Carter GW (1991): Inhibition of leukotriene biosynthesis in the rat peritoneal cavity. *Eur J Pharmacol* 205:259–266

6

Discovery of ACCOLATE™ (ICI 204,219), a Peptide Leukotriene Antagonist for Asthma

Frederick J. Brown

Introduction

In 1940 Kellaway and Trethewie reported that perfusate from antigen-stimulated, resected lung of sensitized guinea pigs produced a slowly developing contraction of isolated guinea pig ileum. The name *slow-reacting substance of anaphylaxis* (SRS-A) was coined to describe the unknown substance(s) mediating this contractile activity. Over the next 30 years a few investigators studied the release of SRS-A from different sources and attempted to purify the highly labile material. Early in the 1970s, scientists from Fisons Ltd. (Augstein et al., 1973) described the discovery of FPL 55712 (Structure 6-A), a relatively potent and competitive antagonist of SRS-A. The compound was identified serendipitously as part of a screening effort at Fisons to identify more potent analogs of their antiasthma drug disodium cromoglycate. While FPL 55712 proved to be relatively nonselective[1] and too metabolically labile to serve as a drug, it became an important pharmacological tool for the evaluation of SRS-A and spurred a resurgence of interest in that substance.

However, the chemical identity of SRS-A remained elusive. A major watershed occurred in the late 1970s when Professor Bengt Samuelsson and co-workers at the Karolinska Institute isolated highly purified SRS-A from ionophore-stimulated leukocytes and identified it as a thioether-linked cysteine-containing derivative of 5-hydroxy-7,9,11,14-eicosatetraenoic acid (Murphy et al., 1979). The definitive, stereochemical proof of the structure of leukotriene C_4 (LTC_4), one of the components of SRS-A, was achieved by total synthesis in the laboratories of Professor Elias Corey at

The Search for Anti-Inflammatory Drugs
Vincent J. Merluzzi and Julian Adams, Editors
© Birkhäuser Boston 1995

FPL 55712
inhibition: 62% @ 4 uM
selectivity: 20

STRUCTURE 6-A

Harvard (Corey et al., 1980). It was shown that there were other structurally-related components of SRS-A, and that all these compounds were derived from arachidonic acid via the enzymatic action of 5-lipoxygenase. The name leukotriene (LT) was proposed to reflect the leukocyte origins of the molecules and the characteristic conjugated triene component of the chemical structure. The leukotrienes have been categorized into two families: those containing amino acid moieties conjugated at carbon 6 (the peptide leukotrienes), and those characterized by the absence of any amino acids. The parent peptide leukotriene, LTC_4, arises from enzyme-mediated addition of the tripeptide glutathione to 5,6-*trans*-5,6-oxido-7,9-*E*-11,14-*Z*-eicosatetraenoic acid (LTA_4). Sequential, enzyme-induced loss of glutamic acid and glycine generates LTD_4 and LTE_4, respectively. The subscripted numbers refer to the total number of double bonds in the molecules, and the letter suffixes to the biogenetic order of synthesis. The peptide leukotrienes are very potent smooth muscle spasmogens. The nonpeptide leukotriene LTB_4 is primarily an attractant and activator of leukocytes (Structure 6-B).

The seminal discoveries of the biogenesis and structures of SRS-A, combined with knowledge of its pronounced biological activities, prompted many pharmaceutical companies to initiate programs aimed at the discovery of leukotriene receptor antagonists or synthesis inhibitors. ZENECA (then ICI) chose in the early 1980s to explore both means of intervention. A US-based pulmonary group pursued receptor antagonists as potential antiasthmatic agents while a UK-based team targeted 5-lipoxygenase inhibitors for therapeutic utility in arthritis and asthma (see Chapter 7 by Crawley et al. in this volume). Both programs were viewed as exciting, but very challenging, high-risk endeavors, with no guarantee that interference with leukotrienes would lead to successful disease therapy. The circumstantial evidence supporting the involvement of peptide leukotrienes in asthma at that time was: (1) the release of SRS-A from sensitized human lung tissue and its presence in airways secretions following an asthmatic attack; (2) the ability of the peptide leukotrienes to elicit a pronounced contraction of human airway

STRUCTURE 6-B

smooth muscle, to increase vascular permeability, and to promote mucus hypersecretion; and (3) the fact that agents blocking the effects of other putative mediators of asthma such as histamine or prostaglandins had not demonstrated clinical utility in asthma.

Initiation of the Leukotriene Antagonist Project

The Pharmaceuticals Group at ICI had identified asthma as an attractive disease target based upon both medical need and market potential. Current data continues to support that view. Asthma is one of the most common chronic illnesses, affecting 3%–5% of the population (Fleming et al., 1987). Despite the availability of various drug therapies such as mediator release inhibitors, xanthines, β_2-agonists, and steroids (Bernstein, 1992), both the prevalence and severity of asthma appear to be increasing (Weiss et al., 1993). The discovery of the leukotrienes opened up a new approach to disease intervention. This confluence of enabling science, medical need, and commercial potential prompted ICI to initiate a leukotriene antagonist program aimed at asthma.

The program commenced with an exploration of synthetic and purification protocols for the preparation of large quantities of the peptide leukotrienes to support pharmacological evaluation and the development of a screening cascade. At this time the average commercial cost of the leukotrienes was approximately $1,000 per milligram. This synthetic undertaking was a major endeavor given the extreme lability of the molecules. Ultimately our processes were refined to the point where the leukotrienes could be produced to high specifications in a very reproducible fashion. The completion of this critical initial phase of the program underpinned our later success by enabling full pharmacological profiling of our subsequent antagonists. The need for leukotrienes continued into the clinical phase, and by the end of the program multi-gram quantities had been supplied by the research team.

Some formidable pharmacologic issues had to be addressed to support the selection of specific receptor targets and testing paradigms. Although there was very little information available on the nature of the leukotriene receptor(s), it was known that the leukotrienes contracted a variety of isolated smooth muscle preparations. However, unlike the prostaglandins, the leukotrienes exhibited rather remarkable tissue and species specificity— guinea pigs were one of the few small animals whose airways responded to leukotrienes. As a generally accepted model for human airway tissue, guinea pig trachea was chosen for functional evaluation of antagonists, accepting the risk that there could be differences between these receptors and those found in human airways. Another major issue was the question of which peptide leukotriene to block. We chose to start in the middle of the cascade using LTD_4 as the agonist, while exploring in detail the issues of receptor heterogeneity.

An assay was established which measured the ability of compounds to inhibit LTD_4-induced contractions of guinea pig tracheal spirals. With a test for functional activity in place, additional chemical resources were dedicated to the project and screening of synthetic compounds commenced. Compounds were initially evaluated at a single concentration and a percentage inhibition of contraction was reported. As a general measure of the selectivity of the compounds for the LT receptor, a similar protocol was established using the nonspecific spasmogen barium chloride as the agonist in place of LTD_4. The more interesting antagonists were profiled in greater detail by using cumulative LTD_4 concentration-response curves to determine dissociation constants (reported here as the negative logarithm: pK_B). Later in the program a receptor-ligand binding assay measuring the displacement of $[^3H]$-LTD_4 from guinea pig lung membranes was added to the screening cascade to facilitate the initial evaluation of new compounds (Aharony et al., 1987).

Our ongoing pharmacological research into leukotriene receptor classification demonstrated that the LTD_4 receptor on guinea pig trachea could be subdivided into high- and low-affinity populations, with LTE_4 acting preferentially at the high-affinity site (Krell et al., 1983). Furthermore, the existence of a distinct receptor for LTC_4 was suggested by the fact that blockade of the metabolic conversion of LTC_4 to LTD_4 by L-serine borate (an inhibitor of gamma-glutamyl transpeptidase) abolished the ability of FPL 55712 to antagonize the contractions of guinea pig trachea mediated by LTC_4 (Snyder et al., 1984). The ability of FPL 55712 to block LTC_4-induced contractions of uninhibited guinea pig trachea was attributed to the rapid metabolism of LTC_4 to the FPL 55712-sensitive LTD_4 and LTE_4 (Aharony et al., 1985). Studies carried out in collaboration with Professor Carl Buckner (who subsequently joined Zeneca) and colleagues at the University of Wisconsin suggested that in human airways all peptide leukotrienes interacted with a single leukotriene receptor, which most closely resembled (but was not identical to) the LTE_4 receptor found in guinea pig trachea (Buckner et al., 1986). In light of this information that the peptide leukotrienes were rapidly converted *in vitro* to LTE_4 and that LTE_4 occupied a single receptor subtype in guinea pig trachea that pharmacologically resembled the one found in human lung, we began using LTE_4 as the agonist in our guinea pig assay instead of LTD_4.

Early in the program a collaboration was established with Professors Roy Patterson and Lewis Smith at Northwestern University to evaluate the airway effects of LTD_4, first in rhesus monkeys (Patterson et al., 1983), subsequently in healthy human volunteers (Kern, et al., 1986; Smith et al., 1987), and then in asthmatic subjects (Smith et al., 1985). To support these studies LTD_4 was produced to very exacting specifications, and an Investigational New Drug application was filed with the Federal Drug Administration to allow administration of this material to humans. These clinical investigations demonstrated the increased sensitivity of asthmatic patients versus nonasthmatic volunteers to inhaled LTD_4. They provided additional support for the hypothesis that peptide leukotrienes are mediators of asthma and established an excellent link to subsequent clinical investigations with our antagonists.

Exploration of Leads

Since no structural information on the leukotriene receptor was available, the design of novel antagonists had to begin from a consideration of the structures of known ligands for the receptor. At the time we commenced this phase of the antagonist program there were only two ligands known to bind to leukotriene receptors—the antagonist FPL 55712 and the leukotrienes themselves. We decided to commit sufficient chemical resources to pursue both leads simultaneously. The project goals were to discover a

competitive, orally active antagonist that was both more potent and more selective than FPL 55712. It was felt that *in vitro* potency equivalent to that of
the natural agonists (1 nM) should be achievable. An aerosol-administered
drug was certainly an option, given our disease target of asthma. However,
we knew that competition in the leukotriene arena was fierce, with most
of our competitors mounting either antagonist- or enzyme-inhibition programs. It was felt that for prophylactic therapy an aerosol-only antagonist
would be commercially vulnerable to subsequent oral agents that promised
greater ease of administration and enhanced patient compliance. Therefore,
we decided that we would target an orally active compound to maximize
versatility in the treatment of asthma and allergic disorders.

With robust synthetic methodology for the leukotrienes in hand, efforts
focused on designing stable molecular surrogates for the labile tetraene
moiety to support an eventual transition from agonists to stable antagonists.
Studies on synthetic leukotriene analogs by several groups had delineated
some of the structural parameters critical for maintaining agonist activity
(reviewed by Krell et al., 1984). Our own work corroborated the importance of the 5(S), 6(R) stereochemistry for biological activity (Tsai et al.,
1982) and suggested that the triene preferred an extended conformation in
solution (Loftus and Bernstein, 1983). Such attributes were considered in
the design of more stable agonists. A series of iterative optimizations led
ultimately to the identification of **1**, an alkylated homocinnamyl derivative
of LTC_4 as a chemically stable agonist (Bernstein et al., 1986). Comparable spasmolytic activity was obtained with the analogous furan **2** (Y.K.
Yee, K.C. Hebbel, and A. Needles, unpublished work from Zeneca). The
contractile activities of both **1** and **2** were antagonized by FPL 55712, and
the 5(S), 6(R) stereochemistry was critical for activity, indicating that these
agonists were acting at leukotriene receptors (Structures **1** and **2**).

1
agonist @ 70 nM

2
agonist @ 70 nM

Simultaneously, work had commenced on exploring structural modifications to the SRS-A antagonist FPL 55712 (Brown et al., 1989). Although
this compound was relatively potent at the leukotriene receptor, it exhibited a variety of additional pharmacologic effects, suffered from a short

biologic half-life, and was orally inactive. Several groups (Appleton et al., 1977; Buckle et al., 1979) had shown that while the left-hand hydroxy-acetophenone portion of the molecule was critical for biological activity, a number of modifications to the right-hand chromone ring were tolerated. We found that the entire right half of FPL 55712 could be substituted by simple thioaliphatic acids of 6–8 atoms in length (e.g., **3**) to provide antagonists with good activity at concentrations of $10\,\mu M$. Replacement of the aliphatic chain by a benzene ring to reduce conformational flexibility led to the para-toluic acid analog **4**, which exhibited potency approaching that of FPL 55712. An investigation of aryl substituents identified a 3-methoxy group as conveying especially good activity. This compound, **5**, was four times more potent and two times more selective than FPL 55712 (Structures **3**, **4**, and **5**).

3
inhibition: 62% @ 10 uM

4
inhibition: 36% @ 4 uM

5
inhibition: 57% @ 1 uM
selectivity: 50
ip: 50 mg/kg @ 60 min

For *in vivo* analysis of the effects of our compounds, pulmonary mechanic parameters were measured in anesthetized guinea pigs. With this assay the efficacy of antagonists at blocking leukotriene-induced changes in pulmonary resistance and dynamic lung compliance could be quantitated. While this assay was quite sensitive and precise, it required surgical preparation of the animals. Envisioning that numerous compounds would need evaluation in the search for an oral agent, we sought a less labor-intensive model for the initial *in vivo* screening. An assay based on blockade of the dyspnea induced by aerosolized LTD_4 in spontaneously breathing, conscious guinea pigs proved particularly robust (Snyder et al., 1988). Compounds could be administered by the route of choice at various time intervals preceding leukotriene challenge. The effectiveness of the antagonists was measured by their ability to delay the onset of dyspnea during a five-minute, continuous exposure to aerosolized LTD_4.[2]

The hydroxyacetophenone antagonist **5** blocked LTD_4-induced dyspnea in conscious guinea pigs at an intraperitoneal dose of 50 mg/kg administered 60 minutes prior to the aerosolized leukotriene challenge. While promising progress had been made with antagonist **5**, it was considerably less active than the leukotrienes themselves, and further improvements were sought.

From a structure-activity standpoint, we viewed the leukotrienes as molecules comprised of three structural domains: the peptide fragment, the lipophilic tetraene portion, and the C(1)–C(6) acid-bearing segment. It had been postulated that the leukotriene receptor contained a hydrophobic pocket for the tetraene chain, a hydrophilic, anion-binding site for the peptide unit, and a polar but not necessarily anion-binding site associated with the C-1 acid (Krell et al., 1984). One of our working hypotheses (Brown et al., 1989) on the possible structural similarities between the hydroxyacetophenones (i.e., FPL 55712 and 5) and the leukotrienes speculated that both series bound to the same portion of the leukotriene receptor with the aryl acids of the hydroxyacetophenones occupying a portion of the region that accommodated the peptidic segment of the leukotrienes. To test the hypothesis we replaced the peptide portion of our stable leukotriene analogs (e.g., 1) with the 3-methoxy-4-methylbenzoic acid moiety from the hydroxyacetophenone 5 (Bernstein et al., 1988). This ploy was successful in converting agonist 1 into an antagonist, 6 (Snyder and Bernstein, 1988). Slight modifications to the homocinnamyl tetraene mimetic demonstrated the presence of a very delicate balance between agonist and antagonist properties within this series.[3] For example, the analogous ether 7, at a concentration of 10 μM, exhibited contractile activity equivalent to that elicited by 8 nM LTE$_4$ (Structures 6 and 7).

6
inhibition: 31% @ 5 uM

7
agonist @ 10 uM

To move away from this propensity for agonism, further structural simplifications to the leukotriene analogs were sought. It was found that removal of the sulfur-linked peptide fragment of 1 (or its benzoic acid surrogate in 6 and 7) and the C5-hydroxyl group provided weak antagonists that did not exhibit any indications of partial agonism. The phenyl hexyl ether (from 7) mimetic for the tetraene portion of the leukotrienes was retained, and a series of structural modifications to the right half of the molecule were investigated. The greatly simplified thioaliphatic acid 8 emerged as one of the better antagonists. These lipid-acid analogs of the leukotrienes were less potent antagonists than compounds such as 6 which mimicked the peptide portion of the LTs as well as the lipophilic and acidic arms. However, the consistent lack of agonism in the series prompted further work.

Variation of the aliphatic chain length did not provide additional receptor affinity, but the use of a benzene ring as a conformationally restrained replacement for the thioaliphatic moiety increased potency, as in the benzyl ether **9**. Structure-activity explorations demonstrated that the para relationship between the two ether moieties, as well as the ortho positioning of the carboxylic acid functionality, provided optimal potency. For example, to obtain comparable inhibition with the meta/para regioisomer **10**, the concentration of the antagonist had to be increased to 50 μM. (Structures **8**, **9**, and **10**; the critical geometrical relationship between the lipophilic and acidic moieties is highlighted in Structure **9**.)

8
inhibition: 56% @ 50 uM

9
inhibition: 40% @ 5 uM
selectivity: 10

10
inhibition: 59% @ 50 uM

Discovery of a Novel Lead

Although great structural simplification had been achieved with these despeptide, lipid-acid mimetics of the leukotrienes, the benzyl ether antagonist **9** had lost four orders of magnitude of receptor affinity compared to LTC$_4$ and was five times less potent and less selective than our best antagonist, hydroxyacetophenone **5**. One approach to enhance the potency of **9** was to introduce further structural rigidity about the benzylic ether linkage by building a bridge between the benzylic carbon atom and the ortho position of that benzene ring (Brown et al., 1990). Thus, a bicyclic framework was sought to which both a lipophilic chain and a benzoic acid moiety could be readily appended to generate molecules of the general form **11**. Indazoles were chosen as an appropriate template for elaboration. The commercial availability of nitroindazoles prompted the use of an amide functionality as a more synthetically accessible means (instead of the ether moiety in **9**) of attaching the lipophilic group. Eight regioisomeric indazoles were prepared, exploring all possible combinations of the substitution patterns indicated by the asterisks in structure **12** (Structures **11**, **12**).

Only one of the eight indazole isomers showed significant antagonist activity at a concentration of 10 μM. It was somewhat disconcerting to find that isomer **13**, in which the relative juxtapositioning of the lipophilic and acidic functionalities was identical to that found to be optimal in the benzyl ether series (i.e., **9**; compare highlighted portions), was inactive. Surprisingly, the one active indazole was **14**, which corresponded more closely to the structural arrangement found in the less active isomer **10** from the benzyl ether series. It was disappointing that the net outcome of this endeavor to increase potency via a conformational constraint had been a twofold loss of activity (compare **9** and **14**) . However, the puzzling divergence of the structure-activity relationships between the benzyl ether and indazole series generated continued interest in the indazole **14** (Structures **13, 14**).

13
inhibition: not active @ 10 uM

14
inhibition: 60% @ 10 uM

In the course of trying to rationalize the indazole/benzyl ether structure-activity discrepancies, it was perceived that perhaps indazole **14** more closely resembled hydroxyacetophenone **5** than its benzyl ether progenitor **9** (Brown et al., 1990). In the design of compound **5** our hypothesis had been that the acidic moiety (of both it and FPL 55712) was mimicking the peptide portion of the leukotrienes. In contrast, the acidic portion of the benzyl ether series had derived from attempts to simulate the C-1 carboxylic acid of the leukotrienes. Thus, it was postulated that the transition from the benzyl ether series to the indazole series (and from a hexyl ether to a hexanamide lipophilic fragment) had resulted in antagonists that now interacted with the receptor more as lipid-peptide mimics of the leukotrienes rather than lipid-acid mimetics. A reasonable overlap of indazole **14** and hydroxyacetophenone **5** could be achieved by superimposing the carbonyl lone pair electrons and the carboxyl groups (see Figure 6-1). Such a comparison to **5** is not possible with the benzyl ether **9** or any of the other seven indazole isomers (e.g., **13**) which had failed to provide significant activity. This perception of a connection between the hydroxyacetophenone and indazole series rationalized the surprising, singular activity of the one

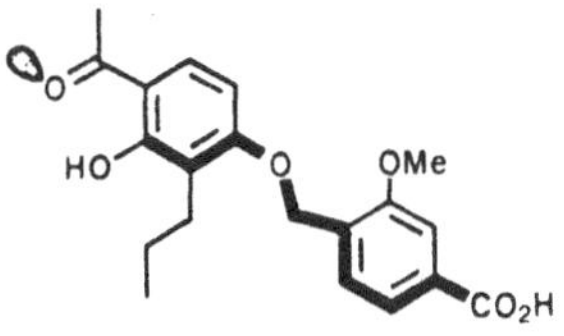

Figure 6-1. Structural similarities between hydroxyacetophenone **5** and indazole **14**.

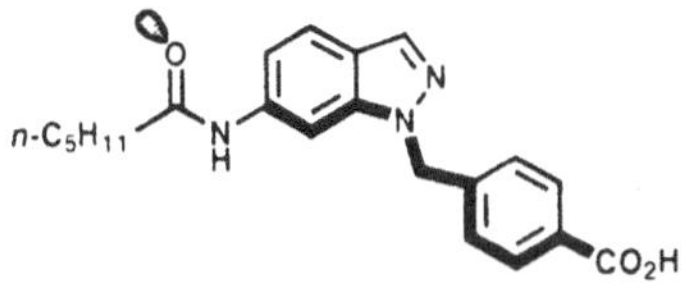

indazole regioisomer and suggested that the beneficial methoxy substituent from **5** should be added to the benzoic acid moiety of **14**.

Since only the N-1 benzylated analog was desired (the N-2 benzylated indazole isomers had been inactive), it made synthetic sense to try substituting an indole for the indazole ring so as to avoid the necessity of separating the N-1 and N-2 indazole alkylation products. Incorporation of the methoxy substituent gave indole **15**, which was approximately twofold more active[4] than indazole **14**. Although indole **15** was still some fivefold less potent than our best antagonist, hydroxyacetophenone **5**, it was surprisingly more selective for the leukotriene receptor. This specificity, combined with the structural novelty of the compound, encouraged further exploration. The decision to pursue compound **15** as a new lead, despite its weak activity, was also influenced by our frustration with the hydroxyacetophenone series. Extensive modifications to **5** (more than 100 derivatives) had failed to provide significant gains in potency (Brown et al., 1989) (Structure **15**).

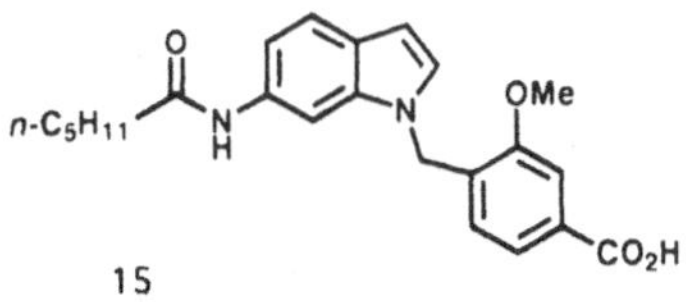

15
inhibition: 70% @ 5 uM
pK_B = 6.0
selectivity: 100

Development of the Indole/Indazole Lead

From a synthetic standpoint, indole 15 was readily dissected into three portions: the lipophilic amide appendage, the toluic acid moiety, and the central heterocyclic ring system. Therefore, the structure-activity investigation of the series was approached in a similar manner, commencing with an exploration of the lipophilic moiety (Brown et al., 1990). An investigation of the amide chain length (ranging from CH_3 to $C_{11}H_{23}$) established that the initial hexanamide in 15 was preferred. A phenylacetamide, 16, was equipotent. The introduction of functionality (e.g., carboxylic acid, amide, ketone, or acetylene) on the hydrocarbon chain of 15 or substituents on the benzene ring of 16 was generally detrimental to activity. These data suggested that the amide appendage was binding to a relatively well defined, hydrophobic pocket of the receptor. Increased potency was obtained with the α-branched amide, 2-ethylhexanamide 17. The activity of this compound was equal to that of our previous best antagonist, hydroxyacetophenone 5, but the barium chloride-assessed selectivity ratio was improved 14-fold. The compound was also orally active at a dose of 233 μmol/kg (100 mg/kg) given one hour prior to LTD_4 challenge. As more potent amide moieties were discovered in the indole series, they were incorporated into the indazoles. The corresponding ethylhexanamido-indazole 18 proved to be an even more active antagonist, finally breaking the long-standing potency record held by compound 5 (Structures 16, 17, 18).

16
inhibition: 71% @ 5 uM
pK_B = 6.0

17 X = CH
inhibition: 59% @ 1 uM
pK_B = 6.6
selectivity: 700
po: 100 mg/kg (233 umol/kg) @ 1 hr
18 X = N
inhibition: 56% @ 0.33 uM
pK_B = 6.9

By this point our interest in the indole/indazole series was thoroughly aroused. Having never surpassed the 1 μM level of activity in the hydroxy-acetophenone series, the rapid progression of selectivity and potency in the new series was particularly exciting. The scope of our investigations widened to also encompass modifications to the central ring system and the acidic moiety. We discovered that the indole ring could be inverted, provided the lipophilic and acidic appendages remained juxtaposed in the same relative 1,4-orientation (Matassa et al., 1990b). Thus, the 3,5-substituted hexanamido-indole **19a** was comparable in activity to the 1,6-substituted indole **15**. Methylation of the indole nitrogen, as in **19b**, increased potency, but larger substituents did not provide any further advantage. Incorporation of the ethylhexanamide moiety, to give **20**, provided increased activity, just as had occurred in the 1,6-substituted series (i.e., **15** versus **17**) (Structures **19, 20**).

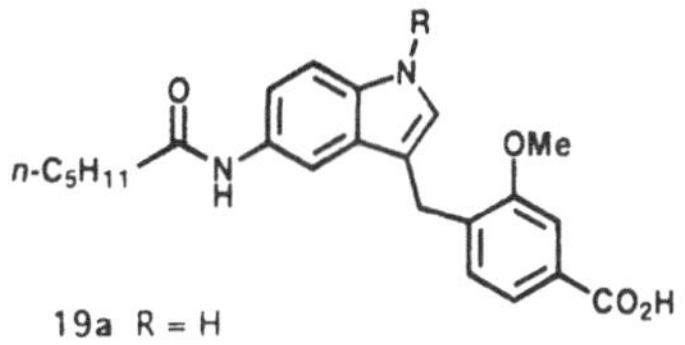

19a R = H
     inhibition: 64% @ 3.3 uM
19b R = Me
     inhibition: 77% @ 1 uM

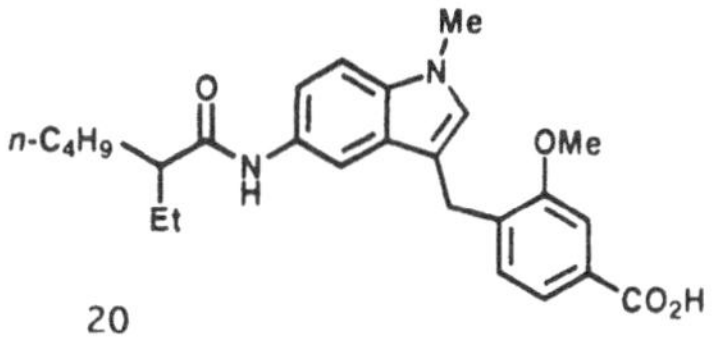

20
inhibition: 70% @ 0.33 uM

Concurrently, alternatives for the acid moiety were under investigation (Yee et al., 1990). A group at Eli Lilly reported that the replacement of an aliphatic carboxylic acid with a tetrazole in a hydroxyacetophenone series of leukotriene antagonists could provide up to 30-fold enhancements in potency (Marshall et al., 1987). The preferred compound, LY171883 (Fleisch et al., 1985), was one of the early leukotriene antagonists to progress to clinical evaluation, but was subsequently withdrawn from clinical trials. In our series of aromatic acid antagonists, tetrazole analogs of hydroxyacetophenone **5** and indole **17** proved to be no more active than the parent carboxylic acids. Other possible bioisosteres for a carboxylic acid include aliphatic or aromatic N-sulfonyl amides (Schaaf, et al., 1979). The methylsulfonimide **21a** did not provide any benefits over the corresponding carboxylic acid **17**. However, a phenylsulfonimide moiety imparted a dramatic 100-fold

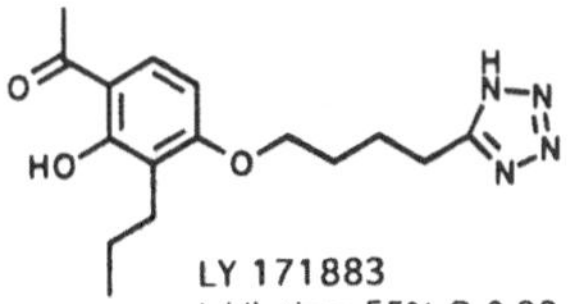

LY 171883
inhibition: 55% @ 0.33 uM
pK$_B$ = 7.2

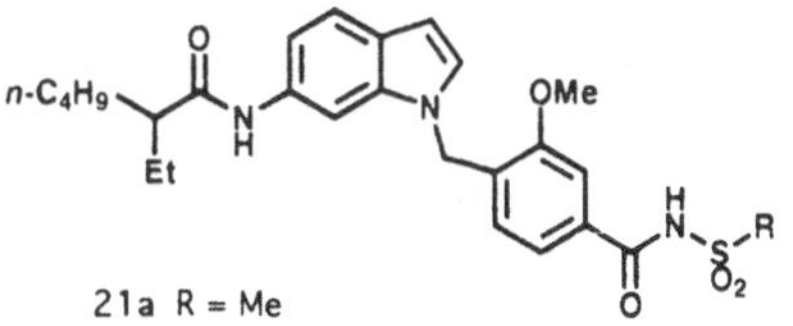

21a R = Me
     inhibition: 60% @ 1 uM
21b R = Ph
     inhibition: 27% @ 10 nM
     pK$_B$ = 8.6
     selectivity: >10,000
     po: 17 mg/kg (30 umol/kg) @ 3 hr

improvement in activity, generating compound **21b**, which was active at a concentration of 10 nM and had a selectivity ratio of over 10,000. A substantial improvement in oral potency and duration of activity also accrued with the phenylsulfonimide; this compound was active at a dose of 30 μmol/kg (17 mg/kg) given three hours prior to LTD$_4$ challenge.

Further structure-activity work (Yee et al., 1990) confirmed our supposition that the 3-methoxy-4-methylbenzoyl fragment, derived from hydroxy-acetophenone **5**, might remain a preferred linking moiety for the indole-phenylsulfonimides.[5] Various perturbations to the sulfonimide (RCONH-SO$_2$Ph) group (e.g., RCONHS$\underline{O}$Ph, R$\underline{CH_2}$NHSO$_2$Ph, and R$\underline{SO_2}$NH$\underline{C}$OPh) decreased potency by more than 30-fold, indicating the critical acidic and geometrical properties conveyed by the sulfonimide functionality. More lipophilic aliphatic homologs of methylsulfonimide **21a** failed to provide the level of activity exhibited by the phenylsulfonimide, suggesting that the sulfonyl phenyl ring occupied a specific aromatic binding site at the receptor.

Additional enhancements to the lipophilic part of the molecule were simultaneously being discovered in the 1,6-substituted indole/carboxylic acid series (Brown et al., 1990). An extensive exploration of acylamino groups culminated in the finding that the β-branched cyclopentylacetamide **22a** was 10 times more potent than the ethylhexanamide **17**. The cyclopentyl-carbamate **22b** was equally good, but the analogous urea **22c** was weaker. With the cyclopentyl acetamide and carbamate, potency had been increased 60-fold from that of the original hexanamide **15**. From the structure-activity relationships of these and other analogs we postulated that the portion of the receptor accommodating the acylamino appendages preferred lipophilic moieties with tetrahedral geometry *beta* to the carbonyl (e.g., compare cyclopentyl acetamide **22a** with phenylacetamide **16**). By this time a number of alternatives to the amide (CONH) link between the lipophilic appendage and the indole ring had been investigated. These alternatives—OC(O), NHC(O), CH$_2$NH, CH$_2$O—were inferior to the amide. For example, the cyclopentylethyl indole ether **23** was 100 times less potent than the cyclopentylacetamide **22a**. Given this latter result, it is interesting to speculate that if the hexyl ether moiety from progenitor **9** had been retained in the indazoles **12** (rather than the amide linkage) whether all of the initial indazole compounds **12** might have been inactive and therefore missed as a lead (Structures **22**, **23**).

The next question was whether the enhanced receptor recognition provided by the cyclopentyl moiety could be combined with the benefits of the phenylsulfonimide group to generate even more potent antagonists. It was very gratifying to find that the increased potency of the cyclopentyl

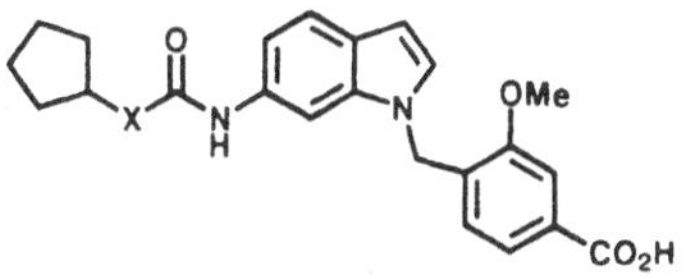

22a X = CH$_2$
     inhibition: 34% @ 0.1 uM
     pK$_B$ = 7.7
22b X = O
     inhibition: 61% @ 0.1 uM
     pK$_B$ = 7.8
22c X = NH
     inhibition: 73% @ 1 uM

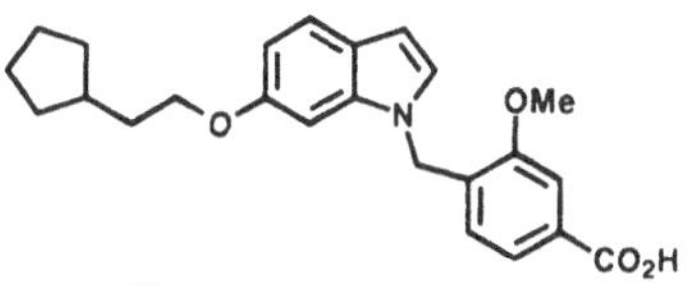

23
inhibition: 75% @ 10 uM

acetamide and carbamate groups did translate well to the phenylsulfon-imides—**24a** and **24b** were approximately 10 times more active than the ethylhexanamide-phenylsulfonimide **21b**. Interestingly, the cyclopenty-lurea, which had not been as effective in the carboxylic acid series (**22c**), was surprisingly equipotent with the acetamide and carbamate in the phenyl-sulfonimide series (**24c**). With these antagonists the subnanomolar recep-tor affinity exhibited by the leukotrienes themselves had been equaled. The carbamate and acetamide were orally active at 30 μmol/kg (15 mg/kg) (Structure **24**).

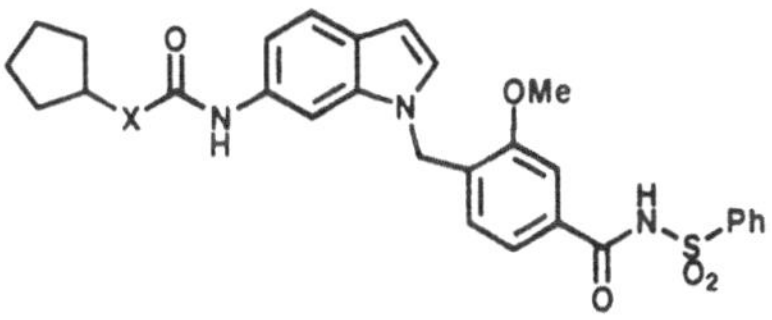

24a X = CH$_2$
     pK$_B$ = 9.6
     po: 30 umol/kg
24b X = O
     pK$_B$ = 9.5
     po: 30 umol/kg
24c X = NH
     pK$_B$ = 9.8

During this period of lead development it was rare for a month to pass without the emergence of a new compound expressing even better potency than the previous best. The various indole/indazole sulfonimide/carboxylic acid series exhibited an astonishing degree of structure-activity concor-dance, which helped the exploration process. We discovered that a molec-ular fragment found to be beneficial in one series could usually be incor-porated in a different series with the same beneficial effect, for example, the successful transfer of the ethylhexanamide effect from indole **17** to indazole **18**. The extension of the cyclopentyl acylamino moieties from the carboxylic acid (**22**) to the phenylsulfonimide (**24**) series is another example of the good concordance. Likewise, the cyclopentyl acylamino and phenylsulfonimide groups could be introduced into 3,5-substituted (inverted) indoles (**25**) and indazoles (**26**) to give potent, orally active antagonists. This ability to mix and match different molecular fragments from the different structural regions of the molecule enabled us to quickly generate a three-by-four matrix (cyclopentyl acylamino groups with 1,6- and 3,5-substituted indoles and indazoles) of phenylsulfonimide LT antagonists with subnanomolar dissociation constants (Structures **25, 26**).

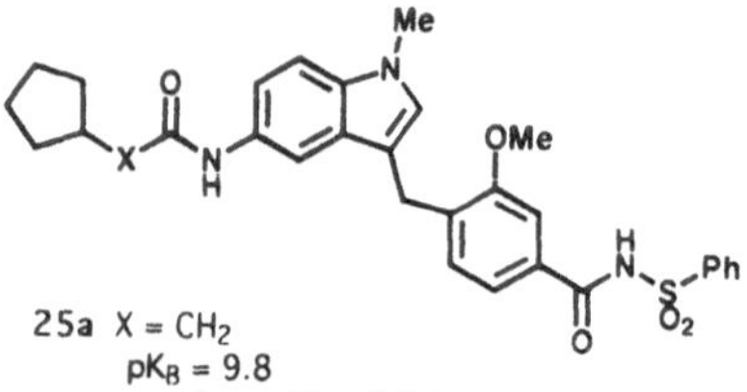

25a X = CH$_2$
   pK$_B$ = 9.8
   po: 1 umol/kg @ 3 hr
25b X = O
   pK$_B$ = 9.7
   po: 3 umol/kg @ 3 hr
25c X = NH
   pK$_B$ = 9.8
   po: 30 umol/kg @ 3 hr

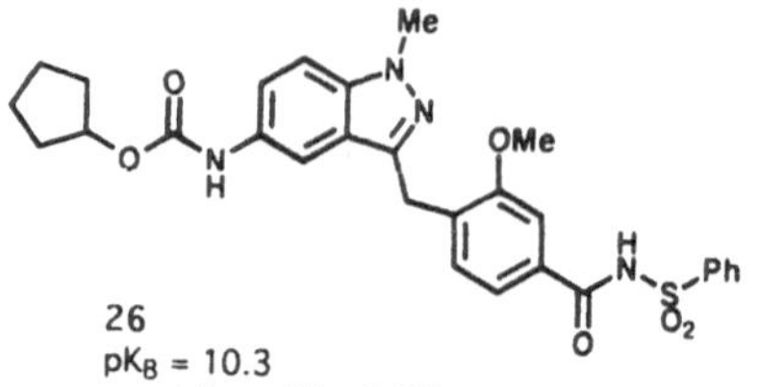

26
pK$_B$ = 10.3
po: 10 umol/kg @ 3 hr

Further explorations showed that the matrix could be expanded to include other bicyclic ring systems in addition to indoles and indazoles (Matassa et al., 1990a). For example, benzothiophene **27** and benzoxazine **28** were both potent antagonists. Such examples strengthened the postulate that the central ring system was serving primarily as a scaffold to fix the lipophilic and acidic appendages in an advantageous geometric orientation for receptor recognition. We focused primarily on the indoles and indazoles due to their greater potency (Structures **27, 28**).

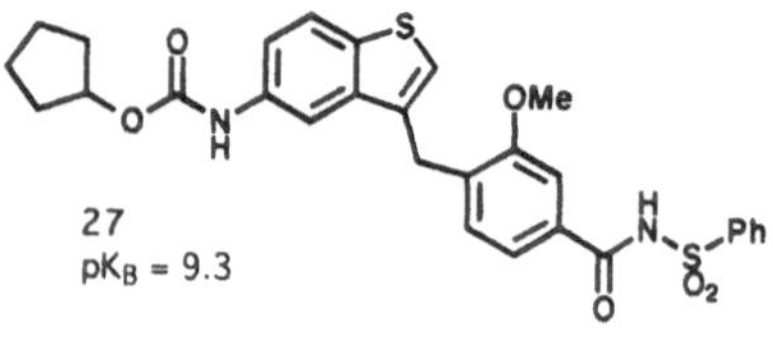

27
pK$_B$ = 9.3

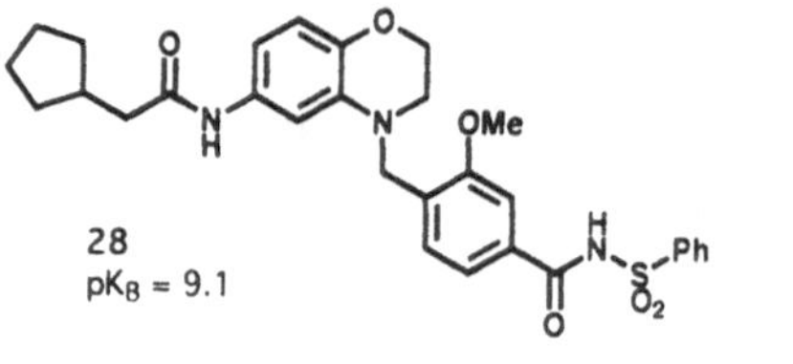

28
pK$_B$ = 9.1

Selection of a Candidate for Development

The indazole-phenylsulfonimides (both the 1,6- and 3,5-substituted regioisomers) were more potent on guinea pig trachea than the corresponding indoles (e.g., compare **26** and **25b** or **29** and **24b**). As the most potent (pK_B = 10.3) leukotriene antagonist known at the time, indazole **29**, ICI 198,615, was profiled extensively. It was at least 1000 times more potent than LY171883 (Aharony et al., 1987) and demonstrated a minimal selectivity ratio of 6300 when assessed opposite a wide range of nonleukotriene receptors (Snyder et al., 1987). The compound blocked antigen-induced contractions of sensitized guinea pig tracheal strips (Redkar-Brown et al., 1989). Tritium-labeled ICI 198,615 was prepared as a metabolically stable, highly selective ligand for investigations of leukotriene receptors (Aharony et al., 1988). ICI 198,615 was orally active at a dose of 10 μmol/kg (6 mg/kg) and exhibited a pharmacologic half-life in excess of 16 hours following oral dosing in the conscious guinea pig model (Krell et al., 1987). It antagonized leukotriene-induced increases in cutaneous vascular permeability in the guinea pig and reversed ovalbumin-induced increases of pulmonary resistance in passively sensitized guinea pigs (Krell et al., 1989) (Structure **29**).

29: ICI 198,615
pK_B = 10.3
po: 10 umol/kg @ 3 hr

In the dyspnea model, ICI 198,615 was active via intravenous administration at doses 333 times lower than that required for oral activity. This difference between the oral-dosing (po) and intravenous-dosing (iv) ED_{50} values suggested that oral bioavailability might be limited. Indeed, pharmacokinetic studies with ICI 198,615 demonstrated less than 0.3% oral bioavailability in dogs and very low blood levels in rats following oral dosing (Yee et al., 1990). These *in vivo* results can be contrasted with those for the related inverted indole **25a**. Although this compound was less active *in vitro* than the indazole, it was more potent orally. Moreover, the po/iv ratio for **25a** was approximately 25. These indications of better oral bioavailability were confirmed in the rat, where **25a** was shown to be 46% bioavailable with a plasma AUC^6 of 8.7 μg/hr per milliliter following an oral dose of 1 mg/kg.

By this point in the program oral activity was the primary concern, and evaluation of oral bioavailability became a critical part of the testing cascade. Low bioavailability in preclinical studies can translate into variable bioavailability in humans, which confounds dose-ranging and proof of efficacy trials. Moreover, we desired good bioavailability figures (greater than 20%) in the rat and dog to prove adequate exposure in toxicological studies. Since pharmacokinetic measurements of blood levels of drug following iv and po dosing were the most resource-limiting step in the cascade, a more facile estimate of bioavailability was sought. We found that for these particular series of compounds the po/iv effective-dose ratios from the conscious guinea pig dyspnea model were generally good predictors of HPLC-assessed oral bioavailability in the rat. Compounds with po/iv ED_{50} ratios lower than 100, and preferably below 30, in the dyspnea model generally demonstrated good bioavailability in the rat. Thus, the attainment of a good po/iv ratio in the guinea pig became a prerequisite for further evaluation (e.g., studies on duration of activity) of a new compound.

As noted above, we had speculated that our acylamino sulfonimide antagonists were binding at those portions of the receptor normally accommodating the lipophilic tetraene and ionic peptide segments of the leukotriene molecule. Thus, these compounds were only mimicking two of the three structural arms of the agonists. We postulated that the addition of a third polar appendage to the heterocyclic ring might enhance potency by accessing that part of the receptor responsible for binding the C-1 terminus of the leukotrienes. Indeed, the incorporation of an N-methylpropionamide[7] substituent at C-3 of the 1,6-substituted indole series (**30**) provided a 12-fold increase in activity over that of its unsubstituted parent **24a** (Brown et al., 1992). Although the oral activity was also improved, the magnitude of the increase (threefold) was not commensurate with the enhancement of the *in vitro* activity. However, with mimetics for all three arms of the leukotrienes now incorporated, this trisubstituted indole did provide further support for our hypothesis of how these antagonists might resemble the natural agonists at the receptor (Structure **30**).

30
pK_B = 10.7
po: 10 umol/kg @ 3 hr

Concurrent with the efforts to find good combinations of acylamino fragments, central templates, and phenylsulfonimides, modifications to the sulfonyl phenyl ring were being investigated (Yee, et al., 1988). Studies in the ethylhexanamido-indole series demonstrated that many substituents in various positions on the phenyl ring were well tolerated. An ortho-methyl group provided approximately a twofold increase in receptor affinity (compare **31** to **21b**). This moderate increase in activity with an ortho-tolylsulfonimide was corroborated in the cyclopentylacetamido-indole series (compare **32a** to **24a**). Other ortho substituents (e.g., Cl, NH_2, OMe) on the sulfonyl phenyl ring were also capable of providing especially potent antagonists. The more striking attribute of the tolylsulfonimide in **32a** was that it provided a dramatic 30-fold improvement in oral potency over that of the phenylsulfonimide **24a**, bringing the effective oral dose down from 30 μmol/kg to 1 μmol/kg. The analogous 2-chlorophenylsulfonimide **32b** also provided enhanced oral activity, although the effect was less dramatic (tenfold) in this particular case (Structures **31, 32**).

31
pK_B = 8.9

32a X = CH_2 ; R = Me
     pK_B = 9.9
     po = 1 umol/kg @ 3 hr
32b X = CH_2 ; R = Cl
     pK_B = 9.8
     po = 3 umol/kg @ 3hr
32c X = NH ; R = Me
     pK_B = 10.0
     po = 3 umol/kg @ 3hr

Once again we found that it was possible to transfer these structural modifications to the other related series and obtain similar beneficial effects. For example, the cyclopentylurea-tolylsulfonimide **32c** provided good oral activity. Likewise, the 3,5-substituted indazole-chlorophenylsulfonimide **33a** was more potent both *in vitro* and *in vivo* than an analogous phenyl-sulfonimide **26**. Indazole **33a** is one of the most potent known antagonists of the leukotrienes. These findings expanded our matrix of better leukotriene antagonists by adding a third dimension: in addition to mixing and matching the three cyclopentyl acylamino fragments with the four indole and indazole templates, ortho-substituted phenylsulfonimides (particularly the ortho-tolyl and 2-chlorophenyl sulfonimides) could also be utilized. In fact, the matrix was even larger. These antagonists were intrinsically so potent that other peripheral substituents were readily accepted. Compound **33b** illustrates that additional cyclic acylamino groups, in this particular case a cyclohexylacetamide, could be utilized. Indazole **33c** exemplifies the use of other small alkyl groups at N-1 and alternative ortho-substituents on the sulfonyl phenyl ring. Analogous substitutions were also feasible in the indole series (Structure **33**).

33a n = 0 ; X = CH$_2$; R = Me ; Y = Cl
        pK$_B$ = 10.8
        po: 1 umol/kg @ 3 hr
33b n = 1 ; X = CH$_2$; R = Me ; Y = Cl
        pK$_B$ = 10.8
        po: 3 umol/kg @ 3 hr
33c n = 0 ; X = O ; R = n-Pr ; Y = Br
        pK$_B$ = 10.0
        po: 3 umol/kg @ 3 hr

Unfortunately, the indazoles continued to suffer from low bioavailability, despite their robust oral profiles. Like the 1,6-substituted indazole **29**, the inverted indazoles **26** and **33a–c** had high po/iv ratios (133, 300, 417, and 295, respectively). Furthermore, **33a**, for example, could not be detected[8] in the blood of rats after an oral dose of 1 mg/kg. The high po/iv ratios recurred for a number of other indazole analogs, ultimately forcing them to be dropped from further consideration in favor of the indoles. The 3,5-substituted indole series provided particularly attractive profiles. The three cyclopentyl acylamino congeners **34a–c** in the tolylsulfonimide series are representative examples. In contrast to the indazoles, these three indoles exhibited low (less than 25) po/iv ratios in the guinea pig model and good oral bioavailability (30% or greater) in rats. The three indoles all had plasma AUC values in excess of 7.1 μg/hr/ml after 1 mg/kg oral doses in rats. In general, the inverted indoles exhibited two- to fivefold better oral activity than the 1,6-substituted series (e.g., compare **34a** and **34b** to **32a** and **32c**). Since a correlation between oral potency and lipophilicity had been observed for these molecules (e.g., the 30-fold improvement resulting from addition of an ortho-methyl group to the phenylsulfonimide), we added a C-3 methyl substituent into the 1,6-indole template to make the series isomeric with the inverted indoles. This ploy was successful (e.g., **35**) in elevating the oral activity of the 1,6-indoles to a level comparable with that of the best compounds (Structures **34, 35**).

34a X = CH$_2$
   pK$_B$ = 9.6
   po: 0.35 umol/kg @ 3 hr
34b X = NH
   pK$_B$ = 9.3
   po: 1.6 umol/kg @ 3 hr
34c X = O : ICI 204,219 ; ACCOLATE™
   pK$_B$ = 9.7
   po: 0.52 umol/kg (0.3 mg/kg) @ 3 hr

35
pK$_B$ = 9.3
po: 0.3 umol/kg @ 3 hr

ACCOLATE™ (ICI 204,219)

From this set of potent, orally active indoles, **34c** (ICI 204,219; ACCOLATE™ is a trademark, the property of Zeneca, for 4-(5-cyclopentyloxycarbonyl-amino- 1 -methylindol- 3 -ylmethyl)- 3 -methoxy- *N* - *o* -tolylsulfonylbenz-amide) was chosen on the basis of its superb overall profile for clinical evaluation (Krell et al., 1990). Accolate inhibited the binding of [^{3}H]-LTD$_4$ to guinea pig lung parenchymal membranes with a K$_i$ of 0.3 nM. In this assay the compound was 2-, 1873-, and 2582-fold more potent than LTD$_4$, LY171883, and FPL 55712, respectively. It inhibited the binding of [^{3}H]-ICI 198,615 to human lung parenchyma with a K$_i$ of 3.7 nM. On isolated guinea pig trachea the compound exhibited competitive antagonism against LTE$_4$-induced contractions, giving a concentration-independent pK$_B$ value of 9.6. It antagonized LTC$_4$-, LTD$_4$-, and LTE$_4$-induced contractions of human isolated intralobar airways with an average, agonist-independent pK$_B$ of 8.4 (Buckner et al., 1990). Accolate was highly selective for leukotriene receptors: at a concentration of 10 μM it showed no affinity for a variety of other receptors. The single exception was the EP$_1$ receptor in guinea pig ileum, where the compound exhibited a pK$_B$ of 5.6 (10,000 times less potent than on the leukotriene receptor). In the dyspnea model the compound was active via several routes of administration (the ED$_{50}$ values for administra-tion of the compound at the designated time interval prior to LTD$_4$ challenge are given): oral (0.52 μmol/kg; 180 minutes), intravenous (0.046 μmol/kg; 30 minutes), and aerosol (5.1 micromolar solution aerosolized for 5 min-utes; 10 minutes). The pharmacodynamic half-lives by these routes of delivery were 815, 85, and 109 minutes, respectively. Accolate was able to reverse ovalbumin-induced bronchoconstriction in passively sensitized, anesthetized guinea pigs, suggesting a potential therapeutic as well as pro-phylactic role for the compound. It also blocked LTD$_4$-induced increases in cutaneous vascular permeability in the guinea pig, demonstrating that nonpulmonary responses to leukotrienes are susceptible to the compound. Accolate demonstrated high oral bioavailability in rat (68%) and dog (67%) (Matassa et al., 1990). The synthesis and properties of the drug have been reviewed recently (Bernstein, 1994).

As has been illustrated here, Accolate is representative of a broad structural class of leukotriene antagonists characterized by a bicyclic ring carrying an acylamino chain and a 3-methoxy-4-methylbenzoyl aryl sulfonimide. While Accolate is one of the more advanced leukotriene antagonists in clinical trials, there are a number of compounds from other companies in various stages of development (Shaw and Krell, 1991). It is possible to discern three general categories of leukotriene antagonists based upon their structures: hydroxyacetophenones, those derived directly from the leukotrienes, and quinoline-based molecules (Figure 6-2). Examples from the hydroxyacetophenone series are YM-16638 and Ro 23-3544. The influence of the leukotriene structure in the design of antagonists is perhaps most clearly evident in LY 170680 and SKF 104353. MK-679 and WY-48,252 are examples from the quinoline series. Accolate does not fit into any of these three categories; it arose from a structurally unique class of LT antagonists discovered at ICI (now Zeneca).

Clinical Studies

Accolate has successfully completed Phase II clinical trials and is the first leukotriene antagonist to have entered Phase III studies in the US. Such trials are also underway in Europe. The drug appears to be well tolerated in asthmatic patients.

Early clinical studies built upon our previous investigations of the bronchoconstrictive effects imparted by aerosol administration of LTD_4 to normal volunteers (*vide supra*). In one of the first trials with Accolate, a single, 40 mg, oral dose provided more than a 100-fold shift of the LTD_4 dose-response curve, and significant antagonism (fivefold shift) persisted for at least 24 hours (Smith et al., 1990). This study demonstrated that Accolate had LTD_4 antagonist effects in humans.

With this encouraging finding in hand, a number of small-scale, double-blind, placebo-controlled trials were undertaken to address specific questions. Having established activity opposite a leukotriene challenge, the compound was evaluated against inhaled allergen provocation in asthmatics. A 40 mg oral dose of Accolate significantly decreased both early and late phase bronchoconstriction (Taylor et al., 1991; Findlay et al., 1992). Following a 20 mg oral dose, the drug was also active versus exercise-induced bronchoconstriction in asthmatic patients (Finnerty et al., 1992).

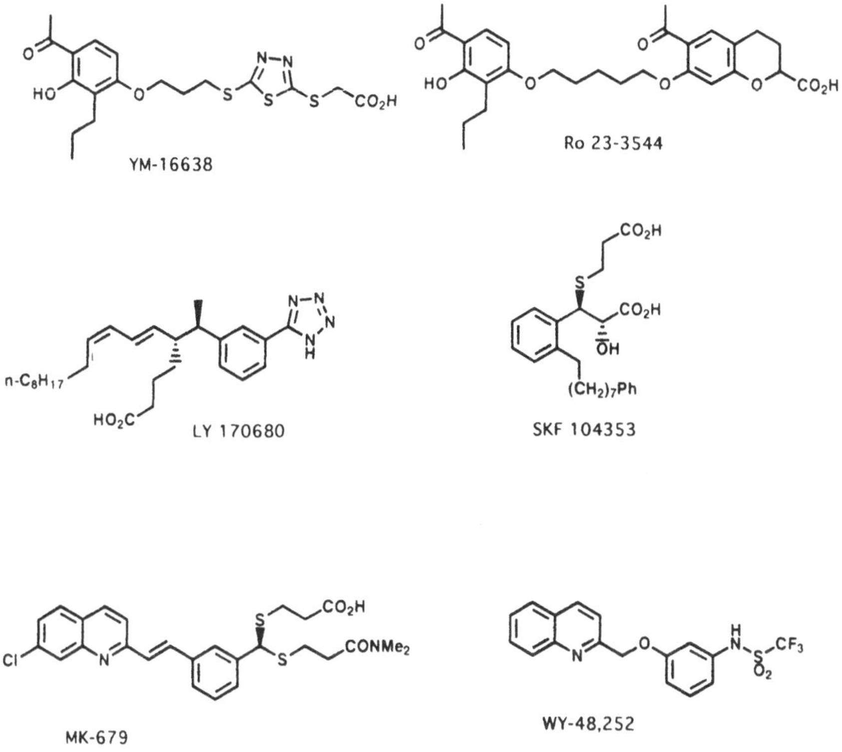

Figure 6-2. Other advanced leukotriene antagonists.

Several studies demonstrated the efficacy of aerosolized Accolate against allergen- or exercise-induced responses in atopic subjects (Makker et al., 1993; O'Shaughnessy et al., 1993).

Larger trials have demonstrated the efficacy of Accolate in improving pulmonary parameters in patients. A six-week, multicenter, dose-ranging study in 276 patients with moderate asthma demonstrated dose-dependent efficacy. Oral Accolate at 20 mg bid provided improvements in several pulmonary indices compared to placebo (Spector et al., 1993, 1994). These positive clinical findings are finally corroborating the long-held hypothesis that the peptide leukotrienes are contributors to the pathogenesis of asthma. Leukotriene antagonists are poised to represent the first novel therapeutic approach to the treatment of asthma in over two decades.

Summary

At the inception of our antagonist program only two ligands were known for the LT receptor—the endogenous agonists and FPL 55712. Given the lack of any structural information about the receptor, these two leads represented the only rational starting points for drug design. We progressed in parallel two series of compounds emanating from the two different starting points. Sequential modifications led to hydroxyacetophenone 5 from the FPL 55712 series and benzyl ether 9 from the leukotriene-based approach. Attempts to increase the potency of 9 via the introduction of conformational constraints generated a less active antagonist, indazole 14. Curiosity over a puzzling structure-activity discrepancy between the indazoles and their benzyl ether progenitors resulted in the perception of a structural similarity between two compounds from very different origins—the hydroxyacetophenone 5 and the LT-derived lipid-acid mimetic 14. This perception provided the incentive to bring together into one molecule structural components from the FPL 55712 and LT series. The hybridization of the two series generated the weak, but surprisingly selective, antagonist, indole 15, which established the indole/indazole compounds as a structurally novel lead.

As with any inventive endeavor, the process leading to the discovery and subsequent recognition of this new lead was dependent upon many varied factors for its success. Certainly, creative scientific design and dedicated effort from a multidisciplinary team were essential elements of the discovery process. But serendipity and good fortune also intervened from time to time. The fragility of the discovery process can be highlighted in this case by acknowledging how easy it would have been to have either missed or failed to capitalize upon the indazole lead. What if only the one indazole regioisomer 13 predicted to be optimal from the structure-activity relationships of the benzyl ether series 9 had been prepared instead of the matrix of eight regioisomers? Given the poor activity of ether 23, what if the ether linkage from 9 had been retained in the indazoles 12 instead of the synthetically-prompted amide functionality? What if less frustration had been encountered in the hydroxyacetophenone series—would there have been the same commitment to the pursuit of inherently weaker antagonists? Perhaps it is fortunate that we will never know the answers to these questions!

For the purposes of exploring structure-activity relationships, indole 15 was dissected into three segments which were investigated simultaneously. Progress toward more potent analogs was greatly accelerated by the discovery that beneficial structural modifications found in any one segment could be matched with changes found in another segment to generate hybrid molecules with enhanced activity. These were very heady days in the project. Every month the team would discover a new modification or a new combination of existing modifications that would shatter the existing potency record. The discovery of equipotent alternatives for each of the acylamino, central template, and acidic regions of the molecule generated an exceptionally large family of antagonists. A relatively high through-put *in vivo* assay was used to define an orally active subset and assess the oral to intravenous potency ratio. Compounds with low po/iv ratios were submitted for pharmacokinetic studies in rats to generate bioavailability data. From this winnowing process, Accolate was selected as an attractive candidate for clinical evaluation.

It is only now, more than 50 years after the discovery of SRS-A and after over a decade of intensive worldwide research following the elucidation of the leukotriene structures, that the importance of these mediators in the pathogenesis of asthma is being proven, as Accolate and other leukotriene antagonists and inhibitors prove successful in clinical trials. It is hoped that the second half of this decade will see the entry of such drugs into our armamentarium for asthma therapy.

A considerable number of very talented scientists contributed to Zeneca's leukotriene antagonist project, and the success of the program is due entirely to their dedicated efforts. Their names are provided in the cited references.

NOTES

[1] The biological activity reported herein was assessed by measuring the ability of the antagonists to inhibit contractions of guinea pig tracheal strips induced by 8 nM LTD_4. Data are presented as the percent inhibition of contraction provided by the given concentration of antagonist. As a general indicator of selectivity for leukotriene-receptor antagonism we measured the concentration of antagonist required to inhibit contractions of guinea pig trachea induced by the non-specific spasmogen barium chloride. The selectivity ratio of 20 for FPL 55712 indicates that at 20 times its leukotriene-inhibitory dose of 4 μM the compound also antagonizes the contractions induced by 1.5 mM barium chloride.

[2] Activity is reported as the dose sufficient to provide significant protection against the onset of LTD_4-induced dyspnea. The route (ip or po) and time interval prior to LT challenge for administration of the antagonists are also reported.

[3] The activities of this and all subsequent antagonists were measured against LTE_4 (instead of LTD_4) because of the demonstrated selectivity of LTE_4 for one of the two LTD_4 receptors in guinea pig trachea and the apparent pharmacological similarity of this receptor to the single human airway peptide-leukotriene receptor.

[4] The pK_B value is the negative logarithm of the dissociation constant of the antagonist, as determined by the construction of cumulative concentration-response curves using guinea pig tracheal strips with LTE_4 as the agonist.

[5] A component of the design rationale for the indole/indazole series had derived from a perceived structural similarity to the hydroxyacetophenone series (*vide*

supra). In this context it is interesting to note that incorporation of the phenylsulfonimide moiety into hydroxyacetophenone 5 was detrimental to *in vitro* activity—the compound was inactive at a concentration of $10 \mu M$ (Brown et al., 1989). Thus, the two series probably bind to the receptor in different ways.

[6] AUC: integrated area under the blood concentration versus time curve for the period 0–24 hours post-dose.

[7] It had been shown that the C-1 carboxylic acid of LTD_4 could be replaced by amide functionalities without seriously compromising the agonist activity (see references in Krell et al., 1984).

[8] The analytical limit of detection for the HPLC methodology was approximately 30 ng/ml.

REFERENCES

Aharony D, Dobson P, Krell RD (1985): *In vitro* metabolism of (^{3}H)-peptide leukotrienes in human and ferret lung: A comparison with the guinea pig. *Biochem Biophys Res Commun* 131:892–898

Aharony D, Falcone RC, Krell RD (1987): Inhibition of ^{3}H-leukotriene D_4 binding to guinea pig lung receptors by the novel leukotriene antagonist ICI 198,615. *J Pharm Exp Ther* 243:921–926

Aharony D, Falcone RC, Yee YK, Hesp B, Giles RE, Krell RD (1988): Biochemical and pharmacological characterization of the binding of the selective peptide-leukotriene antagonist ^{3}H-ICI 198,615, to leukotriene D_4 receptors in guinea-pig lung membranes. *Ann NY Acad Sci* 524:162–180

Appleton RA, Bantick JR, Chamberlain TR, Hardern DN, Lee TB, Pratt AD (1977): Antagonists of slow reacting substance of anaphylaxis. Synthesis of a series of chromone-2-carboxylic acids. *J Med Chem* 20:371–379

Augstein J, Farmer JB, Lee, TB, Sheard P, Tattershall ML (1973): Selective inhibitor of slow reacting substance of anaphylaxis. *Nature (London) New Biol* 245: 215–217

Bernstein PR (1992): Antiasthmatic agents. *Kirk-Othmer Encyclopedia of Chemical Technology*, 4th ed., Vol. 2. New York: John Wiley & Sons

Bernstein PR (1994): ACCOLATE™. *Drugs of the Future* 19:217–220

Bernstein PR, Snyder DW, Adams EJ, Krell RD, Vacek EP, Willard AK (1986): Chemically stable homocinnamyl analogues of the leukotrienes: Synthesis and preliminary biological evaluation. *J Med Chem* 29:2477–2483

Bernstein PR, Vacek EP, Adams EJ, Snyder DW, Krell RD (1988): Synthesis and pharmacological characterization of a series of leukotriene analogues with antagonist and agonist activities. *J Med Chem* 31:692–696

Brown FJ, Bernstein PR, Cronk LA, Dosset DL, Hebbel KC, Maduskuie TP Jr, Shapiro HS, Vacek EP, Yee YK, Willard AK, Krell RD, Snyder DW (1989): Hydroxyacetophenone-derived antagonists of the peptidoleukotrienes. *J Med Chem* 32:807–826

Brown FJ, Cronk LA, Aharony D, Snyder DW (1992): 1,3,6-Trisubstituted indoles as peptidoleukotriene antagonists: Benefits of a second, polar, pyrrole substituent. *J Med Chem* 35:2419–2439

Brown FJ, Yee YK, Cronk LA, Hebbel KC, Krell RD, Snyder DW (1990): Evolution of a series of peptidoleukotriene antagonists: Synthesis and structure-activity relationships of 1,6-disubstituted indoles and indazoles. *J Med Chem* 33:1771–1781

Buckle DR, Outred DJ, Ross JW, Smith H, Smith RJ, Spicer BA, Gasson BC (1979): Aryloxyalkyloxy- and aralkyloxy-4-hydroxy-3-nitrocoumarins which inhibit histamine release in the rat and also antagonize the effects of a slow reacting substance of anaphylaxis. *J Med Chem* 22:158–168

Buckner CK, Fedyna JS, Robertson JL, Will JA, England DM, Krell RD, Saban R (1990): An examination of the influence of the epithelium on contractile responses to peptidoleukotrienes and blockade by ICI 204,219 in isolated guinea pig trachea and human intralobar airways. *J Pharm Exp Ther* 252:77–85

Buckner CK, Krell RD, Laravuso RB, Coursin DB, Bernstein PR, Will JA (1986): Pharmacological evidence that human intralobar airways do not contain different receptors that mediate contractions to leukotriene C_4 and leukotriene D_4. *J Pharm Exp Ther* 237:558–562

Corey EJ, Clark DA, Goto G, Marfat A, Mioskowski C, Samuelsson B, Hammarström S (1980): Stereospecific total synthesis of a "slow reacting substance" of anaphylaxis, leukotriene C-1. *J Am Chem Soc* 102:1436–1439

Findlay SR, Bardern JM, Easley CB, Glass MJ (1992): Effect of the oral leukotriene antagonist, ICI 204,219, on antigen-induced bronchoconstriction in subjects with asthma. *J Allergy Clin Immunol* 89:1040–1045

Finnerty JP, Wood-Baker R, Thomson H, Holgate ST (1992): Role of leukotrienes in exercise-induced asthma: Inhibitory effect of ICI 204219, a potent leukotriene D_4 receptor antagonist. *Am Rev Respir Dis* 145:746–749

Fleisch JH, Rinkema LE, Haisch KD, Swanson-Bean D, Goodson T, Ho PPK, Marshall WS (1985): LY 171883, 1-[2-hydroxy-3-propyl-4-[4-(1H-tetrazol-5-yl)butoxy]phenyl]ethanone, an orally active leukotriene D_4 antagonist. *J Pharm Exp Ther* 233:148–157

Fleming DM, Crombie DL (1987): Prevalence of asthma and hay fever in England and Wales. *Br Med J* 294:279–283

Kellaway CH, Trethewie ER (1940): The liberation of a slow-reacting smooth muscle-stimulating substance in anaphylaxis. *Quart J Exp Physiol Cogn Med Sci* 30:121–145

Kern R, Smith LJ, Patterson R, Krell RD, Bernstein PR (1986): Characterization of the airway response to inhaled leukotriene D_4 in normal subjects. *Am Rev Respir Dis* 133:1127–1132

Krell RD, Aharony D, Buckner, C, Keith RA, Kusner EJ, Snyder DW, Bernstein PR, Matassa VG, Yee YK, Brown FJ, Hesp B, Giles RE (1990): The preclinical pharmacology of ICI 204,219—a peptide leukotriene antagonist. *Am Rev Resp Dis* 141:978–987

Krell RD, Brown FJ, Willard AK, Giles RE (1984): Pharmacologic antagonism of the leukotrienes. In: *The Leukotrienes, Chemistry and Biology*, Chakrin LW, Bailey DM, eds. Orlando FL: Academic Press

Krell RD, Giles RE, Yee YK, Snyder DW (1987): *In vivo* pharmacology of ICI 198,615: A novel, potent and selective peptide leukotriene antagonist. *J Pharm Exp Ther* 243:557–564

Krell RD, Kusner EJ, Aharony D, Giles RE (1989): Biochemical and pharmacological characterization of ICI 198,615: A peptide leukotriene receptor antagonist. *Eur J Pharmacol* 159:73–81

Krell RD, Tsai B-S, Berdoulay A, Barone M, Giles RG (1983): Heterogeneity of leukotriene receptors in guinea-pig trachea. *Prostaglandins* 25:171–178

Loftus P, Bernstein PR (1983): A study of some leukotriene A_4 and D_4 analogues by proton nuclear magnetic resonance spectroscopy. *J Org Chem* 48:40–44

Makker HK, Lau LC, Thomson HW, Binks SM, Holgate ST (1993): The protective effects of inhaled leukotriene D_4 receptor antagonist ICI 204,219 against exercise-induced asthma. *Am Rev Respir Dis* 147:1413–1418

Marshall WS, Goodson T, Cullinan GJ, Swanson-Bean D, Haisch KD, Rinkema LE, Fleisch JH (1987): Leukotriene receptor antagonists. 1. Synthesis and structure-activity relationships of alkoxyacetophenone derivatives. *J Med Chem* 30:682–689

Matassa VG, Brown FJ, Bernstein PR, Shapiro HS, Maduskuie TP Jr, Cronk LA, Vacek EP, Yee YK, Snyder DW, Krell RD, Lerman CL, Maloney JJ (1990a): Synthesis and *in vitro* LTD_4 antagonist activity of bicyclic and monocyclic cyclopentylurethane and cylopentylacetamide N-arylsulfonyl amides. *J Med Chem* 33:2621–2629

Matassa VG, Maduskuie TP Jr, Shapiro HS, Hesp B, Snyder DW, Aharony D, Krell RD, Keith RA (1990b): Evolution of a series of peptidoleukotriene antagonists: Synthesis and structure/activity relationships of 1,3,5-substituted indoles and indazoles. *J Med Chem* 33:1781–1790

Murphy RC, Hammarström S, Samuelsson B (1979): Leukotriene C: A slow-reacting substance from murine mastocytoma cell. *Proc Natl Acad Sci USA* 76:4275–4279

O'Shaughnessy KM, Taylor IK, O'Connor B, O'Connell F, Thomson H, Dollery CT (1993): Potent leukotriene D_4 receptor antagonist ICI 204,219 given by the inhaled route inhibits the early but not the late phase of allergen-induced bronchoconstriction. *Am Rev Respir Dis* 147:1431–1435

Patterson R, Harris KE, Smith LJ, Greenberger PA, Shaughnessy MA, Bernstein PR, Krell RD (1983): Airway response to leukotriene D_4 in rhesus monkeys. *Int Arch Allergy Appl Immun* 71:156–160

Redkar-Brown DG, Aharony D (1989): Inhibition of antigen-induced contraction of guinea pig trachea by ICI 198,615. *Eur J Pharmacol* 165:113–121

Schaaf TK, Hess H-J (1979): Synthesis and biological activity of carboxyl-terminus modified prostaglandin analogues. *J Med Chem* 22:1340–1346

Shaw A, Krell RD (1991): Peptide leukotrienes: Current status of research. *J Med Chem* 34:1235–1242

Smith LJ, Geller S, Ebright L, Glass M, Thyrum PT (1990): Inhibition of leukotriene D_4-induced bronchoconstriction in normal subjects by the oral LTD_4 receptor antagonist ICI 204,219. *Am Rev Respir Dis* 141:988–992

Smith LJ, Greenberger, PA, Patterson R, Krell RD, Bernstein PR (1985): The effect of inhaled leukotriene D_4 in humans. *Am Rev Resp Dis* 131:368–372

Smith LJ, Kern R, Patterson R, Krell RD, Bernstein PR (1987): Mechanism of leukotriene D_4-induced bronchoconstriction in normal subjects. *J Allergy Clin Immun* 80:340–347

Snyder DW, Bernstein PR (1988): U19052 (ICIAm): A novel leukotriene analog which antagonizes LTC_4, LTD_4, and LTE_4. *Prostaglandins* 35:903–915

Snyder DW, Giles RE, Keith RA, Yee YK, Krell RD (1987): *In vitro* pharmacology of ICI 198,615: A novel, potent and selective peptide leukotriene antagonist. *J Pharm Exp Ther* 243:548–556

Snyder DW, Krell RD (1984): Pharmacological evidence for a distinct leukotriene C_4 receptor in guinea pig trachea. *J Pharm Exp Ther* 231:616–622

Snyder DW, Liberati NJ, McCarthy MM (1988): Conscious guinea-pig aerosol model for evaluation of peptide leukotriene antagonists. *J Pharmacol Methods* 19:219–231

Spector SL, Smith LJ, Glass M, and the Accolate™ Trialists Group (1993): Dose-ranging study of oral ICI 204,219, a selective leukotriene antagonist, in asthma. *Clin Res* 41:199A

Spector SL, Smith LJ, Glass M, and the Accolate™ Trialists Group (1994): Effects of 6 weeks of therapy with oral doses of ICI 204,219, a leukotriene D_4 receptor antagonist, in subjects with bronchial asthma. *Amer J Respir Crit Care Med*, 150:618.

Taylor IK, O'Shaughnessy KM, Fuller RW, Dollery CT (1991): Effect of cysteinyl-leukotriene receptor antagonist ICI 204,219 on allergen-induced bronchocon-striction and airway hyperreactivity in atopic subjects. *Lancet* 337:690–694

Tsai BS, Bernstein PR, Macia RA, Conaty J, Krell RD (1982): Comparative po-tency and pharmacology of isomers of leukotriene D_4 on guinea pig trachea: Requirement for a 5(S)6(R) configuration. *Prostaglandins* 23:489–506

Weiss KB, Gergen PJ, Wagener DK (1993): Breathing better or wheezing worse? The changing epidemiology of asthma morbidity and mortality. *Ann Rev Publ Health* 14:491–513

Yee YK, Bernstein PR, Adams EJ, Brown FJ, Cronk LA, Hebbel KC, Vacek EP, Krell RD, Snyder DW (1990): A novel series of selective leukotriene antagonists: Exploration and optimization of the acidic region in 1,6-disubstituted indoles and indazoles. *J Med Chem* 33:2437–2451

Yee YK, Brown FJ, Hebbel KC, Cronk LA, Snyder DW, Krell RD (1988): Structure-activity relationships based on the peptide leukotriene receptor anatgonist, ICI 198,615, enhancement of potency. *Ann NY Acad Sci* 524:458–461

7

Discovery of ZD2138, a Potent, Selective, Well-Tolerated, Nonredox Inhibitor of the Enzyme 5-Lipoxygenase

G.C. Crawley, S.J. Foster, R.M. McMillan, and E.R.H. Walker

Introduction

Leukotrienes are a family of biologically active lipids formed by oxidative metabolism of arachidonic acid. Leukotriene B_4 (LTB_4) is a potent chemotactic agent and induces inflammatory reactions in animals and in humans. The peptido-leukotrienes, leukotrienes C_4, D_4, and E_4, have powerful spasmogenic actions in vascular and bronchial tissue and are the major mediators responsible for slow reacting substance of anaphylaxis (SRS-A) activity.

The discovery that leukotriene synthesis is catalyzed by 5-lipoxygenase (5LO) was recognized as an exciting opportunity for drug discovery. The target of the Zeneca 5LO inhibitor program was the development of well-tolerated, potent, orally active inhibitors of the enzyme, which possessed sufficient selectivity to avoid inhibition of the related cyclooxygenase (CO) enzyme system. In view of the fact that no validated 5LO-sensitive animal models were available at the outset of the work, the Team pursued the strategy of seeking compounds with optimized *in vitro* 5LO inhibitory properties followed by evaluation of characterized inhibitors in animal models. This chapter outlines the progress and issues associated with profiling potential inhibitors, the development of appropriate animal models, the design of selective enzyme inhibitors, and the selection of the Zeneca development candidate, ZD2138.

The Search for Anti-Inflammatory Drugs
Vincent J. Merluzzi and Julian Adams, Editors
© Birkhäuser Boston 1995

It was anticipated that drug design would be aided considerably by analogies with other widely studied lipoxygenases: for example, soybean 15-lipoxygenase was one of the first enzymes to be prepared in pure crystalline form. Thus, rapid progress in obtaining effective drugs was anticipated. However, discovery of inhibitors of 5LO was confounded by the anomalous behavior of the enzyme in *in vitro* systems. This is illustrated below by consideration of three assumptions that represent conventional wisdom in most drug discovery programs for enzyme inhibitors.

The first assumption is that the selectivity of inhibitors should be determined using isolated enzyme preparations. At the time of the elucidation of the 5LO pathway, modern molecular biological techniques for cloning and expression were less widely available than today and most researchers opted to employ purified or semipurified preparations of 5LO. Publications appeared on a number of broken cell assays for 5LO using leukocytes from different sources. It became clear that the enzyme was isolated from the soluble fraction; routinely the high-speed supernatant (100,000 g) from differential centrifugation was used as the enzyme source. Since CO, the related enzyme of arachidonate metabolism, which catalyzes formation of prostaglandins and thromboxanes, was known to be microsomal in origin (100,000 g pellet), it was straightforward to separate the two enzymes for selectivity testing.

Table 7-1. Inhibition of 5LO and CO by NDGA in different systems.

System	5LO*	CO*
Cell-free fractions (RBL-1)	0.5	>100
Intact rat PMN	0.2	2.5
Murine macrophages	0.6	2.5

* IC_{50} (μM)

Initial studies with isolated 5LO revealed that the enzyme was exquisitely sensitive to inhibition by a variety of known compounds that participate in redox reactions. This was not entirely unexpected since the accepted mechanism of all lipoxygenases involved an iron-catalyzed redox cycle as described below. More surprising was the observation that a number of such frank antioxidants exhibited impressive selectivity in isolated enzyme preparations. This is illustrated by nordihydroguaiaretic acid (NDGA) which profiled as a submicromolar inhibitor of 5LO but did not inhibit CO significantly at concentrations up to 100 μM *in vitro* (Table 7-1).

The profile of NDGA was much less impressive in cellular systems. Although the compound retained 5LO inhibitory potency in both PMN and macrophages, this was accompanied by inhibition of synthesis of CO products. Similar observations were obtained in this system using a variety of other redox inhibitors including flavonoids and the Takeda compound, AA861. Other groups reported comparable data with other compounds with similar mechanisms. The conclusion from these studies was that selectivity data obtained in isolated enzyme systems could be misleading.

The second assumption is that mechanistic and kinetic studies in isolated enzyme preparations are critical for the design of enzyme inhibitors. One interpretation of the data described above is that the major problem was the nonspecific nature of the inhibitors employed and that there would be better relationships between cell-free and intact cell systems when more selective inhibitors became available. To detect such compounds, mechanistic data would be required and therefore much research activity was directed at developing kinetic assays of 5LO. Cytosolic enzyme preparations were employed, and a continuous spectrophotometric assay was developed by our colleagues at Zeneca US to allow measurement of initial rates (Aharony and Stein, 1986). A complex pattern emerged. Enzyme reactions were characterized by the presence of lag phases prior to catalysis that, by analogy with other lipoxygenases, was assumed to represent autoactivation of 5LO by low levels of initial hydroperoxide product. Reactions also terminated prematurely, that is, prior to the depletion of substrate. For example, reduction of assay temperature or dilution of enzyme resulted not only in slowing of reaction rates but also in a marked reduction in the extent of the reaction. Addition of further substrate failed to induce more catalysis indicating that product-catalyzed inactivation had occurred.

A final problem with developing a kinetic assay for 5LO was that there was a marked reduction in initial rates at concentrations of substrate ($> 15 \mu$M) that were only slightly higher than its K_m for catalysis (approximately 5μM). In contrast, this was not observed with another 5LO substrate, eicosapentaenoic acid, which exhibited similar affinity for the enzyme. One explanation of this phenomenon was that arachidonic acid interacted with an additional site on the enzyme that resulted in substrate inhibition whereas eicosapentaenoic acid interacted only with the active site. An alternative explanation was that the reduction in initial rates was a consequence of poor solubility of arachidonic acid. Support for this hypothesis was obtained in a spectrophotometric assay of substrate solubility in buffer; arachidonic acid precipitated from solution at the concentrations at which the apparent substrate inhibition was observed whereas eicosapentaenoic acid did not (McMillan et al., 1989). Whatever the precise explanation for the phe-

Table 7-2. Potency of 5LO inhibitors in cell-free and intact cell assays.

No.	Structure	Class	Isolated enzyme*	Intact cells*
1 ZM198143	OH / NHCONH₂	Redox	60.0	1.40
2 ZM207968	(structure)	Redox	30.0	3.20
3 PF-5901	(structure)	Nonredox	10.0	0.06
4 ZM211965	(structure)	Nonredox	0.10	0.01

* IC$_{50}$ (μM)

nomenon, it was clear that a kinetic assay using arachidonic acid was not feasible, and even with eicosapentaenoic acid there remained problems of long lag phases and product inactivation.

The third assumption is that potency determination of enzyme inhibitors should be performed using isolated enzyme preparations; potency will decrease in more complex intact cell systems. As a result of anomalies such as those described above, there were concerns about the use of isolated enzyme preparations for primary screening of putative 5LO inhibitors. Our approach was to evaluate compounds in parallel in a cytosolic enzyme assay (guinea pig neutrophils) and in intact cells (murine macrophages). Again there were surprising data: a large number of structurally diverse compounds exhibited greater potency in the cellular assay than in the isolated enzyme system. Representative examples are shown in Table 7-2. The pattern was observed with inhibitors derived from both redox and nonredox series. This was not a result of species differences since similar observations were obtained when human leukocytes were employed for both the cellular and isolated enzyme assays.

A number of possible explanations were considered. For example, could the increased potency in cells be indicative of active transport of compounds? This was unlikely since there was a broad range of structures that showed the effect. Another possibility was that the compounds were inhibiting leukotriene synthesis in cells by an alternative mechanism (e.g., phospholipase A_2 inhibition, cytotoxicity). This was excluded since each of the compounds illustrated in Table 7-2 inhibited leukotriene synthesis selectively, that is, either without inhibiting synthesis of CO products or doing so only at much higher doses.

An explanation for the anomalies became apparent when it was discovered by researchers at Merck-Frosst that leukotriene synthesis in intact cells required translocation of 5LO from the cytosol to the nuclear membrane where it is in proximity to 5LO-activating protein (FLAP), which appears to present substrate to the enzyme. Thus, the environments of the enzyme and of the substrate in the cell-free assay are not representative of that within cells, and it is therefore not surprising that there are such discrepancies between the two assay systems.

As a consequence of these data, the testing cascade for 5LO inhibitors shown in Figure 7-1 was adopted.

The Mechanism of 5-Lipoxygenase

In designing selective inhibitors of 5LO, the team was influenced by consideration of the putative structure and mechanism of the enzyme although at the time that the program was initiated, detailed structural information about the enzyme family was limited. Thus, arachidonic acid was known to be oxygenated at positions 5, 12, and 15 by mammalian 5-, 12-, and 15-lipoxygenase, respectively, as well as at position 15 by the related plant enzyme, soybean lipoxygenase, and it was well established that lipoxygenases catalyze the hydroperoxidation of cis-1,4-pentadienyl systems. Soybean lipoxygenase was known to contain a non-heme iron atom (Sloane et al., 1990) and therefore by analogy the mammalian enzymes were also believed to be iron-containing. However, the amino acid residues responsible for iron chelation, catalysis, substrate binding, and substrate specificity were unknown, and no three-dimensional structural information was available. Although complex, the kinetics of lipoxygenases (Aharony and Stein, 1986) shed some light on the enzyme mechanism. On addition of substrate, three distinct kinetic phases are observed: a lag phase, during which the rate of product generation slowly increases to a steady state; a steady state phase, which obeys Michaelis Menten kinetics, and an irreversible enzyme inactivation phase, perhaps due to radical-induced damage. Indeed, incubation of the plant enzyme with substrate in the absence of oxygen generates

ISOLATED ENZYME ASSAY
- radiochemical (end point)
- spectrophotometric (kinetic)

⇓

CELLULAR EICOSANOID SYNTHESIS *IN VITRO*
- zymosan-stimulated macrophages
- LTC_4 and PGE_2 synthesis

⇓

WHOLE BLOOD EICOSANOID SYNTHESIS *IN VITRO*
- A23187-stimulated blood (human, rat, dog)
- LTB_4 and TxB_2 synthesis

⇓

EICOSANOID SYNTHESIS *IN* OR *EX VIVO*
- A23187-stimulated blood (rat, dog)
- zymosan-inflamed air pouch (rat)
- LTB_4 and cyclo-oxygenase products

⇓

SYSTEM-DEPENDENT ANIMAL MODELS
- arachidonic acid inflammation
- allergic bronchospasm

Figure 7-1. Screening cascade for discovery of 5-lipoxygenase inhibitors.

substrate-derived radical species which can be trapped with appropriate spin-trapping agents (Gibian and Galaway, 1977).

Gibian and Galaway rationalized these observations via a hypothetical mechanism for soybean lipoxygenase, modification of which (Figure 7-2) led to a putative mechanism for 5LO that was used to guide inhibitor design at Zeneca.

Briefly, this envisages that the resting state of the enzyme contains iron in the Fe^{2+} oxidation state. Enzyme activation occurs when the iron is oxidized to the active Fe^{3+} state by traces of hydroperoxides present in or formed from added substrate. The time taken to activate the enzyme explains the lag phase observed in kinetic studies. It is then envisaged that

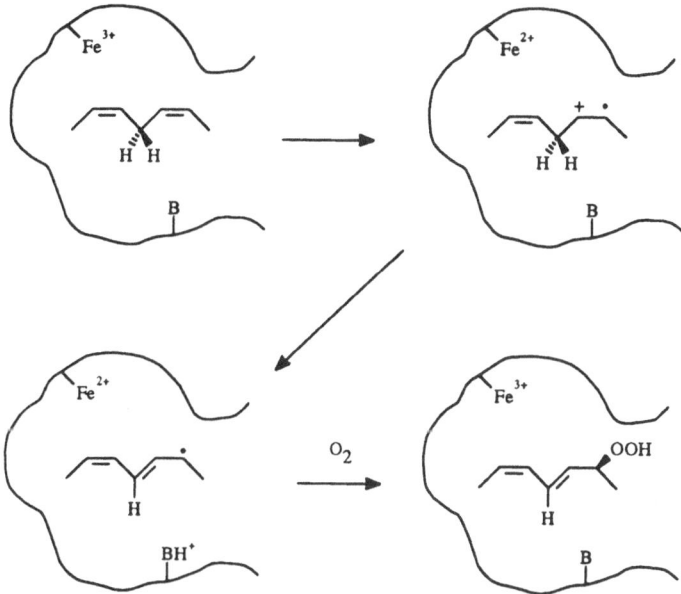

Figure 7-2. Hypothetical mechanism of 5-lipoxygenase.

Fe^{3+} specifically oxidizes the 5,8 diene system to generate a pentadienyl radical and a proton, with iron being reduced to the Fe^{2+} form. The radical, while located at the enzyme active site, stereospecifically reacts with oxygen at C-5 to generate the 5S-hydroperoxy radical. This undergoes electron transfer and protonation to regenerate the Fe^{3+} form of the enzyme together with the hydroperoxide product, 5-HPETE, which may diffuse from the active site.

Table 7-3 summarizes the target biochemical profile the Team used in designing selective 5-lipoxygenase inhibitors. The key aspects of the profile on which the Team focused were potency and selectivity in the *in vitro* and *ex vivo* systems. Thus, as the program evolved, the Team moved away from a specific potency target in the isolated enzyme system, and structure-activity analyzes were concentrated on cellular and whole blood assays.

As discussed below, medicinal chemistry focused on optimizing bio-chemical efficacy both *in vitro* and *ex vivo*. Once compounds were obtained with the target biochemical profile they were evaluated in systems models of inflammation. Selected compounds were also evaluated in cyclooxygenase-dependent *in vivo* models (e.g., carrageenan-induced inflammation, gastro-intestinal tolerance); in these cases the objective was to seek confirmation

Table 7-3. Target biochemical profile for selective 5-lipoxygenase inhibitors.

Biological system	5LO*	CO*
Isolated enzyme	Activity	Not measured routinely
Cellular assay	$<10^{-7}$ M	$>10^{-4}$ M
Whole blood *in vitro*	10^{-7} M to 10^{-6} M	$>10^{-4}$ M
Whole blood *ex vivo*	<1 mg/Kg	>100 mg/Kg

* IC_{50} or ED_{50}

that none of the pharmacology of the agents could be ascribed to cyclooxygenase inhibition by the parent compound or a metabolite.

The Design of Inhibitors of 5-Lipoxygenase

The hypothetical mechanism described in Figure 7-2 was used as a framework to rationalize the inhibitory mechanisms of known 5LO inhibitors and to aid design of new ones. Medicinal chemistry approaches were based on *de novo* design, modification of published inhibitors, and the evaluation of compounds from the Zeneca compound collection which were selected to explore design hypotheses.

Three broad approaches were explored to identify inhibitors of the enzyme; these can be classified as redox inhibitors, iron ligand inhibitors, and nonredox active site directed inhibitors.

Redox Inhibitors

5LO and the related arachidonic acid metabolizing enzyme, cyclooxygenase, are susceptible to inhibition by compounds with low redox potentials. The prototypes of this class of inhibitors are NDGA, phenidone, and the Wellcome compound, BW755C. All these compounds can readily function as one-electron reducing agents (Duniec et al., 1983). Redox inhibitors, as these compounds are loosely termed, potentially can interact at a number of points in the enzyme mechanism described above. Thus, for example, they may reduce iron from the active Fe^{3+} to the inactive Fe^{2+} or reduce one of the radical intermediates leaving the enzyme in the inactive Fe^{2+} state. Despite the attractions of high *in vitro* activity displayed by this class of compounds, there were concerns about the ability of redox agents to interact with other biological redox systems, and one toxic consequence of these interactions is the formation of methemoglobin, a side-effect that had been observed in *in vivo* studies with BW755C (Fort et al., 1984).

At Zeneca, a range of catechol and aminophenol analogs was evaluated in order to explore the activity and SAR of these series and to determine whether they could provide compounds that met our target profile. Compounds were prepared that incorporated the catechol unit into a lipophilic chain that might mimic some portions of arachidonic acid (e.g., Structure **6**, Table 7-4).

Table 7-4. Catechol 5LO inhibitors.

No.	Structure	Isolated enzyme IC$_{50}$ (μM)	Rat whole blood % inhibition at 1 μM LTB$_4$	PGE$_2$
5		0.55	50	8
6		0.25	45	36

Typically these compounds were of similar potency to NDGA (Structure **5**) as *in vitro* inhibitors of the RBL enzyme system and of LTB$_4$ generation in blood. As a class, catechols also inhibited CO and additionally failed to show any activity when administered orally, and therefore were not pursued further.

Exploration of aminophenol series generated compounds with more interesting profiles of activity (Table 7-5). In particular, hydroxydiphenylamines possessed much greater intrinsic potency than the catechols and not only inhibited LTB$_4$ generation in whole blood *in vitro* but, in contrast to catechols, were orally active in the rat when administered at high doses (100 mg/kg). Ureidophenols such as ZM198143 (Structure **1**) also shared these characteristics.

Taken together, catechols and aminophenols show poor selectivity for 5LO relative to CO and, despite displaying potent activity *in vitro*, are weak or inactive enzyme inhibitors *in vivo* when dosed orally. Structure-activity relationships of compounds in this type are generally diffuse. The most notable feature is that, although redox properties are essential for activity, lipophilicity plays a dominant role in determining potency (Hammond et al., 1989; Hlasta et al., 1991). In view of the known propensity for redox

Table 7-5. Amino- and ureidophenol 5LO inhibitors.

No.	Analog	Isolated enzyme IC$_{50}$ (μM)	In vitro % inhibition* LTB$_4$	PGE$_2$	Ex vivo % inhibition** LTB$_4$	PGE$_2$
7		0.15	96	43	<50	<50
8		0.03	80	75	75	56
9		0.16	83	30	70	19
1 ZM198143		59	54	0	75	45

* inhibition at 3 μM in rat whole blood
** inhibition in rat whole blood 1 hr after an oral dose of 100 mg/kg

processes in proteins to occur over large donor/acceptor distances, agents of this type can potentially inhibit the enzyme without the need to form strong enzyme-inhibitor interactions and this could account for the modest selectivity for 5LO and diffuse structure activity relationships. In addition, they may readily interact with other biological redox systems and these effects could explain the formation of methemoglobin, a side-effect that has plagued many redox inhibitors.

Cyclic voltammetry indicates that aminophenols and catechols possess redox potentials in the 0.2 to 0.6 v range and can function as powerful one-electron reducing agents. In an attempt to design more selective 5-lipoxygenase inhibitors, we focused on the preparation of agents with relatively weak redox properties, which we hoped would be capable of inhibiting 5LO without disrupting CO or other essential oxidative processes. This approach lead to the interesting discovery that indazolinones that possess weaker reducing properties (0.7–0.9 v) are effective inhibitors of the enzyme with some members of the series showing high selectivity for the 5LO pathway (Bruneau et al., 1991).

The unsubstituted parent indazolinone (Table 7-6) inhibited 5LO with an IC_{50} of $2\mu M$ but showed poor selectivity and no oral activity at 100 mg/kg. Substitution at N-2 with alkyl or aralkyl groups provided derivatives with improved profiles. For example, the benzyl derivative **11** was more potent, more selective and displayed oral activity at 20 mg/kg. While this compound showed an interesting profile, further increases in both selectivity and oral activity were required to meet the team's target profile. Simple substitution of the benzyl ring failed to modulate activity but replacement of the N-2 benzyl group by a naphthylmethyl moiety markedly varied activity. The 1-naphthyl analog, **13**, showed potent inhibition of LTB_4 generation with a good selectivity ratio IC_{50} $(PGE_2)/IC_{50}(LTB_4)$ of 160. Its close 2-naphthyl analog, **12**, showed similar LTB_4 inhibitory potency *in vitro* but was a more potent inhibitor of PGE_2 generation, resulting in a diminished selectivity ratio of 18. The pure enantiomers, (R)- and (S)-**14** of the methylated 2-naphthyl analog also showed different profiles. These data support the view that, in contrast to catechols and aminophenols, small structural changes modulate the potency and selectivity of indazolinones that possess identical lipophilicity and redox properties. In contrast to aralkyl derivatives, the 3-pyridylmethyl analog, ZM207968 (**2**), was notable in combining potent inhibition of LTB_4 generation *in vitro*, a high level of selectivity and potent oral activity. ZM207968 inhibited LTB_4 generation in blood with an ED_{50} of 3 mg/kg while not inhibiting PGE_2 production at doses as high as 300 mg/kg.

The structure-activity relationships for selectivity among pyridine-containing indazolinones were reminiscent of those for thromboxane synthetase inhibition by pyridine derivatives. It was found that in whole blood ZM207968 was a more potent inhibitor of thromboxane B2 synthesis (IC_{50} 50 μM) than PGE_2 synthesis (IC_{50} 470 μM). Thus, the impressive LTB_4/PGE_2 selectivity of ZM207968 reflects a balance between two effects: (1) reduced leukotriene synthesis as a result of 5LO inhibition, and (2) elevated PGE_2 levels as a consequence of diversion of the pivotal endoperoxide intermediate to prostaglandin products during thromboxane synthetase inhibition (Bruneau et al., 1991).

Despite displaying high selectivity for the 5LO pathway, ZM207968 caused transient methemoglobin formation when dosed orally in the dog, almost certainly as a consequence of its redox properties, and this precluded its development for clinical use. By incorporation of a 4 carbamoyl substituent, e.g., **15**, which further reduces redox properties, it was possible to prevent methemoglobin formation without significantly reducing 5LO inhibitory potency (Bruneau et al., 1991). Thus, methemoglobin formation appears to be more sensitive to redox potential than 5LO inhibitory activ-

Table 7-6. Indazolinone 5LO inhibitors.

No.	Structure	Human whole blood IC$_{50}$ (μM) LTB$_4$	PGE$_2$	Rat oral activity ED$_{50}$ for LTB$_4$ inhibition at 1 h LTB$_4$ (mg/Kg)	Redox Potential (volts)
10		2	3	>100	0.71
11		0.36	18	20	0.69
12		0.97	18	>100	
13		0.61	100	>100	0.73
14		R 1.4 S 11	400 48	400 48	>100 0.69
2 ZM207968		1.5	460	3	0.71
15		1.5	>250	30	0.9

ity. This dissociation of 5LO inhibition and methemoglobin induction was encouraging and supported the hypothesis that redox potential can modulate selectivity. Unfortunately, the 4 carbamoyl analogs did not share the oral potency of the parent compound. In addition, the risk of metabolic conversion *in vivo* to potent methemoglobin inducers was considered too high to justify further development of such compounds.

Iron Ligand Inhibitors

An alternative approach to the rational design of 5LO inhibitors is via compounds that contain a functional group capable of forming a ligand interaction with the iron atom thought to be present at the active site of the enzyme. One of the most powerful metal ligand groups is the hydroxamic acid moiety, and one of the first reports of compounds from this class was from Corey, who described the *in vitro* 5LO inhibitory properties of arachidonyl hydroxamic acid and its N-alkyl derivatives (Corey et al., 1984). To explore this approach, a range of alkyl and aryl hydroxamic acids was prepared that contained alkyl and aryl groups designed to mimic the C_{10-20} portion of arachidonic acid and a functionalized alkyl moiety attached to the nitrogen atom that might mimic the C_{1-5} portion (Table 7-7). Interestingly, in both the alkyl and aryl series, N-substituents containing ester and nitrile groups were more potent than the corresponding acids, and no convincing acidic binding could be identified with these compounds.

With the exception of sterically demanding 2-substituted aryl derivative **18**, a wide range of lipophilic groups was acceptable, with many analogs showing good potency as inhibitors of LTB_4 generation in whole blood. However, as in the redox series, structure-activity relationships were diffuse, with lipophilicity having a dominant influence on activity, and consequently there was little evidence of direct interactions between the inhibitors and the enzyme. In order to probe whether these compounds interacted in a specific way with the enzyme, a number of chiral inhibitors were prepared in pure enantiomeric forms. In each case, pairs of enantiomers (e.g., (*R*)- and (*S*)-**20**) displayed essentially the same level of enzyme inhibitory activity. A further concern over this approach was that most analogs had limited selectivity for leukotriene versus prostaglandin inhibition in whole blood.

In view of the doubts that these compounds were forming specific interactions with the enzyme, it was concluded that this series was unlikely to lead to compounds with the desired level of selectivity and therefore the area was not pursued further. Recent work has established that, in addition to their metal liganding properties, hydroxymates also possess weak redox properties, which are likely to be enhanced in the presence of iron, and there is increasing evidence that the 5LO inhibitory activity of this class of compound may be mediated in whole or in part by these effects (Riendeau et al., 1991). Limited selectivity and lack of enantioselectivity were also observed by the Abbott and Wellcome groups, although they successfully translated the *in vitro* potency of hydroxymates and N-hydroxyureas to orally active compounds which were evaluated clinically (Summers et al., 1988; Carter et al., 1991; Jackson et al., 1988).

Nonredox Inhibitors

The foregoing discussion illustrates the evolution of the Zeneca Team's

Table 7-7. Hydroxamic acid 5LO inhibitors.

No.	Structure	Human whole blood IC$_{50}$ (μM)
16		0.31
17		0.1
18		>10
19		0.25
20		R 0.16 S 0.3

concerns over the 5LO inhibitor profile of redox and iron ligand inhibitors. The Team always had reservations about the level of selectivity of such agents and their potential to provide the oral potency and safety required of a clinical candidate. Thus, although for a period the majority of our medicinal chemistry resource was deployed on redox and iron ligand inhibitors, we always maintained some effort deployed on other approaches. The target for these was a compound that acted at the active site of the enzyme and was devoid of redox and iron-chelating properties.

Ideally, to design agents to act at the active site of an enzyme, some knowledge of the structure of the enzyme is required, but at the stage in the program at which this work was initiated, none was available. We therefore used the hypothetical 5LO mechanism described earlier and outlined in Figure 7-2 to predict the functionality present at the active site of the enzyme and to use this information to design inhibitors that could fit this model. This mechanism implies that an iron atom and a basic group are present in the active site, and it was principally these groups with which our inhibitors were designed to interact. Potential inhibitors were designed by incorporating hydrogen bond-donor and iron-interacting groups as well as lipophilic groups to mimic the lipophilic tail of the substrate. Two ideas that were explored extensively are illustrated in Figure 7-3.

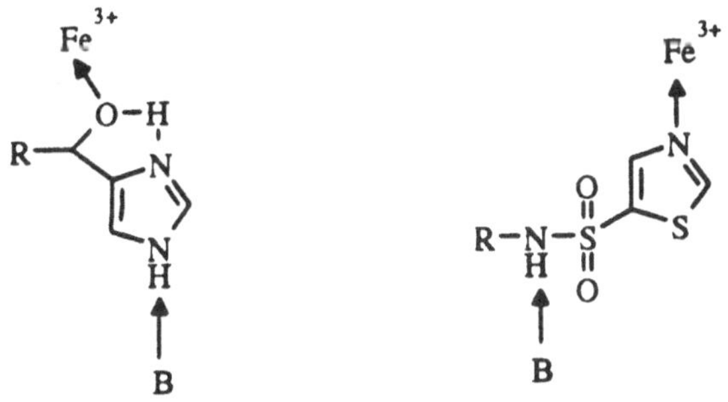

Figure 7-3. Approaches to nonredox 5LO inhibitors.

None of our approaches to nonredox inhibitors yielded active compounds in the isolated enzyme assay based on a radiochemical end point, which was the primary screen used at the time. This screen used 5LO from rat basophilic leukemia cells and measured total product after enzyme inactivation. The assay was very effective at detecting the activity of redox inhibitors but we were concerned that in this assay compounds could inhibit the initial rate, but not total product, and would appear inactive. Using the spectrophotometric assay described earlier, we found activity in two of our nonredox approaches, (hydroxyalkyl)imidazoles and the sulfonamidothiazole, exemplified in Structures **21** and **22** (Figure 7-4).

For a time, these two series were developed concurrently. With the sulfonamidothiazole series it proved easier initially to find activity in rat whole blood *in vitro*, but potency could not be improved much below 10 μM in this assay (Table 7-8).

In the case of the (hydroxyalkyl)imidazoles, activity in whole blood proved elusive. As part of a wider investigation of the SAR of (hydroxyalkyl)imidazoles, we explored various other heterocycles and found that imidazole could be replaced by 2-thiazolyl (**25**, Table 7-9). As a result of

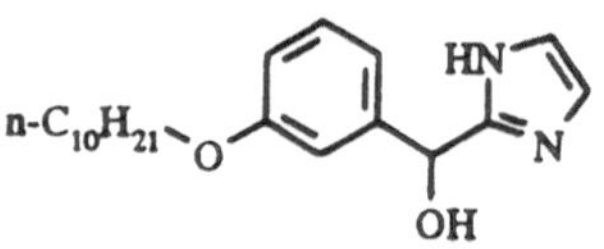

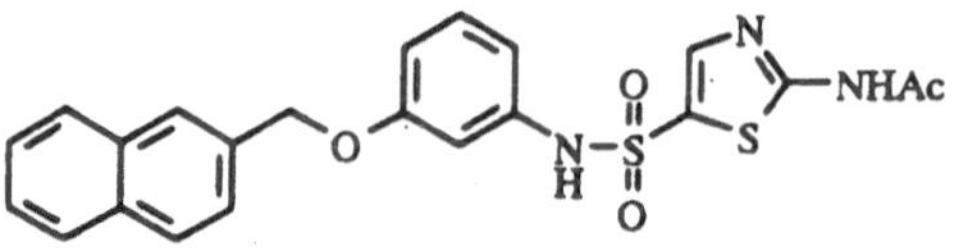

21

11 μM

22

4.6 μM

Figure 7-4. Lead 5LO nonredox inhibitors; inhibition of guinea pig 5LO.

Table 7-8. Sulphonamidothiazole 5LO inhibitors.

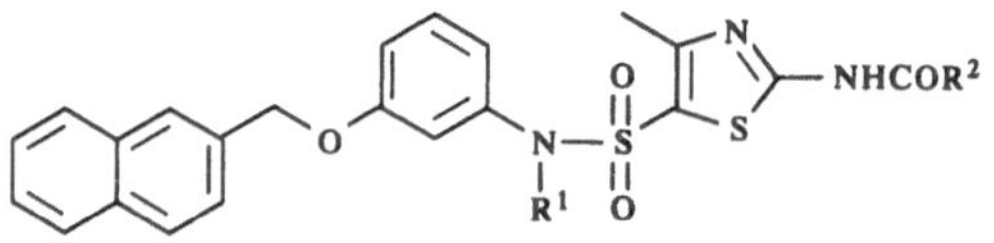

No.	R^1	R^2	Guinea pig 5LO IC_{50} (μM)	Rat blood 5LO IC_{50} (μM)
23	H	NHMe	0.6	10
24	Me	OEt	0.6	8

this change, activity in the isolated enzyme assay was retained, and we also observed weak inhibition in blood. In moving from imidazole to thiazole, we had removed an H-bond donor group, and we speculated that the hydroxyl was now fulfilling that role. An analog containing methoxy in place of hydroxy was prepared with the expectation, on the basis of this hypothesis, that it would have lower activity. Surprisingly, not only was activity retained, but compound **26** exhibited increased potency in the whole blood systems. Further methylation at the benzylic carbon, which has the effect of removing some conformational freedom, produced further improvement in potency so that in the (methoxyalkyl)thiazole ZM211965 (**4**) we had a potent inhibitor in various *in vitro* systems.

Based on their structures, redox properties were not anticipated, and cyclic voltammetry measurements indicated no significant redox potential in either series. In view of the greater potency of the (methoxyalkyl)thiazoles in blood, resources were deployed on this series of nonredox inhibitors.

Selectivity of ZM211965 for inhibition of the 5LO pathway compared

Table 7-9. Enhancement of 5LO inhibition in blood by (methoxyalkyl)thiazoles.

No.	R^1	R^2	Guinea pig 5LO IC_{50} (μM)	Rat whole blood 5LO IC_{50} (μM)
25	H	H	5.1	30
26	H	Me	12	6
4	Et	Me	0.1	0.5

to the CO pathway was determined in zymosan-stimulated macrophages and ionophore-stimulated human whole blood. ZM211965 inhibited LTC$_4$ synthesis in macrophages and LTB$_4$ synthesis in human blood with IC$_{50}$s of 8 nM and 0.4 μM respectively but did not inhibit formation of CO products up to 50 μM in macrophages or 100 μM in human blood. These data represented selectivities in excess of 250 in blood and 5000 in macrophages. Subsequently, we found ZM211965 to be a weak inhibitor of LT synthesis in rat blood *ex vivo* from an oral dose, the first demonstration of oral activity in a nonredox series. Based on these data and the growing concerns with the redox indazolinone series, we transferred all our medicinal chemistry behind the (methoxyalkyl)thiazole series.

In contrast to the diffuse nature of the SAR of redox and iron ligand series, it quickly became apparent that the SAR of (methoxyalkyl)thiazoles was far more distinct; small changes between essentially iso-lipophilic structures often produced large changes in *in vitro* potency. We will illustrate this here with one example (Table 7-10).

The (methoxyalkyl)thiazoles are chiral, and this gave us the opportunity to determine whether the separate enantiomers display different activities, arguably the definitive test for specificity of interaction with a protein target. The activity of the indane analog, Structure **27** (Table 7-10), was found to reside almost exclusively in the (+)-isomer with the (−)-isomer essentially inactive (Bird et al., 1991). This was true not only for cell-free 5LO but also for plasma protein-free cultures of macrophages as well as human whole blood. More recently, enantioselectivity has been demonstrated in the open chain derivative, Structure **28** (Lambert-van der Brempt et al., 1994). The high enantioselectivity of the (methoxyalkyl)thiazoles and a related series (see below) is unique, as far as we are aware, in the 5LO area.

The demonstration of distinct and convincing SAR gave us confidence

Table 7-10. Enantioselective inhibition of 5LO by (methoxyalkyl)thiazoles.

27   **28**

No.	RBL 5LO IC$_{50}$ (μM)	Mouse macro- phage 5LO IC$_{50}$ (μM)	Human whole blood 5LO IC$_{50}$ (μM)
($\pm$)-**27**	0.8	0.014	0.7
(+)-**27**	0.13	<0.03	0.5
(−)-**27**	>20	1.5	>40
($\pm$)-**28**			0.73
(R)-(−)-**28**		0.51	6.5, >10
(S)-(+)-**28**		0.0007	0.54

that the (methoxyalkyl)thiazole skeleton could be used as a framework for the rational design of more potent inhibitors.

ZM211965 showed modest oral activity in *ex vivo* rat blood with ED$_{50}$s of 10 and >100 mg/kg, 1 and 3 hrs, respectively, after dosing. Despite extensive efforts, no analog in the (methoxyalkyl)thiazole series could be found with oral potency that matched the target we were seeking for a clinical candidate. At this stage, we undertook a wider exploration of the SAR of (methoxyalkyl)thiazoles, and this led by the steps outlined in Figure 7-5 to 4-methoxytetrahydropyrans, a second series of nonredox inhibitors.

The parent compound of this series, ZM218287 (Structure **29**), was more potent than ZM211965 in all *in vitro* tests (Table 7-11).

In progressing from ZM211965 to ZM218287, the ethyl and thiazolyl groups of the former had been replaced by a tetrahydropyran ring in the latter. We wondered at first whether the 4-methoxytetrahydropyrans rep- resented a completely new series, but since then, we have generated con- siderable SAR data which run parallel in each series, implying that the two series are related. Some of this SAR is presented in Table 7-12.

For example, in each series, substituting a methyl group at the 2-position of the central phenyl ring retained potency at a similar level (entry 2) whereas introducing a methyl group at the 6-position reduced potency by

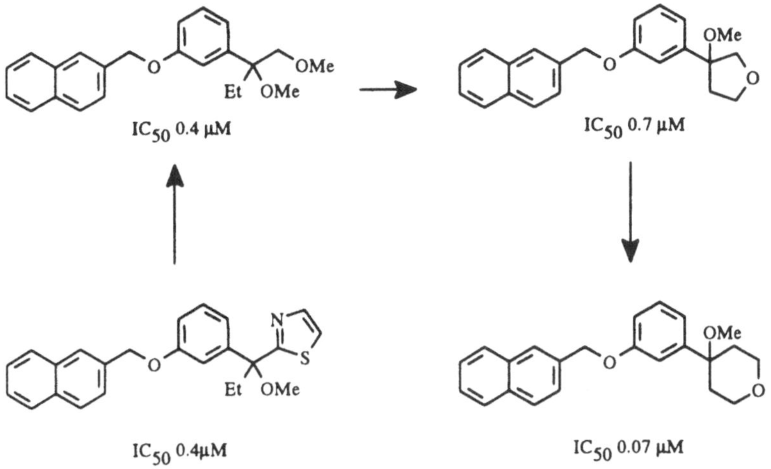

Figure 7-5. Evolution of 4-methoxytetrahydropyran series from (methoxyalkyl)thiazole series: IC_{50}s of LT biosynthesis in human whole blood.

Table 7-11. Inhibition of 5LO *in vitro* by ZM218287.

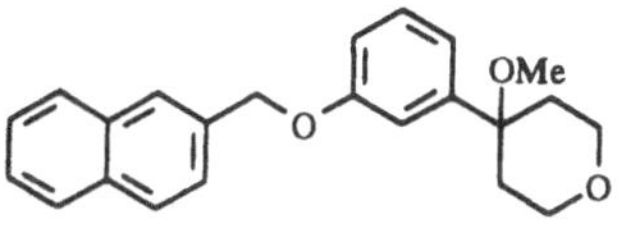

29

No.	RBL 5LO IC_{50} (μM)	Mouse macrophage 5LO IC_{50} (μM)	Human whole blood 5LO IC_{50} (μM)
29 ZM218287	0.1	0.0005	0.07

at least 100-fold (entry 3). In both series, a para arrangement of groups in the central phenyl ring markedly reduced potency (entry 4) which was restored by reintroduction of an ether substituent at position 3 (entry 5). Results such as these indicated that the two series are related and implied that each bound at the same locus on 5LO with several contact points in common.

At this stage, the medicinal chemistry resource on the project was increased, and we became highly focused on finding a clinical candidate from within the 4-methoxytetrahydropyran series. This necessitated streamlin-

Table 7-12. Comparative SAR of (methoxyalkyl)thiazole and 4-methoxytetra-hydropyran series of 5LO inhibitors.

Entry	Position of naphthylmethoxy group in central phenyl ring	R	Human whole blood 5LO IC$_{50}$ (μM)	
			X = [thiazole, Me/Et, OMe]	X = [tetrahydropyran, OMe]
1	3	H	0.5 (Me); 0.4 (Et)	0.07
2	3	2-Me	0.8 (Me)	0.6
3	3	6-Me	~ 40 (Me)	~ 40
4	4	H	~ 40 (Et)	19
5	4	3-OMe	0.5 (Et)	0.6

ing the testing cascade to the essential assays to achieve our objective. As the crucial role of FLAP in the cellular mechanism of 5LO emerged (Miller et al., 1990), we had become increasingly concerned over the relevance of our broken cell 5LO assay, in which FLAP was absent, to LT biosynthesis in whole cells. We found that there was a reasonable correlation between potency of inhibition of 5LO in the isolated enzyme assay and potency of inhibition of LT biosynthesis in zymosan-stimulated macrophages (Figure 7-6A). However, inhibitors were consistently more potent in whole cells than against the enzyme. This observation could have been interpreted as indicating that our nonredox inhibitors had activity in addition to 5LO inhibition that was contributing to inhibition of LT biosynthesis. However, this was regarded as unlikely as we found a very similar correlation with indazolinone redox inhibitors (Figure 7-6B). In view of the good correlation between cell-free and cellular assays and the fact that the cellular assay was providing more relevant potency data, the enzyme assay was dropped from the cascade. The human whole blood test was used as the sole *in vitro* test driving medicinal chemistry, with occasional compounds evaluated in macrophages. The reasons for our reliance on the blood assay were that

A

IC50 (ZSM) vs IC50 (5LO) FOR ZENECA NON-REDOX INHIBITORS

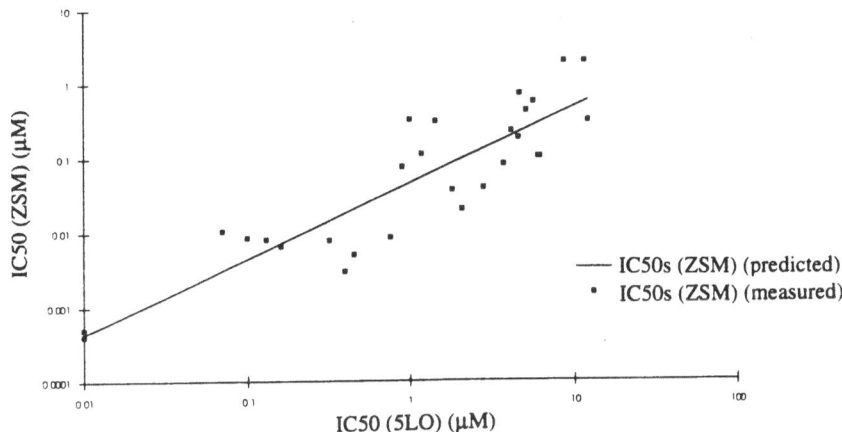

Experimental data fit by linear regression analysis to $\log(IC_{50}.ZSM) = 1.01 \log(IC_{50}.5LO) -$
1.34, $n = 28$, $r^2 = 0.78$, $F_{1,26} = 90.9$ ($p < 0.01$)

B

IC50 (ZSM) vs IC50 (5LO) FOR INDAZOLINONE REDOX INHIBITORS

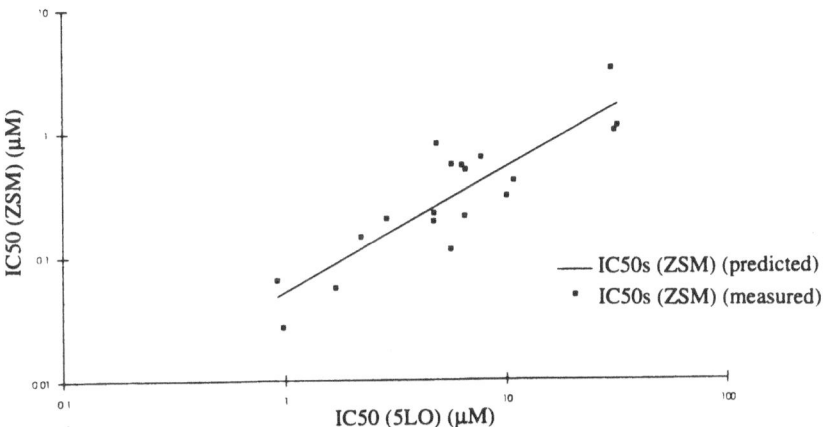

Experimental data fit by linear regression analysis to $\log(IC_{50}.ZSM) = 1.00 \log(IC_{50}.5LO) -$
1.29, $n = 19$, $r^2 = 0.77$, $F_{1,18} = 56.56$ ($p < 0.01$)

Figure 7-6. Correlation between 5LO enzyme and zymosan-stimulated mouse macrophage assays: inhibition by redox and nonredox 5LO inhibitors.

it gave IC$_{50}$s in which the effects of plasma protein binding were already taken into account, and these IC$_{50}$s also indicated what plasma concentrations needed to be achieved in order to inhibit LT biosynthesis *in vivo*. For evaluation of oral potency, the *ex vivo* rat blood and zymosan-inflamed air pouch tests were run routinely to assess inhibition of LT biosynthesis in blood and inflammatory exudate.

Despite improvement of 5LO inhibitory potency *in vitro*, ZM218287 was not significantly more potent orally than ZM211965. Bioavailabilities in the dog from lactose formulations of ZM211965 and ZM218287 were 9% and 3%, respectively. To help understand the reasons for the poor bioavailability of ZM218287, ^{14}C-labelled material was prepared with radiolabel in the methoxy group. This was the easiest site synthetically at which to introduce a label, although we recognized that it was also a metabolically unstable position. Nevertheless, we found that the majority of the label from ^{14}C-ZM218287 was excreted to the same extent in rat feces, irrespective of whether it was dosed orally (as a solution in soybean oil) or intravenously. A further study with ZM218287, dosed orally in soybean oil, to a portal-vein-cannulated dog resulted in bioavailabilities in the portal vein of 90%, but only 26% in the jugular vein. These studies indicated that ZM218287 was poorly absorbed from a solid formulation but well absorbed when in solution. However, once absorbed it experienced extensive elimination from the bile, probably through first-pass metabolism. In support of the latter conclusion, we found using ^{19}F NMR that bile-duct-cannulated dogs dosed the difluoro analog of ZM218287, Structure **30** (Figure 7-7), produced a number of metabolites in bile but few in urine.

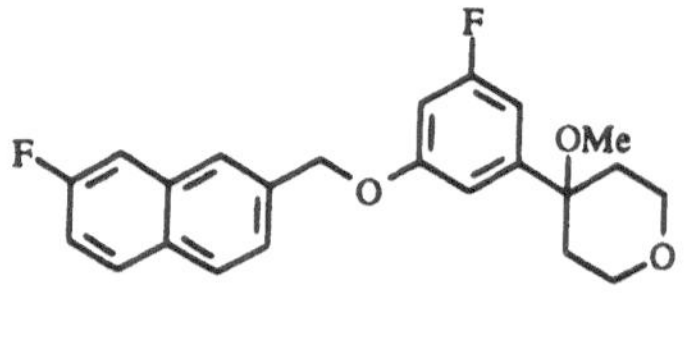

30

Figure 7-7. Difluoro-analog of ZM218287.

These studies indicated that the reasons for poor oral bioavailability of ZM218287 were low aqueous solubility and rapid metabolism, and our medicinal chemistry was designed to address both these points. With regard to reducing metabolism, we introduced blocking groups, or functionality that was anticipated to be metabolically more robust, into the ZM218287 structure as indicated by the arrows in Figure 7-8.

The low aqueous solubility of ZM218287 (measured at 0.08 μM) was a consequence of its high lipophilicity, and in order to find more soluble

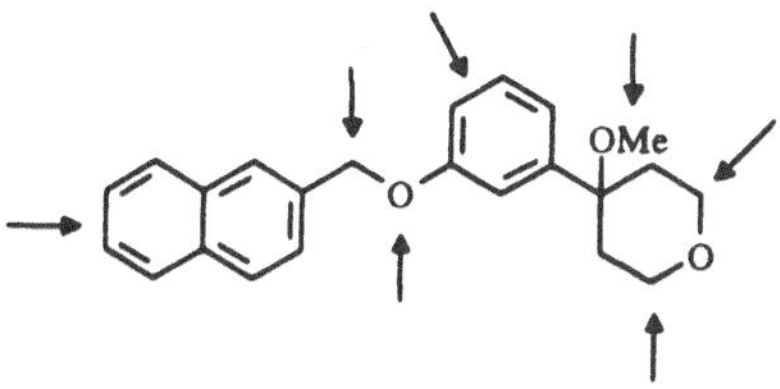

Figure 7-8. Approaches to blocking metabolism in 4-methoxytetrahydropyran series.

inhibitors, we sought, in congeners of ZM218287, to reduce lipophilicity without sacrificing potency. At the outset, it was not clear to us that this target could be achieved because several reports as well as our previous experience had indicated a dependence of 5LO inhibitory potency on lipophilicity. The naphthyl fragment contributes most to the n-octanol-water partition coefficient (P) of ZM218287 (log P 5.1) and one approach was to replace the naphthyl fragment by less lipophilic heterocycles. These would also have the potential to form extra interactions with the enzyme which could compensate for any loss of potency arising from reduced lipophilicity. To test this strategy, the naphthyl group was replaced by aza- and oxo-heterocycles to probe for hydrogen bond donor-acceptor interactions. Examples of the heterocycles investigated are shown in Table 7-13.

Typically, quinoxalines (**32**) and quinazolines reduced lipophilicity by 1.65 to 1.95 log P units compared to ZM218287 but were some fourfold less potent in human whole blood. On the other hand, the N-methyl 2-quinolone **33** was less lipophilic by 1.5 log P units but showed a small but significant increase in potency over ZM218287. The quinoxalinone **34** demonstrated most clearly that we had achieved our objective of reducing lipophilicity (Δ log P $-$ 2.2) without compromising potency *in vitro*. The IC$_{50}$s of ZM218287 and **33** in plasma-free cultures of mouse macrophages, 0.5 nM and 3 nM respectively, when compared with their IC$_{50}$s in human whole blood, suggested that the enhanced potency of **33** in blood arose from reduced plasma protein binding, relative to ZM218287.

Of interest is **31**, which contains a 2-quinolylmethoxy fragment. This group appears to be a pharmacophore for inhibition of LT biosynthesis through FLAP antagonism. Structure **33** does not antagonize FLAP (J. Evans, personal communication, September 20, 1992) and the observation that 2-quinolyl was inferior to N-methyl 2-quinolone in this series is indicative, perhaps, that the SAR of 5LO inhibition and FLAP antagonism are distinct.

The reason behind making these heterocyclic derivatives was to improve oral potency, and in this endeavor, our efforts were well rewarded. In fact,

Table 7-13. Enhancement of 5LO oral potency of 4-methoxytetrahydropyran series.

No.	Het	Human whole blood 5LO IC$_{50}$ (μM)	Rat inflam-matory exudate ED$_{50}$ (mg/Kg po)	log P	Aqueous solubility μM
31		0.2	> 1.5		
32		0.3	30	3.45	50
33		0.02	0.3	3.6	4
34		0.05	1.5	2.9*	71**
29	ZM218287	0.07	10	5.1*	0.08

* estimated
** aqueous solubility (S) estimated using $\log S = -1.32 \log P - 0.017(Tm - 25) + 7.4$

these were exciting times, for as soon as we started making compounds containing the N-methyl 2-quinolone system, for the first time since the initiation of the 5LO project, we began regularly hitting our target for oral potency (Crawley et al., 1992). Structure **33** had an ED$_{50}$ for inhibition of LT synthesis in rat blood *ex vivo* and in rat inflammatory exudate *in vivo* of less than 1 mg/kg when measured 3 hr after oral dosing in either test. Structure **33** is 40-fold more soluble than ZM218287, but solubility increases alone were insufficient to explain the increased oral potency since **32** and **34**, which were still potent inhibitors of LT synthesis and more soluble than **33**, were less potent orally.

We have shown that (methoxyalkyl)thiazoles and 4-methoxytetrahydropyrans are related and that the former series exhibited enantioselective inhibition of 5LO. So far, all the 4-methoxytetrahydropyrans described here have been achiral, and thus it is not possible to demonstrate enantioselective interactions. However, we have explored chiral tetrahydropyrans bearing methyl substitution, and in 4-methoxy-2-methyl-tetrahydropyrans, enantioselectivity is observed (Crawley et al., 1993). For example, the enantiomers shown in Table 7-14 demonstrated separations of IC$_{50}$s in macrophages and human whole blood. Although of interest from an SAR perspective, chirality can slow down drug development significantly, and since we wanted an expeditious development, we concentrated on achiral compounds.

Table 7-14. Enantioselective inhibition of LT biosynthesis in human whole blood of chiral 4-methoxy-2-methyl-tetrahydropyrans.

35			**36**

No.	Stereochemistry	Mouse macrophage 5LO IC$_{50}$ (nM)	Human whole blood 5LO IC$_{50}$ (μM)
35	2S, 4R	0.4	0.02
36	2R, 4S	9	0.7

From both the approaches aimed specifically at reducing metabolism and those addressing solubility, a number of orally potent compounds had by this stage been identified. In order to select a clinical candidate, a short list of compounds from both approaches, was drawn up for head-to-head comparisons by oral administration in rat blood *ex vivo* and rat inflammatory exudate *in vivo*. The short-listed compounds are shown in Figure 7-9.

The results of these experiments showed that **33** was consistently the most potent of this group of compounds at the 3 hr time point (Figure 7-10) as well as at 6 and 10 hr post dose (data not shown). Interestingly, **39**, the des-fluoro-analog of **33**, was significantly less active. The fluoro-substituent was introduced into the central phenyl ring to block a potential metabolic site; we have no evidence that metabolism occurs there, but clearly the fluorine atom influences the oral potency of **33**.

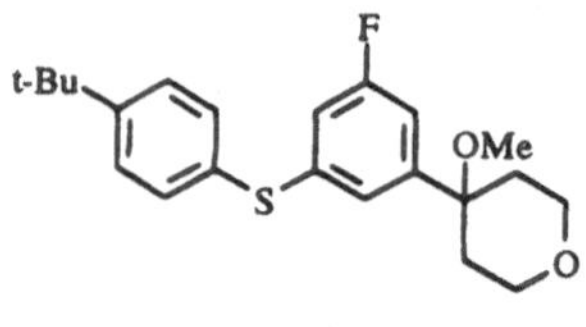

37 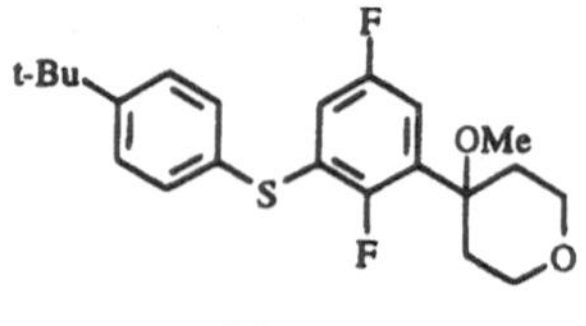   38

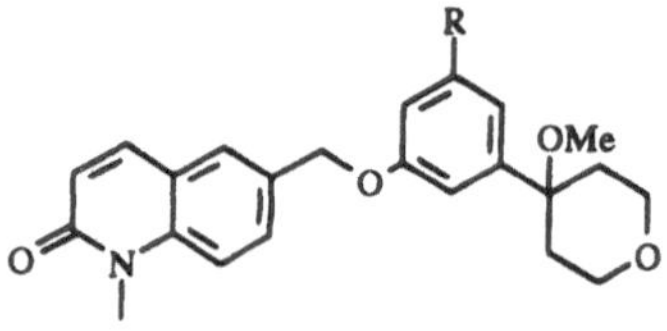

39, R = H

33, R = F

Figure 7-9. Short-listed drug development candidates.

Further studies showed that **33** inhibited *ex vivo* LTB$_4$ synthesis in rat blood in a dose-dependent manner with IC$_{50}$s at 3 and 10 hr of 0.9 and 4 mg/kg po respectively. Synthesis of LTB$_4$ in zymosan-inflamed air pouch exudate was inhibited with IC$_{50}$s of 0.3 and 2 mg/kg obtained at 3 and 10 hr after oral dosing. Whereas inhibition in rat blood from a 10 mg/kg oral dose of **33** had reversed by 24 hr, oral administration of 3 and 10 mg/kg doses to dogs produced maximal inhibition for 9 hr and at least 31 hr, respectively. These data were consistent with the observed half-lives in rat and dog of 2 hr and 6 hr, respectively, and we were optimistic that **33** might provide us with our target of 24-hr suppression of leukotriene synthesis with once-daily dosing. The compound therefore met our target for oral potency and **33** entered development in February 1990 as ZD2138.

We compared ZD2138 with the LT synthesis inhibitor competitor compounds that had been studied clinically, zileuton and MK-886 (McMillan et al., 1992). ZD2138 was 25–100-fold more potent than either of these compounds in whole blood *in vitro*, depending on the species (Table 7-15), and was also more selective than zileuton. In dog and human blood, no inhibition of thromboxane B$_2$(TxB$_2$) was observed with ZD2138 at the highest concentration (500 μM) tested, which corresponds to selectivity ratios for 5LO compared to CO in excess of 20,000. In rat blood, some inhibition of TxB$_2$ was seen at 500 μM so that the selectivity ratio was 4000 in this species.

Following oral administration, ZD2138 was a more potent inhibitor of LT synthesis in the rat than either of the competitor compounds (Table 7-16). In the dog, zileuton had comparable potency to ZD2138, but it was

Inhibition of LT Synthesis in Rat Blood *Ex vivo*

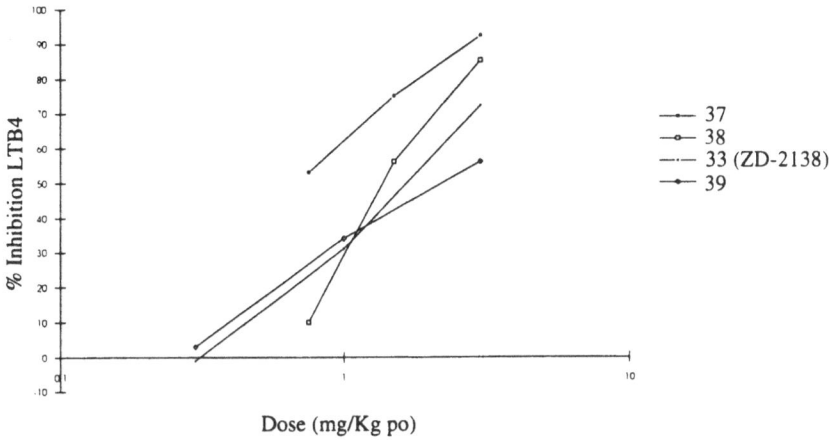

Inhibition of LT Synthesis in Rat Inflammatory Exudate *In vivo*

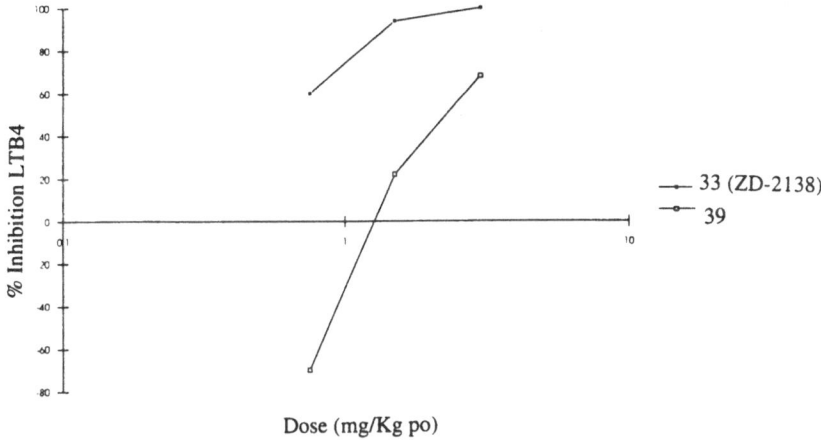

Figure 7-10. Comparison of 5LO oral potency of short-listed compounds.

known that the half-life of zileuton in humans was much shorter than in dogs. Based on these data, we believed that ZD2138 had the potential to be more potent in humans than either zileuton or MK-886 and, in the case of the former, to be more selective.

The one-month toxicity evaluation raised no concerns, and ZD2138 proceeded to a volunteer study (Yates et al., 1992). This comprised single doses rising to 1000 mg per volunteer, and at all doses, ZD2138 was well tolerated. Blood was taken to determine plasma concentrations and to

Table 7-15. Inhibition of eicosanoid generation in blood *in vitro* by LT biosynthesis inhibitors.

Compound	Human		Rat		Dog	
	LTB_4	TxB_2	LTB_4	TxB_2	LTB_4	TxB_2
ZD2138	0.024	>500	0.033	156	0.02	>500
zileuton	2.6	40	2.3	>100	0.56	51
MK-886	0.95	>100	0.11	200		

Table 7-16. Inhibition of LTB_4 generation in rat blood *ex vivo* and Zymosan-inflamed exudate *in vivo* by LT biosynthesis inhibitors.

Compound	Rat blood *ex vivo* (ED_{50})		Rat inflammatory exudate *in vivo* (ED_{50})	
	3 hr	10 hr	3 hr	10 hr
ZD2138	0.9	4	0.3	2
zileuton	5	20	3.4	
MK-886	2	2.5	2	

measure inhibition of LT synthesis in blood following *ex vivo* stimulation with A-23187. It was found that at doses of 350 mg and above, LT synthesis was still completely inhibited in blood taken 24 hr post-dose. In addition, the degree of inhibition achieved correlated well with plasma concentrations of ZD2138. ZD2138 was more potent in humans than in rat or dog and this was explained by the half-life in humans, which was estimated to be 12–16 hr.

The conclusion from this study was that we would produce complete suppression of LT synthesis in patients with a once-per-day dose of 350 mg po, which compared very favorably with the predicted daily dose of 600 mg four times daily for zileuton (Table 7-17).

Table 7-17. Half-lives and predicted clinical doses of ZD2138 and zileuton

Compound	Elimination half-life (hrs)			Predicted clinical dose
	Rat	Dog	Human	
ZD2138	1–2	5–6	12–16	350 mg uid
zileuton	2.3	7.5	2–3	600 mg qid

Pharmacological Efficacy of 5-Lipoxygenase Inhibitors

Anti-Inflammatory Activity

At the outset of the 5LO program, one of the objectives of the team was to establish the role of LTB_4 in inflammatory reactions and thus to determine whether inhibitors of 5LO would be effective anti-inflammatory agents. Our approach to this was to try and answer three key questions:

(1) Does LTB_4 produce inflammatory symptoms?
(2) Is LTB_4 produced at sites of inflammation?
(3) Are agents that block LTB_4 formation effective in appropriate animal models?

To address the first question, natural LTB_4 purified from calcium ionophore (A23187)-stimulated pig leukocytes was used to confirm early literature reports that LTB_4 was a potent activator and chemotactic agent for PMNL *in vitro*. When synthetic LTB_4 became available, it proved to be an invaluable tool allowing investigation of *in vivo* chemotactic activity to be extended. Thus, we were able to confirm that LTB_4 was a potent chemotactic agent and produced a PMNL-dependent edema when co-injected with the vasodilator PGE_2 into rabbit dermis. These observations, together with literature reports that LTB_4 and the peptidoleukotrienes produced inflammatory symptoms when injected intradermally in humans, encouraged the view that inhibitors of leukotriene synthesis might be anti-inflammatory.

Early work to identify LTB_4 at sites of inflammation was confounded by insensitive assay systems even though Davidson had reported LTB_4 in rheumatoid synovial fluid (Davidson et al., 1982). A major breakthrough was made when sensitive and specific RIAs were developed. A group from the Wellcome Research Laboratories were among the first to develop such an assay, which they used to demonstrate the release of LTB_4 in inflammatory exudates obtained following the subcutaneous implantation of saline- or carrageenan-soaked polyester sponges in rats (Simmons et al., 1983). Using an LTB_4 RIA developed at Zeneca we were able to confirm these observations in the rat (Foster et al., 1986) and extend the observations to the rabbit (Aked and Foster, 1987).

Addressing the third question was a dilemma since no selective 5LO inhibitors or appropriate animal models were available. Thus, it appeared that validating a potentially relevant inflammation model would have to await the discovery of an efficacious selective 5LO inhibitor. However, Higgs (Higgs et al., 1979) had reported exciting anti-inflammatory activity with the dual arachidonate CO and 5LO inhibitor BW755C. This compound not only inhibited carrageenan-induced paw edema in the rat but also had a greater effect on leukocyte migration than indomethacin, an activity that was suggested to be due to inhibition of the production of the chemotactic hydroxy acid, 5-HETE. In subsequent studies, Salmon (Salmon et al., 1983) demonstrated that BW755C reduced LTB_4 concentration and leukocyte numbers in exudate derived from the implantation of carrageenan-impregnated sponges in the rat. These observations led them to speculate that inhibition of LTB_4 may, in part, account for the lower number of exudate leukocytes, since LTB_4 was a potent chemotactic agent.

These data led us to investigate the subcutaneous sponge implant model of acute inflammation as a suitable test for evaluating the anti-inflammatory efficacy of 5LO inhibitors. We were able to demonstrate that elevated concentrations of LTB_4 preceded the infiltration of leukocytes into inflammatory exudate (Foster et al., 1986), an observation that was also reported at the time in rat air pouches challenged with a variety of stimuli. In view of this temporal relationship it was tempting to speculate that LTB_4 was the key chemotactic factor. If this was the case, then inhibition of its synthesis would be expected to cause reduced leukocyte recruitment.

When rats were treated with an efficacious dose of BW755C one hour prior to sponge implantation, a reduction in both exudate LTB_4 and leukocyte content was demonstrated for 6 hours following implantation. However, phenidone, an analog of BW755C, which was at least equipotent with BW755C as an inhibitor of LTB_4 generation, was much less effective at inhibiting leukocyte recruitment. These observations suggested that BW755C exerted its effect on leukocyte recruitment by a mechanism that was independent of its inhibitory effect on LTB_4 synthesis. Further experiments with BW755C, phenidone, and the CO inhibitors indomethacin and fluorbiprofen demonstrated that the inhibition of leukocyte migration in the sponge implant model correlated better with inhibition of prostaglandin biosynthesis than LTB_4. In fact, inhibition of leukocyte recruitment by CO inhibitors had been previously reported, although the mechanism was unknown. The suitability of the rat as a species for evaluating the potential anti-inflammatory efficacy of 5LO inhibitors was questioned by the observation that sponges that had been soaked in LTB_4 produced only a modest (twofold) recruitment of leukocytes in contrast to carrageenan (11-fold). Moreover, the threshold concentration of LTB_4 required to cause modest increase in leukocyte recruitment was 10-fold higher than the maximal levels detected in inflammatory exudate, an observation supported by histological assessment following intradermal LTB_4 administration. Thus, the levels of LTB_4 detected in inflammatory exudate were insufficient to cause leukocyte recruitment. The failure of LTB_4 to induce leukocyte infiltration in the rat was clarified by the observations of Kreisle (Kreisle et al., 1985), who demonstrated that rat PMNL lack a specific, high-affinity LTB$_4$ binding site which is present in human PMNL and plays a role in the chemotactic response to LTB$_4$. We concluded, therefore, that the rat sponge implant model was unsuitable for demonstrating the potential anti-inflammatory efficacy of 5LO inhibitors, and more generally that the rat was inappropriate for investigating the role of LTB$_4$ in acute inflammation.

Experience with the rat raised the issue of which species was most appropriate for investigating leukotriene-dependent models of inflammation. Since LTB$_4$ had been shown to be potently chemotactic towards PMNL in rabbit, and because technology for investigating dermal inflammation was well established, we reasoned that rabbit might be an appropriate species. A further issue to address was what would be an appropriate inflammatory stimulus. Topical application of arachidonic acid to human skin under occlusion had been reported to induce erythema and leukocyte accumulation, while high concentrations applied to the ears of mice or injected intradermally in rabbit caused inflammatory edema. These studies suggested that arachidonic acid, the precursor of inflammatory eicosanoids, might be an appropriate inflammatory stimulus. We demonstrated that intradermal injection of arachidonic acid into rabbit dermis caused a concentration-dependent increase in plasma extravasation and was two- to threefold more effective than two related C$_{20}$ fatty acids, 5,8,11,14,17-eicosapentenoic acid and 8,11,14-eicosatrienoic acid, while the C$_{18}$ fatty acids oleic, linoleic, and linolenic acids were inactive (Aked et al., 1986). Since the C$_{18}$ fatty acids are not metabolized to prostaglandins or leukotrienes, the induction of plasma extravasation appeared to be selective for the fatty acids that are precursors of these eicosanoids. Histological examination of arachidonic acid injected skin sites revealed an infiltration of PMNL that was related to the concentration of injected arachidonic acid. Further studies demonstrated that the plasma extravasation induced by arachidonic acid was dependent on circulating PMNL.

Wedmore and Williams (Wedmore and Williams, 1981) had previously demonstrated PMNL-dependent plasma exudation following co-injection of LTB$_4$ and PGE$_2$ into rabbit dermis. This led us to speculate that the arachidonic acid-induced plasma and PMNL extravasation were a consequence of the metabolism of arachidonic acid to vasodilatory prostaglandins and chemotactic LTB$_4$. Demonstration that inhibition of plasma extravasation and PMNL infiltration by the dual cyclooxygenase/lipoxygenase inhibitors BW755C and phenidone were consistent with this view. In subsequent studies we demonstrated elevated concentration of LTB$_4$ and PGE$_2$ in arachidonic acid-injected skin sites. Co-injection of LTB$_4$ and PGE$_2$ at concentrations detected in the inflamed site could induce plasma extravasation. This arachidonic acid-induced inflammation was mediated by formation of leukotrienes and prostaglandins, and we concluded that this was an appropriate model for evaluating the anti-inflammatory efficacy of inhibitors of arachidonic acid metabolism, including 5LO inhibitors.

Initial pharmacological studies in the model employed redox inhibitors. One such inhibitor was the indazolinone ZM207968, which was approximately 300 times more potent as a 5LO than CO inhibitor. Coadministration of ZM207968 with arachidonic acid into rabbit dermis potently inhibited both plasma extravasation and PMNL infiltration (Foster et al., 1990). Encouragingly, the anti-inflammatory activity of other intradermally administered indazolinones was related to their potency as inhibitors of LTB_4 generation by ionophore-stimulated blood (Table 7-16). These results suggested that the anti-inflammatory action of the indazolinones was a consequence of their inhibitory effect on LTB_4 biosynthesis and supported the view that selective 5LO inhibitors would be anti-inflammatory. However, there was a concern that since the indazolinones were weak redox agents, their anti-inflammatory effect might be due to a nonspecific redox mechanism. This was inconsistent with our results, since indazolinones having similar redox potential to ZM207968 showed widely different anti-inflammatory activity in the model (Table 7-18).

Although these results were encouraging, definitive proof of the utility of the model had to await the discovery of selective nonredox 5LO inhibitors. This was provided by the (methoxyalkyl)thiazoles and later methoxytetrahydropyrans, two series of potent, highly selective nonredox 5LO inhibitors. As previously observed with the indazolines, a good correlation was seen between the anti-inflammatory potency of these nonredox 5LO inhibitors and their potency versus leukotriene production by ionophore-challenged blood (Table 7-19).

Discovery of ZM216800, which is a chiral, potent, selective inhibitor of 5LO, provided further evidence for the utility of the model. Resolution of the compound into its enantiomers revealed that the (+)-enantiomer was at least 70 times more potent an inhibitor of 5LO than the (−)-enantiomer. When the enantiomers were evaluated against arachidonic acid-induced skin inflammation, only the (+)-enantiomer produced anti-inflammatory activity (McMillan et al., 1990). Taken together, these results demonstrated that inhibition of 5LO *in vivo* translated into anti-inflammatory activity in a leukotriene-mediated model of inflammation.

Of more recent interest are the observations of Teixeira and Hellewell (Teixeira and Hellewell, 1994) who demonstrated that intradermal injection of arachidonic acid into guinea pig dermis induced the infiltration of ^{111}In-eosinophils, a process that could be abrogated by co-injection of the

Table 7-18. Comparison of the anti-inflammatory activity of several indazolinones with their potency as inhibitors of leukotriene biosynthesis.

No.	Structure	Plasma extravasation* IC_{30} (nmol per site)	In vitro blood[♯] IC_{50} (μM)	Redox potential (V)
11		15.6 ± 6.3 (4)	0.4 ± 0.2 (6)	0.69
2		40.0 ± 5.3 (4)	1.5 ± 0.6 (6)	0.71
40		>170.0 (3)	>50.0 (3)	0.81
41		>125.0 (3)	>100.0 (3)	0.78

Number of experiments shown in parentheses.
* Plasma extravasation was assessed as described (Aked et al., 1986) following co-injection of compounds or vehicle with arachidonic acid into rabbit dermis.
[♯] 5-lipoxygenase was assessed by measuring LTB_4 production by A23187-stimulated rat blood as described (Foster et al., 1990).

selective nonredox 5LO inhibitor ZM230487 with the arachidonic acid. In addition, ZM230487 also inhibited the accumulation of ^{111}In-eosinophils induced by a passive cutaneous anaphylaxis reaction in guinea pig skin. These data suggest that 5LO product(s), play a role in allergic inflammation in the guinea pig and make this species also suitable for investigating the effects of inhibitors of leukotriene synthesis.

Although arachidonic acid-inflamed rabbit skin proved to be an excellent model for evaluating compounds topically, oral evaluation of Zeneca's early 5LO inhibitors in the rabbit proved to be difficult due to the large size of the animals and, therefore, the large amounts of compound that needed to be dosed. These difficulties led us to consider alternative species for the evaluation of oral activity.

Several pharmaceutical companies had reported modulation of arachidonic acid-induced mouse ear edema by CO and 5LO inhibitors (Carlson et al., 1985; Young et al., 1984). Also, a group from Merck (Opas et al., 1985) had reported elevated concentration of LTC_4 and PGE_2 in extracts

Table 7-19. Comparison of the anti-inflammatory activity of selective nonredox 5-lipoxygenase inhibitors with their potency as inhibitors of leukotriene biosynthesis.

No.	Structure	Inhibition of plasma extravasation* ID_{30} (μg per site)	Inhibition of 5-lipoxygenase[#] IC_{50} (μM)
42		0.25	0.04
29		0.48	0.06
43		1.60	0.31
4		5.70	0.30
44		8.00	0.49
45		140.00	>40.0

Values shown are the means of two separate experiments.

* Plasma extravasation was assessed as described (Aked et al., 1986) following co-injection of compounds or vehicle with arachidonic acid into rabbit dermis.

[#] 5-lipoxygenase was assessed by measuring LTB_4 production by A23187-stimulated human blood as described (Foster et al., 1990).

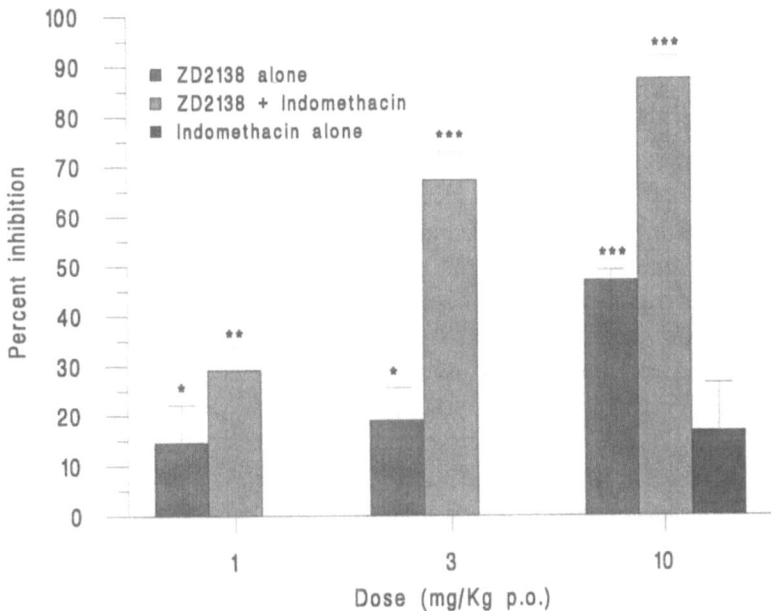

Figure 7-11. Synergistic anti-inflammatory effect of ZD2138 and indomethacin on arachidonic acid-induced mouse ear edema.

from arachidonic acid-inflamed mouse ear and suggested that these products, by acting together on the microvasculature, were responsible for the edema. These reports led us to investigate the suitability of this model for evaluating the oral anti-inflammatory efficacy of our nonredox 5LO inhibitors. As shown in Figure 7-11, orally administered ZD2138 produced dose-dependent inhibition of arachidonic acid-induced mouse ear edema. When ZD2138 was co-administered with the CO inhibitor, indomethacin, synergistic anti-inflammatory activity was observed (Figure 7-11). Synergistic anti-inflammatory activity was also observed when ZD2138 was combined with other CO inhibitors (Foster et al., 1992). This raised the issue of the likely mechanism of this synergistic effect. Williams (Williams and Peck, 1977; Williams, 1979) proposed a hypothesis that vascular changes associated with acute inflammation are the results of vasodilation induced by prostaglandins of the E series together with increased vascular permeability mediated by another component. Subsequently, Wedmore and Williams (Wedmore and Williams, 1981) demonstrated a synergistic proinflammatory interaction between the vasodilatory prostanoids and chemotactic agents including LTB$_4$. We noted that co-administration of leukotrienes and prostaglandins could lead to synergistic pro-inflammatory effects, and we reasoned that the anti-inflammatory synergy is a consequence of inhibition of both classes of eicosanoid. It has been suggested (Higgs et al., 1979) that dual inhibitors of arachidonate CO and 5LO would have a more comprehensive anti-inflammatory effect than selective 5LO inhibitors. The discovery of an orally effective mixed inhibitor in a single molecule has proved elusive to date, but such mixed inhibition could also be achieved by co-administration of a selective 5LO inhibitor and CO inhibitor. It is possible that combined treatment of certain inflammatory conditions with a selective 5LO inhibitor and CO inhibitor may have greater anti-inflammatory efficacy than treatment with either agent alone.

In addition to their potential efficacy in inflammatory conditions, one of the early concerns over the clinical utility of 5LO inhibitors was whether they would cause gastrointestinal toxicity similar to the NSAIDs. On theoretical grounds this seemed unlikely, but it was nevertheless an important issue. It was known that treatment of rats with moderate doses of CO inhibitors caused gastrointestinal toxicity within a few days. In marked contrast, long term treatment of rats or dogs (6 months) with high doses of selective 5LO inhibitors demonstrated that inhibition of 5LO was not associated with gastrointestinal toxicity. In fact there was a body of literature evidence suggesting that 5LO products may play a role in protecting the gastric mucosa from injury induced by various agents including CO inhibitors. In support of this, we demonstrated that the selective 5LO inhibitor, ZM207968, protected the rat gastric mucosa from acute ethanol- or indomethacin-induced damage (Foster et al., 1989). However, since ZM207968 was a weak redox agent, it was possible that a nonspecific redox mechanism might have been responsible for its gastrointestinal cytoprotective effect. Subsequently we demonstrated that nonredox 5LO inhibitors, ZD2138 and ZM230487, also protected the rat gastric mucosa from acute indomethacin-induced damage (Figure 7-12). In addition, oral administration of ZD2138 or ZM230487 (10 mg/kg bid) to rat inhibited gastrointestinal toxicity (assessed histologically), increase in the plasma concentration of the acute phase protein α1-acid glycoprotein, and weight loss induced by subacute administration of indomethacin (4.5 mg/kg/day for 5 days). These results suggest that ZD2138 and possibly other 5LO inhibitors may have clinical utility for protecting the gastrointestinal tract from CO inhibitor-induced damage. Moreover, treatment of inflammatory conditions with 5LO inhibitor/CO inhibitor combinations would be expected to be much less toxic to the gastrointestinal tract than CO inhibitors alone.

Based on its preclinical biochemical and pharmacological profile, ZD2138 was entered into development. The compound is well tolerated in humans and a single oral dose of 350 mg ZD2138 is sufficient to abrogate

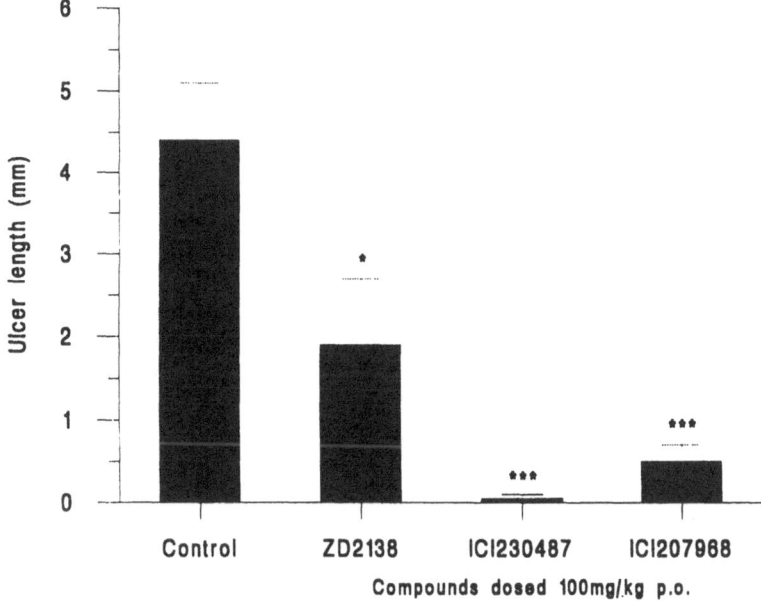

Figure 7-12. Protective effect of 5-lipoxygenase inhibitors versus versus acute indomethacin-induced gastric ulceration in rat.

completely *ex vivo* LTB_4 production in blood for 24 hours. ZD2138 is currently undergoing clinical trials in rheumatoid arthritis and asthma. These
studies will establish the relationship between systemic 5LO inhibition and
clinical efficacy at the target organ.

5-Lipoxygenase Inhibitors and the Lung

In addition to chronic inflammatory conditions such as rheumatoid arthritis, psoriasis, and inflammatory bowel disease, a major therapeutic target
for 5LO inhibitors is asthma. Preclinical development of 5LO inhibitors
for the treatment of asthma was hampered by the lack of suitable *in vivo*
models. In 1983, Anderson (Anderson et al., 1983) demonstrated that
antigen-induced anaphylaxis in actively sensitized guinea pigs pretreated
with a cocktail of indomethacin, pyrilamine, and propranolol was characterized by a delayed onset, slowly developing bronchoconstriction indicative of a slow-reacting substance of anaphylaxis (SRS-A) response
(Anderson et al., 1983). These workers demonstrated that SRS-A appeared in the plasma of the animals prior to the onset of antigen-induced
bronchoconstrictor response of sensitized animals. The slowly developing
antigen-induced bronchoconstriction was inhibited by the SRS-A antagonist, FPL55712, and the SRS-A synthesis inhibitors BW755C, phenidone,

and NDGA. These observations suggested that peptidoleukotrienes mediated the bronchoconstriction and, therefore, that the model might be useful for evaluating the pharmacological efficacy of 5LO inhibitors. We were able to confirm the utility of the model for evaluating the efficacy of both 5LO inhibitors and peptidoleukotriene antagonists versus antigen-induced bronchoconstriction. In a series of studies, elevated concentration of immunoreactive peptidoleukotrienes were detected in lung lavage fluid taken from antigen-challenged actively sensitized guinea pigs. This immunoreactive material was dose-dependently inhibited by phenidone and indazolinone 5LO inhibitors. Demonstrable ED_{50}s of approximately 1 mg/kg were achieved versus antigen-induced bronchoconstriction by intravenous administration of a number of indazolinones and the Takeda 5LO inhibitor, AA861. The peptidoleukotriene antagonist, ZM198615, which is structurally related to the Zeneca development compound Accolate (Krell et al., 1990), had an ED_{50} of 0.1 mg/kg. Later experiments with the 5LO inhibitor ZD2138 demonstrated it to be equipotent with ZM198615. Further characterization of the effect of ZD2138 on antigen-induced bronchoconstriction monitored pulmonary mechanics changes in response to antigen. ZD2138 (1 mg/kg) antagonized the antigen-induced increase in pulmonary resistance (R_p) and decrease in dynamic compliance (C_{dyn}) but had no effect on methacholine-induced changes in R_p or C_{dyn}. These data supported the development of ZD2138 for the treatment of asthma.

ZD2138 is currently undergoing clinical trials in asthma. Phase II clinical trials have demonstrated that in aspirin-sensitive asthmatic patients, a single oral dose of 350 mg ZD2138 caused bronchodilation and significantly inhibited the fall in FEV1 induced by aspirin ingestion, and this is associated with substantial 5LO inhibition (Nasser et al., 1994a). In contrast, ZD2138 did not attenuate the early or late asthmatic responses following allergen challenge of asthmatic patients (Nasser et al., 1994b). A similar observation has been reported for the 5LO inhibitor zileuton (Hui et al., 1991), which is also undergoing clinical trials in asthma.

REFERENCES

Aharony D, Stein RL (1986): Kinetic mechanism of guinea pig neutrophil 5-lipoxygenase. *J Biol Chem* 261:11512–11519

Aked D, Foster SJ (1987): Leukotriene B_4 and acute inflammation in the rabbit. *Br J Pharmacol* 90:235P

Aked D, Foster SJ, Howarth A, McCormick ME, Potts HC (1986): The inflammatory response of rabbit skin to topical arachidonic-acid and its pharmacological modulation. *Br J Pharmacol* 89:431–438

Anderson WH, O'Donnell M, Simko BA, Welton AF (1983): An *in-vivo* model for measuring antigen induced slow reacting substance of anaphylaxis mediated bronchoconstriction and plasma slow reacting substance of anaphylaxis levels in the guinea-pig. *Br J Pharmacol* 78:67–74

Bird TGC, Bruneau P, Crawley GC, Edwards MP, Foster SJ, Girodeau J-M, Kingston JF, McMillan RM (1991): (Methoxyalkyl)Thiazoles: A new series of potent, selective, and orally active 5-lipoxygenase inhibitors displaying high enantioselectivity. *J Med Chem* 34:2176–2186

Bruneau P, Delvare D, Edwards MP, McMillan RM (1991): Indazolinones, a new series of redox-active 5-lipoxygenase inhibitors with built-in selectivity and oral activity. *J Med Chem* 34:1028–1036

Carlson RP, O'Neill-Davies L, Chang J, Lewis AJ (1985): Modulation of mouse ear edema by cyclooxygenase and lipoxygenase inhibitors and other pharmacologic agents. *Agents Actions* 17:197–204

Carter GW, Young PR, Albert DH, Bouska J, Dyer R, Bell RL, Summers JB, Brooks DW (1991): 5-Lipoxygenase inhibitory activity of zileuton. *J Pharm Exp Ther* 256:929–937

Corey EJ, Cashman JR, Kantner SS, Wright SW (1984): Rationally designed, potent competitive inhibitors of leukotriene biosynthesis. *J Am Chem Soc* 106:1503–1504

Crawley GC, Briggs MT, Dowell RI, Edwards PN, Hamilton PM, Kingston JF, Oldham K, Waterson D, Whalley DP (1993): 4-Methoxy-2-methyltetrahydropyrans: Chiral leukotriene biosynthesis inhibitors, related to ICI D2138, which display enantioselectivity. *J Med Chem* 36:295–296

Crawley GC, Dowell RI, Edwards PN, Foster SJ, McMillan RM, Walker ERH, Waterson D, Bird TGC, Bruneau P, Girodeau J-M (1992): Methoxytetrahydropyrans. A new series of selective and orally potent 5-lipoxygenase inhibitors. *J Med Chem* 35:2600–2609

Davidson EM, Rae SA, Smith MJH (1982): Leukotriene B4 in synovial fluid. *J Pharm Pharmac* 34:410

Duniec Z, Robak J, Gryglewski R (1983): Antioxidant properties of some chemicals vs. their influence on cyclooxygenase and lipoxidase activities. *Biochem Pharmacol* 14:2283–2286

Fort FL, Pratt MC, Carter GW, Lewkowski JP, Heyman IA, Cusick PK, Kesterson JW (1984): Heinz bodies, methemoglobinemia, and hemolytic anemia induced in rats by 3-amino-1-(m-(trifluoromethyl) phenyl)-2-pyrazoline. *Fundamental and Applied Toxicology* 4:216–220

Foster SJ, Aked DM, McCormick ME, Potts HC (1989): Cytoprotective properties of ICI207968 a selective 5 lipoxygenase inhibitor on the rat gastrointestinal mucosa. *Br J Pharmacol* 96:38P

Foster SJ, Bruneau P, Walker ERH, McMillan RM (1990): 2 Substituted indazolinones orally active and selective 5-lipoxygenase inhibitors with anti-inflammatory activity. *Br J Pharmacol* 99:113–118

Foster SJ, McCormick ME, Howarth A, Aked D (1986): Leukocyte recruitment in the subcutaneous sponge implant model of acute inflammation in the rat is not mediated by leukotriene B4. *Biochem Pharmacol* 35:1709–1718

Foster SJ, Potts HC (1992): Arachidonic acid-induced mouse ear oedema synergistic inhibition by combined treatment with ICI D2138 a selective 5 lipoxygenase inhibitor and non-steroidal anti-inflammatory agents. *Arthritis Rheumatism* 35:309 (AbsP0148)

Gibian MJ, Galaway RA (1977): Chemical aspects of lipoxygenase reactions. In: *Bio-organic Chemistry*, Vol. 1, van Tamelen EE, ed. New York: Academic Press

Hammond ML, Kopka IE, Zambias RA, Caldwell GC, Boger J, Baker F, Luell S, MacIntyre DE (1989): 2,3-Dihydro-5-benzofuranols as antioxidant-based inhibitors of leukotriene biosynthesis. *J Med Chem* 32:1006–1020

Higgs GA, Flower RJ, Vane JR (1979): A new approach to antiinflammatory drugs. *Biochem Pharmacol* 28:1959–1961

Hlasta DJ, Casey FB, Ferguson EW, Gangell SJ, Heimann MR, Jaeger EP, Kullnig RK, Gordon RJ (1991): 5-Lipoxygenase inhibitors: The synthesis and structure-activity relationships of a series of 1-phenyl-3-pyrazolidinones. *J Med Chem* 34:1560–1570

Hui KP, Taylor IK, Taylor GW, Rubin P, Kesterson J, Barnes NC, Barnes PJ (1991): Effect of a 5 lipoxygenase inhibitor on leukotriene generation and airway responses after allergen challenge in asthmatic patients. *Thorax* 46:184–189

Jackson WP, Islip PJ, Kneen G, Pugh A, Wates PJ (1988): Acetohydroxamic acids as potent selective orally active 5-lipoxygenase inhibitors. *J Med Chem* 31:500–503

Kreisle RA, Parker CW, Griffin GL, Senior RM, Stenton WF (1985): Studies of leukotriene B4-specific binding and function in rat polymorphonuclear leukocytes absence of a chemotactic response. *J Immunol* 134:3356–3363

Krell RD, Aharony D, Buckner C, Keith RA, Kusner EJ, Snyder DW, Bernstein PR, Matassa VG, Yee YK, Brown FJ, Hesp B, Giles RE (1990): The preclinical pharmacology of ICI 204,219. *Am Rev Resp Dis* 141:978–987

Lambert-van der Brempt C, Bruneau P, Lamorlette MA, Foster SJ (1994): Conformational analysis of 5-lipoxygenase inhibitors: Role of the substituents in chiral recognition and on the active conformations of the (methoxyalkyl)thiazole and methoxytetrahydropyran series. *J Med Chem* 37:113–124

McMillan RM, Girodeau J-M, Foster SJ (1990): Selective chiral inhibitors of 5-lipoxygenase with anti-inflammatory activity. *Br J Pharmacol* 101:501–503

McMillan RM, Masters DJ, Vickers CV, Dicken MP, Jacobs VN (1989): Metabolism of unsaturated fatty acids by rbl-1 5-lipoxygenase: Influence of substrate solubility and product inactivation. *Biochem Biophys Acta* 1005:170–176

McMillan RM, Spruce KE, Crawley GC, Walker ERH, Foster SJ (1992): Preclinical pharmacology of ICI D2138, a potent orally-active non-redox inhibitor of 5-lipoxygenase. *Br J Pharmacol* 107:1042–1047

Miller DK, Gillard JW, Vickers PJ, Sadowski S, Leveille C, Mancini JA, Charleson P, Dixon RAF, Ford-Hutchinson AW, Fortin R, Gauthier JY, Rodkey J, Rosen

R, Rouzer C, Sigal IS, Strader CD, Evans JF (1990): Identification and isolation of a membrane protein necessary for leukotriene production. *Nature (Lond)* 343:278–281

Nasser SM, Bell GS, Foster SJ, Spruce K, McMillan RM, Williams AJ, Arm JP, Lee TK (1994a): Effect of the 5-lipoxegenase inhibitor ZD2138 on aspirin-induced asthma. *Thorax* 49:749–756

Nasser SM, Bell GS, Hawksworth RJ, Spruce KE, McMillan R, Williams AJ, Lee TH, Arm JP (1994b): Effect of the 5-lipoxegenase inhibitor ZD2138 on allergen-induced early and late asthmatic responses. *Thorax* 49:743–748

Opas EE, Bonney RJ, Humes JL (1985): Prostaglandin and leukotriene synthesis in mouse ears inflamed by arachidonic-acid. *J Invest Dermatol* 84:253–256

Riendeau D, Falgueyret JP, Guay J, Ueda N, Yamamota S (1991): Pseudoperoxidase activity of 5-lipoxygenase stimulated by potent benzofuranol and N-hydroxyurea inhibitors of the lipoxygenase reaction. *Biochem J* 274:287–292

Salmon JA, Simmons PM, Moncada S (1983): The effects of BW755C and other anti-inflammatory drugs on eicosanoid concentrations and leukocyte accummulation in experimentally-induced acute inflammation. *J Pharm Pharmacol* 35:808–813

Simmons PM, Salmon JA, Moncada S (1983): Release of leukotriene B4 during experimental inflammation. *Biochem Pharmacol* 32:1353–1359

Sloane DL, Browner MF, Dauter Z, Wilson K, Fletterick RJ, Sigal E (1990): Purification and crystallization of 15 lipoxygenase from rabbit reticulocytes. *Biochem Biophys Res Communications* 173:507–513

Summers JB, Gunn BP, Martin JG, Martin MB, Mazdiyasnu H, Stewart AO, Young PR, Bouska JB, Goetze AM, Dyer RD, Brooks DW, Carter GW (1988): Structure-activity analysis of a class of orally active hydroxamic acid inhibitors of leukotriene biosynthesis. *J Med Chem* 31:1960–1964

Teixeira MM, Hellewell PG (1994): Effects of a 5-lipoxygenase inhibitor, ZM-230487, on cutaneous allergic inflammation in the guinea-pig. *Br J Pharmacol* 111:1205–1211

Wedmore CV, Williams TJ (1981): Control of vascular permeability by polymorphonuclear leukocytes in inflammation. *Nature (Lond)* 289:646–650

Williams TJ (1979): Prostaglandin E2, prostaglandin I2 and the vascular changes of inflammation. *Br J Pharmacol* 65:517–524

Williams TJ, Peck MJ (1977): Role of prostaglandin-mediated vasodilatation in inflammation. *Nature (Lond)* 270:530–532

Yates RA, McMillan RM, Ellis SH, Hutchinson M, Culmore EM, Wilkinson DM (1992): A new non-redox 5-lipoxygenase inhibitor ICI D2138 is well tolerated and inhibits leukotriene synthesis in health volunteers. *Am Rev Respir Dis* 145:A745

Young JM, Spires DA, Charles MS, Bedord CJ, Wagner B, Ballaron SJ, De Young LM (1984): The mouse ear inflammatory response to topical arachidonic acid. *J Invest Dermatol* 82:367–371

8

Development of MK 0591: An Orally Active Leukotriene Biosynthesis Inhibitor with a Novel Mechanism of Action

Petpiboon Prasit and Philip J. Vickers

Introduction

Asthma is a chronic inflammatory condition characterized by bronchial hyper-responsiveness and reversible airway obstruction. A number of mediators have been implicated in the pathophysiology of this complex disease, including leukotrienes (LTs) (Ford-Hutchinson, 1989; Chanarin and Johnston, 1994). The biological effects of these arachidonic acid (AA) metabolites, including smooth muscle contraction, vasoconstriction and vascular permeability mimic the pathological changes seen in asthma (Ford-Hutchinson et al., 1980; Samuelsson, 1983; Jones et al., 1989). Furthermore, baseline LT production is higher in asthmatic subjects and is increased significantly during asthmatic attacks (Isono et al., 1985; Lam et al., 1988; Tagari et al., 1990; Knapp et al., 1992; Picado et al., 1992). The pro-inflammatory effects of LTs have also implicated these compounds in inflammatory bowel disease (IBD) (Rask-Madsen et al., 1992). The demonstration that LTs have properties associated with asthma and IBD has stimulated the development of many potential therapeutic agents to block the effects of these compounds.

LT biosynthesis inhibitors represent one of the most interesting new classes of compound for the treatment of asthma and IBD in development at present. Here, we would like to describe one of the approaches to the development of such an entity taken at Merck Frosst, which resulted in the discovery of our clinical candidate MK 0591. During this process,

The Search for Anti-Inflammatory Drugs
Vincent J. Merluzzi and Julian Adams, Editors
© Birkhäuser Boston 1995

elucidation of the mechanism of action of MK 0591 and its predecessor MK-886 provided novel insights into the cellular synthesis of LTs.

The enzyme 5-lipoxygenase (5-LO) catalyzes the first two steps in the synthesis of all LTs from AA, namely the oxidation of AA to 5-hydroperoxyeicosatetraenoic acid (5-HPETE) followed by an LTA_4 synthase activity which converts 5-HPETE to LTA_4 (Rouzer et al., 1986; Shimizu et al., 1986) (Figure 8-1). LTA_4 can be further metabolized to the pro-inflammatory compound LTB_4 or to the bronchoconstrictive peptidoleukotrienes LTC_4, LTD_4, and LTE_4, with the amounts and ratios of these products depending upon the cell type being studied (Maycock et al., 1989). 5-LO has therefore been the focus of intense efforts to develop inhibitors of LT biosynthesis. Various whole cell and isolated enzyme assays have been used to identify a number of potent inhibitors of 5-LO (Fitzsimmons and Rokach, 1989). Some of these compounds, including zileuton (Carter et al., 1991), MK 0591 (Brideau et al., 1992) and D 2138 (McMillan et al., 1992) are currently in clinical development.

The Early Years

Early in our search for inhibitors of 5-LO we proposed that, as AA is a substrate for both 5-LO and cyclooxygenase enzymes, examination of cyclooxygenase inhibitors may reveal compounds with 5-LO inhibitory activity, particularly if these compounds are AA mimetics or active site inhibitors. Indeed, study of the indole-2-alkanoic acids, indomethacin-like molecules, such as 1 yielded compounds possessing moderate activity (IC_{50} $\sim 5\,\mu M$) for the inhibition of LT synthesis in human polymorphonuclear leukocytes (HPMN). This compound also possessed thromboxane receptor antagonist activity. Medicinal chemistry effort was initiated to dissociate these two activities. This effort was successful and led to the discovery that by appropriate manipulation of the indole substitution as well as constraint of the alkanoic side chain, it was possible to derive specific thromboxane receptor antagonists with high potency, such as (-)-L-670,596 (Ford-Hutchinson et al., 1989). Introduction of bulky substituents on the indole nucleus resulted in potent inhibitors of LT biosynthesis which were essentially devoid of thromboxane receptor antagonist or cyclooxygenase activity. Optimization of this activity led to the discovery of MK-886 (Gillard et al., 1989).

MK-886 was a potent inhibitor of LT synthesis in a variety of cell types, including human and rat PMN with IC_{50} values of approximately of 3 nM. However, the compound was essentially inactive in broken cell 5-LO assays or when tested against purified enzyme ($IC_{50} > 10\,\mu M$). In

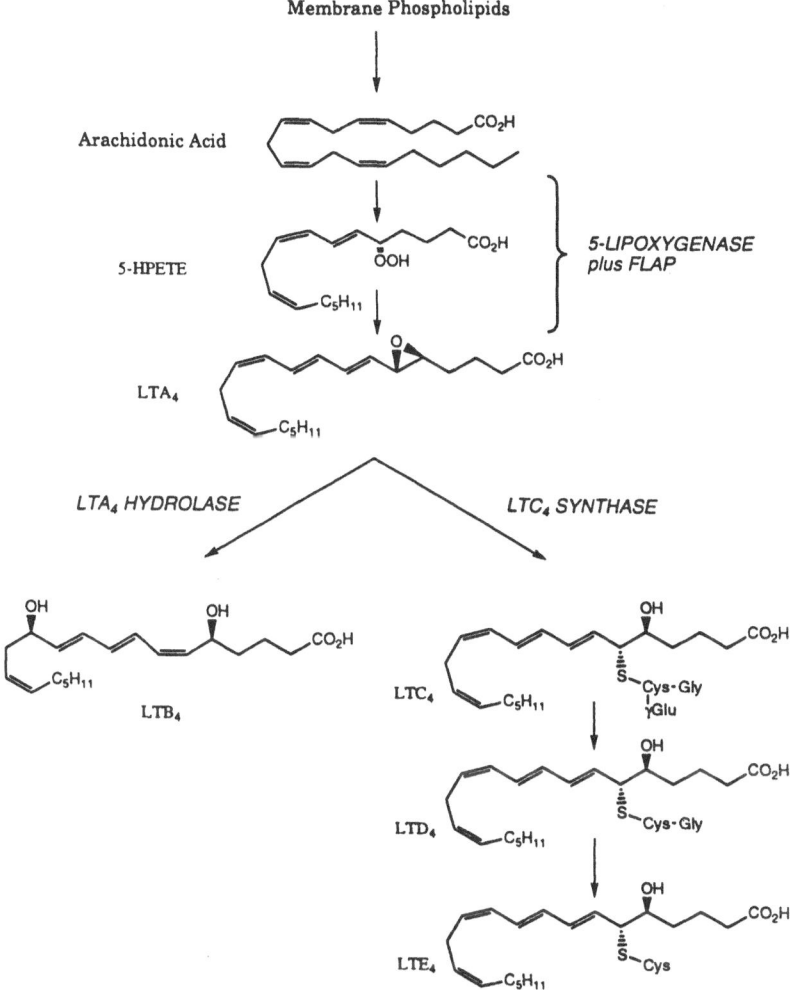

Figure 8-1. Arachidonic acid cascade.

spite of this lack of activity on the enzyme, the compound inhibited LT production in a number of *in vivo* models and attenuated antigen-induced dyspnea in the hyperreactive rat and allergic squirrel monkey at doses of 0.05–1.0 mg/kg po.

The question then remained, what was the mechanism of action of MK-886? To address the mechanism of action of MK-886, two approaches were taken (see Structure 8-A). At Merck Frosst, it was demonstrated that [125I]-L-669,083, a radioiodinated photoaffinity analog of MK-886, specifically bound to an 18 kDa membrane protein from rat and human leukocytes.

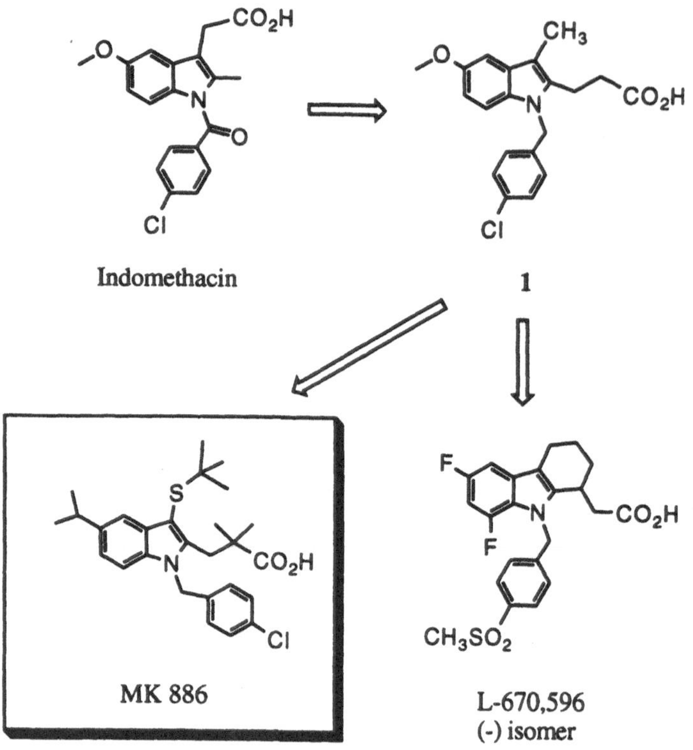

STRUCTURE 8-A

Concurrently with these studies, our colleagues at Merck Research Laboratories in Rahway demonstrated that an 18 kDa membrane protein from leukocytes specifically bound to an affinity matrix to which was attached a structural analogue of MK-886 (Miller et al., 1990) (Structure 8-B).

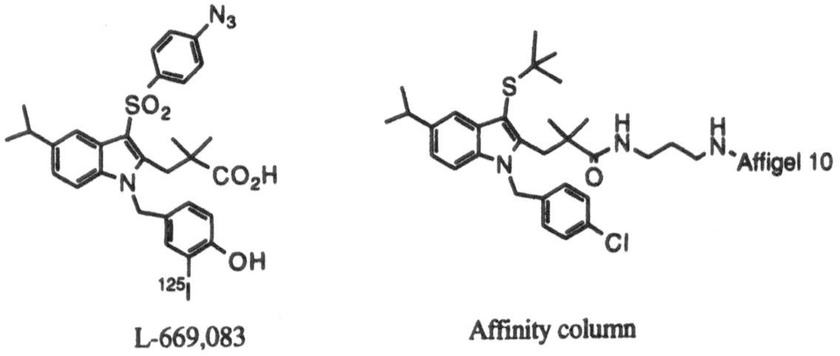

STRUCTURE 8-B

The protein eluted from these columns was subsequently purified to homogeneity by sequential column chromatography, and partial amino acid sequences were obtained. Oligonucleotides based on these sequences were then used to isolate full length cDNA molecules for both the rat and human forms of the protein from RBL-1 and DMSO-differentiated HL-60 cell cDNA libraries, respectively (Dixon et al., 1990). The photoaffinity and affinity column approaches to identify the target of MK-886 converged in immunoprecipitation studies in which antisera that recognized the protein isolated from affinity columns immunoprecipitated the photoaffinity labelled protein.

The primary amino acid sequence of the rat and human 18 kDa proteins deduced from the cDNA sequences was 161 amino acid residues in length, and shared little homology with any other sequence in protein and DNA data banks. Consistent with the biochemical studies, the protein contains three stretches of amino acids of sufficient length and hydrophobicity to potentially represent transmembrane regions (Figure 8-2).

While these studies identified a novel protein that specifically bound MK-886, it was critical to establish whether this protein was absolutely required for cellular LT synthesis. The demonstration of cellular LT synthesis that does not require this protein would render its utility as a therapeutic target questionable. In a series of transfection studies using an osteosarcoma cell line that does not normally express 5-LO or the 18 kDa protein, it was demonstrated that expression of the recombinant forms of either of these proteins alone did not result in cellular LT synthesis in response to the calcium ionophore A23187. In contrast, cells expressing both the 18 kDa protein and 6-LO synthesized LTs in response to A23187, with this synthesis being inhibited by compounds that specifically interacted with either the 18 kDa protein or 5-LO (Dixon et al., 1990). These studies clearly demonstrated that the 18 kDa protein is required for the cellular synthesis of LTs by 5-LO, hence it was termed 5-lipoxygenase-activating protein and is now more commonly referred to in its abbreviated form, FLAP.

Biological Role of FLAP

Cellular LT synthesis in a variety of cell types in response to a number of agents, including A23187, thapsigargin, IgE, and fMetLeuPhe is associated with the translocation of 5-LO from a soluble to a membrane fraction of the cell (Rouzer and Kargman, 1988; Kargman et al., 1991; Wong et al., 1991; Wong et al., 1992). Compounds that specifically interact with FLAP, such as MK-886, inhibit and reverse this specific membrane association of the enzyme (Rouzer et al., 1990). As FLAP is an integral membrane

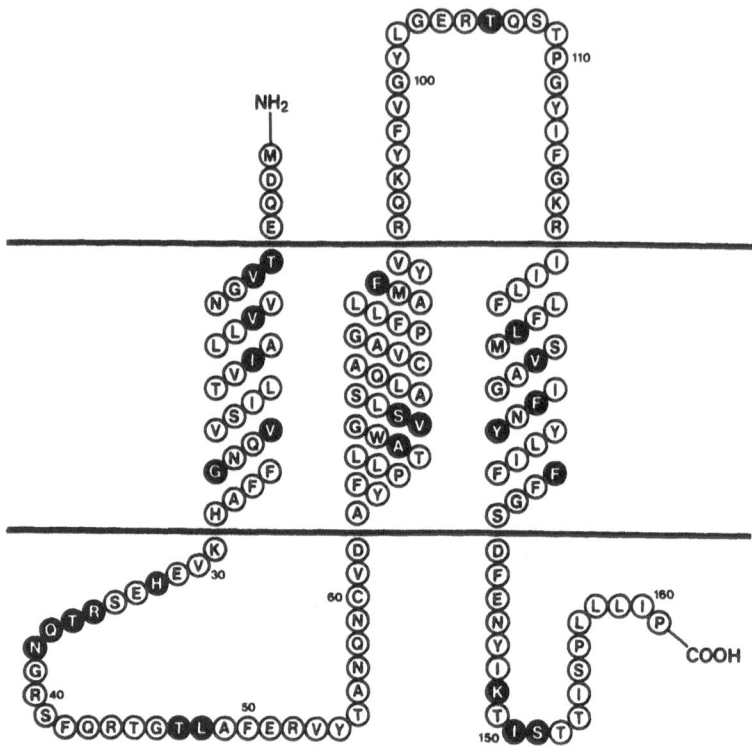

Figure 8-2. Amino acid sequence and proposed membrane topology of human FLAP. The proposed membrane topology of human FLAP based on hydropathy plot analysis (Dixon et al., 1990) is shown. Horizontal lines represent boundaries of a membrane bilayer. The standard single letter amino acid code is used. Highlighted residues differ from human FLAP in one or more of seven mammalian species (Vickers et al., 1992a). Residues 42–62 are critical for the binding of LT biosynthesis inhibitors (Vickers et al., 1992a; Mancini et al., 1994).

protein, these studies suggested that 5-LO may dock with FLAP at the membrane during LT synthesis, presumably to efficiently use AA released from membrane phospholipids by phospholipases.

Studies in osteosarcoma cells (Dixon et al., 1990) and Sf9 insect cells (Abramovitz et al., 1993) in which the cellular activity of recombinant 5-LO was assessed in the absence or presence of recombinant FLAP demonstrated that FLAP stimulated the utilization of AA as a 5-LO substrate. To test whether FLAP specifically bound AA, we synthesized [^{125}I]-L-739,059, a radioiodinated photoaffinity analog of AA (Perrier, 1994). This ligand binds to FLAP and is covalently attached to it upon irradiation with uv light. This binding is inhibited in a concentration-dependent manner by AA and

MK-886 but not by structural analogs of MK-886 which were not effective as inhibitors of LT synthesis (Mancini et al., 1993). The affinity of AA for FLAP in this system was low, with an IC_{50} for the inhibition of $[^{125}I]$-L-739,059 binding of approximately $10 \mu M$. This is similar to the Km for AA reported for human and porcine 5-LO (Lewis and Austen, 1984; Ueda et al., 1986), and may be consistent with a mechanism in which FLAP binds AA with low affinity prior to providing access to this substrate for 5-LO. In the corollary of these studies, we have recently demonstrated that AA competes with the LT biosynthesis inhibitor $[^{125}I]$-L-691,831 for binding to FLAP, with an IC_{50} of approximately $45 \mu M$ (Charleson et al., 1994). This study also demonstrated that the fatty acid binding site on FLAP is not absolutely specific for AA. It can accommodate alternate unsaturated fatty acids, such as 12-HETE, 15-HETE, and eicosapentanoic acid but not fully saturated fatty acids. While the biological significance of this finding is not yet clear, recent studies in human neutrophils have suggested that FLAP may be required for the utilization of 12-HETE and 15-HETE as substrates by 5-LO (Hill et al., 1992) (Structure 8-C).

[^{125}I]-L-739,059   [^{125}I]-L-691,831

STRUCTURE 8-C

In addition to stimulating the oxidation of AA and other fatty acids by 5-LO, in Sf9 cells FLAP increased the ratio of LT to 5-HPETE reaction products compared to cells expressing 5-LO alone (Abramovitz et al., 1993). This suggested that FLAP not only stimulates AA utilization by 5-LO but also specifically increases the LTA_4 synthase activity of the enzyme. This is consistent with the demonstration that the ratio of LT to 5-HPETE reaction products is considerably higher in intact cells (i.e. in the presence of FLAP) following ionophore challenge than that obtained in 5-LO assays using the soluble enzyme from the same cells (Rouzer et al., 1988). The proposed stimulation of the LTA_4 synthase activity of 5-LO by FLAP was confirmed by the demonstration that FLAP stimulates the conversion of exogenously provided 5-HPETE to LTA_4 by 5-LO when the proteins are expressed in Sf9 insect cells (Vickers et al., 1994).

The studies described above demonstrate that LT biosynthesis inhibitors that bind to FLAP compete with AA for binding to the protein, block the utilization of this substrate, and inhibit the membrane association of 5-LO. This suggests that AA and LT biosynthesis inhibitors that specifically bind to FLAP share a common binding site on the protein and that 5-LO may only associate with FLAP when FLAP has AA bound. This model for the cellular synthesis of LTs is shown in Figure 8-3. Our original hypothesis for searching for direct 5-LO inhibitors that were AA mimetics therefore serendipitously resulted in the identification of LT synthesis inhibitors that specifically bind to FLAP.

Clinical Study with MK-886

The result from the clinical asthma study with MK-886 was somewhat disappointing (Bel et al., 1990; Friedman et al., 1993). Eight atopic men participated in a two-part, double-blind, placebo controlled crossover trail. MK-886 was administered in two oral doses of 500 mg and 250 mg, 1 h before and 2 h after antigen inhalation, respectively. Biochemical effects of MK-886 were evaluated by the inhibition of urinary LTE_4 excretion and calcium ionophore-stimulated LTB_4 biosynthesis in whole blood *ex vivo*. MK-886 significantly inhibited the antigen-induced early asthmatic reactions by 58.4% (AUC_{0-3h}) and the late asthmatic reactions by 43.6% (AUC_{3-7h}) when compared to placebo ($p < 0.01$). MK-886 inhibited calcium ionophore-stimulated LTB_4 production in whole blood (54.2%) for up to 6 h post allergen challenge. LTE_4 excretion in urine was inhibited by 51.5% during the early asthmatic reactions by as much as 80% during the late asthmatic reactions. Although the level of inhibition of LT synthesis by MK-886 was disappointing, the data clearly indicated that LTs play a role in allergen-induced asthmatic reactions in humans and suggested that effective LT synthesis inhibitors may provide useful protection against allergen.

From REV 5901 to MK 0591

Clearly our next objective was to develop a compound with superior *in vivo* activity to MK-886. The desired compound had to inhibit LT production essentially completely and work functionally in relevant animal models. Having adequately defined the mechanism of action of MK-886, we sought other structures that could act through a similar mechanism. Since at that point in time a FLAP radioligand binding assay had not been developed, we used the radiolabeled photoaffinity probe [^{125}I]-L-663,083 to search for

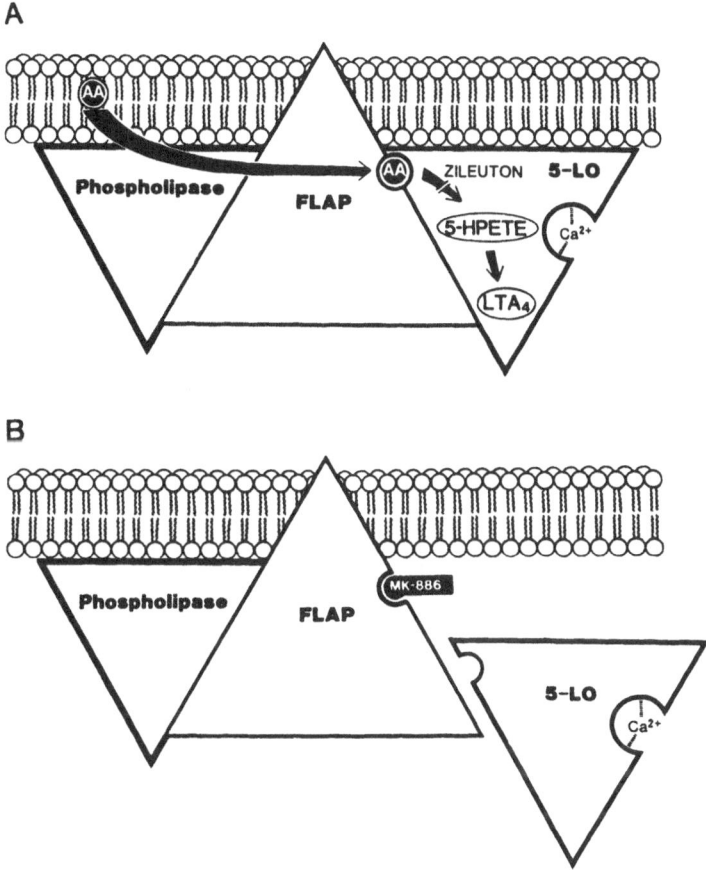

Figure 8-3. A model for cellular LT synthesis. A schematic representation of cellular LTA$_4$ synthesis is shown. FLAP is shown as an integral membrane protein forming a calcium-dependent complex with a phospholipase and 5-LO (A). Arachidonic acid (AA) is released from membrane phospholipids by phospholipase activity and transferred to 5-LO by FLAP for use as a substrate. Zileuton directly inhibits 5-LO as indicated. MK-886 competes with AA for binding to FLAP (B) and prevents the productive membrane association of 5-LO. Immunoelectron microscopy has shown that 5-LO translocates to the nuclear membrane of activated human leukocytes, where FLAP resides (Woods et al., 1993).

other structural classes that would bind to FLAP. Such screening revealed that the Revlon compound, REV-5901, competed with the photoaffinity probe for binding to FLAP in a dose-dependent manner. However this compound had moderate potency for inhibition of LT biosynthesis (IC$_{50}$ in HPMN = 1800 nM). It has since been shown that REV-5901 is less potent in a FLAP radioligand binding assay than MK-886 (Charleson et al., 1992;

Figure 8-4). This quinoline-containing molecule represented an entirely different structural class to MK-886, but like MK-886 it was essentially inactive against isolated 5-LO. Optimization of this lead was initiated, and we quickly discovered that the para-isomer **2** gave a 200-fold boost in potency in the HPMN assay (IC_{50} = 90 nM) (Young and Zamboni, 1987). However, this compound was not orally active and our next goals were to enhance both the intrinsic potency and the oral activity. In addition, it would be desirable to minimize the potential for metabolism at the terminal ω-alkyl site.

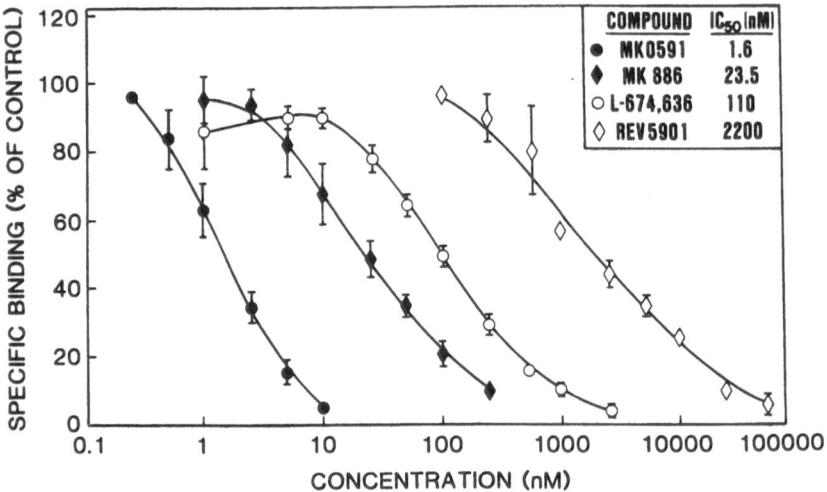

Figure 8-4. Competition analysis of [^{125}I]-L-691,831 binding to human FLAP. The binding of [^{125}I]-L-691,831 to membrane preparations from human leukocytes was analyzed in the presence of the indicated compounds as previously described (Charleson et al., 1992).

Replacement of the n-pentyl side chain of **2** with a 3-phenylpropyl side chain gave the metabolically more stable analog **3**, with a slight enhancement in potency over the initial lead (HPMN IC_{50} = 30nM). The 3-phenylpropyl side chain was found to be superior to both the 2-phenylethyl and the 4-phenylbutyl side chain. Having secured the optimal chain length, we then turned our attention to other parts of the molecule. We rapidly determined that substituents on the quinoline ring usually have a detrimental effect on the potency of the compound and that the nitrogen atom of the quinoline was essential for the activity since the naphthalene analog was virtually inactive. In addition, the methoxy bridge linking the two aromatic units was superior to both methylthio or 1,2-ethyl linkage. The hydroxyl

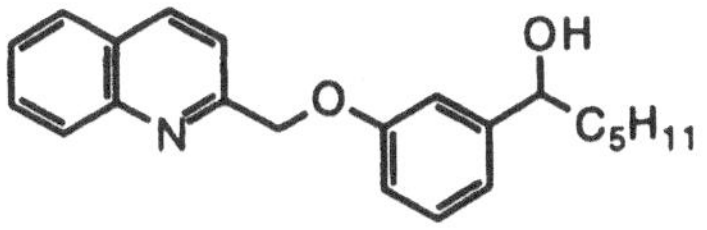

REV 5901

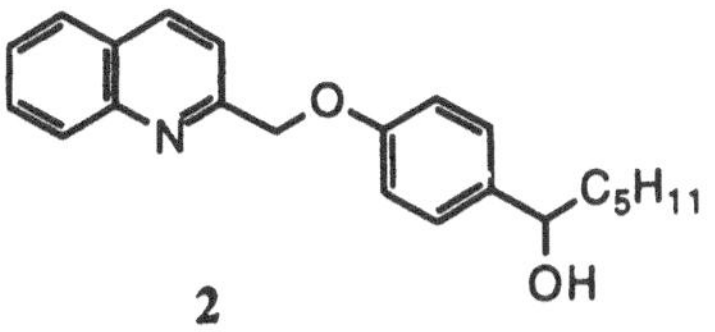

2

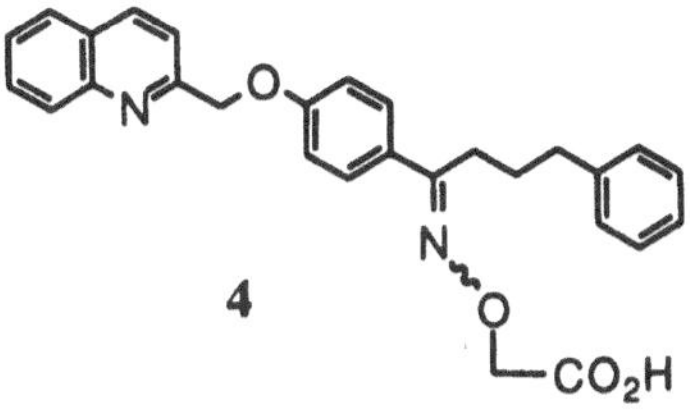

4

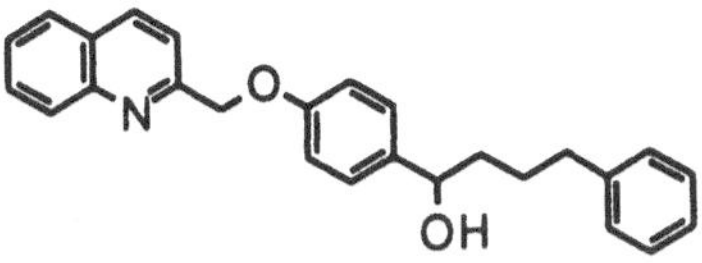

3

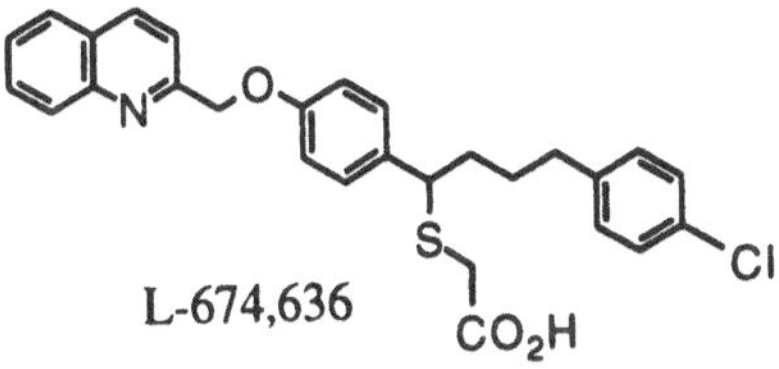

L-674,636

STRUCTURE 8-D

group was not critical for activity since its removal or its replacement with a keto group did not drastically affect activity (Structure 8-D).

Since **3** binds to FLAP in the same manner as MK-886 (Evans et al., 1991), we postulated that the potency of the quinoline series may be enhanced by the addition of an acid moiety such as that in MK-886. This notion was particularly attractive because in addition to potentially picking up an acid-binding domain on FLAP that may enhance the intrinsic potency of the compound, the sodium salt of the compound should also be more bioavailable. Toward this end, a carboxylic acid moiety was appended at various sites. After several disappointing attempts in which the activity was completely lost, a successful acid attachment point was identified. This led to **4** (HPMN IC_{50} = 80nM), a molecule with comparable potency to the initial lead. Encouraged by this result, the (oximinooxy)acetic acid side chain was replaced by a 2-thioacetic acid side chain and further optimization led to L-674,636, which was an extremely potent LT synthesis inhibitor (HPMN IC_{50} = 8 nM). Unlike its predecessor **3**, the sodium salt of L-674,636 was orally active, had excellent bioavailability in rats and monkeys, and was active *in vivo* in a number of animal models (Prasit et al., 1991). The high potency of this compound in different species is consistent with the observation that amino acid sequence of FLAP being highly conserved among all of the species in which it has been determined (Vickers et al., 1992b).

As L-674,636 contains a stereogenic center, it was of interest to compare the activities of the two enantiomers of L-674,636. The two enantiomers were conveniently prepared by separating the R-(−)-2,2,2,-trifluoro-1-(9-anthryl) ethyl ester of L-674,636 and subsequent removal of the chiral auxiliary. The (+)-enantiomer **5** has $[\alpha]_D = +160°$ (c1, CHCl$_3$) and the (−)-enantiomer **6** has $[\alpha]_D = -166°$ (c1, CHCl$_3$) respectively. Surprisingly, potencies of compounds **5** and **6** in the HPMN assay were not significantly different, the (−)-enantiomer **6** being only two times more potent than the (+)-enantiomer **5** (5 nM compared with 10 nM, respectively). This result was particularly intriguing since the two side chains are vastly different (Structure 8-E).

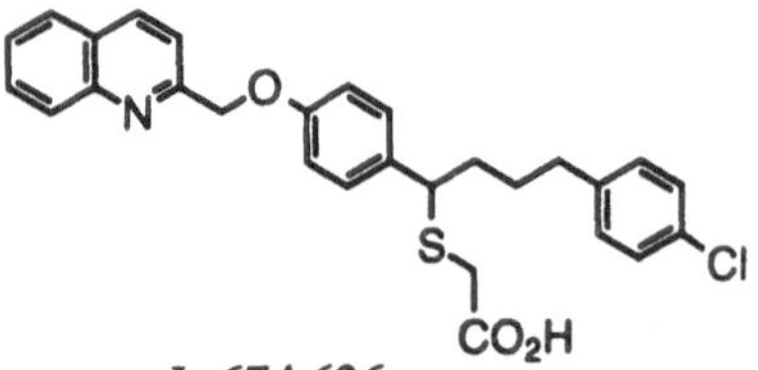

L-674,636
5(+)-enantiomer
6(-)-enantiomer

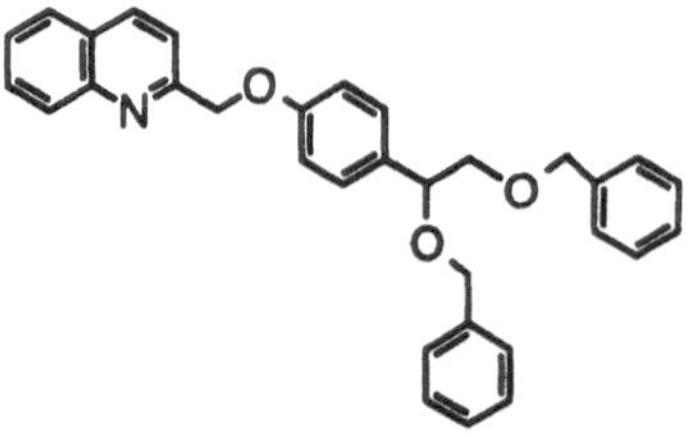

7

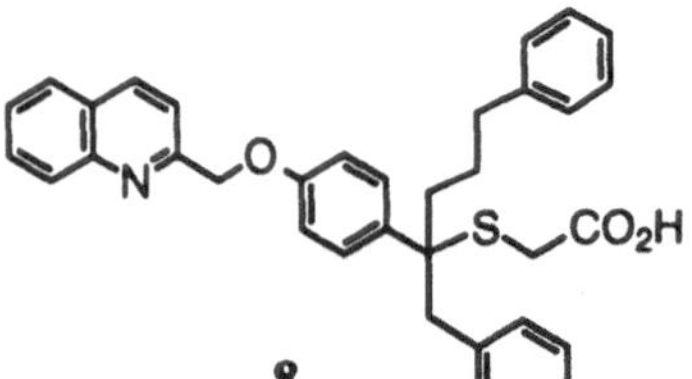

8

STRUCTURE 8-E

At the same time we also looked at the effect of the introduction of hetero-atoms on to the lipophilic side chain to increase the solubility and bioavailability of the series. During this endeavor, it was observed that compound **7**, a synthetic intermediate that contains two lipophilic side chains but lacking the acidic side chain, was surprisingly active (HPMN IC$_{50}$ = 30 nM). This led us to the hypothesis that there may be an additional lipophilic binding site in FLAP for this class of compounds, and hence compound **8** was prepared. Indeed, compound **8** was found to be more active than L-674,636 (4 nM compared with 20 nM, respectively, when the two compounds were tested in parallel HPMN assays). Thus, the addition of another potential binding domain appeared to enhance the potency of the compound. This result may explain why there was not a significant difference in potency between the two enantiomers **5** and **6**. The lipophilic side chain of **5** may bind with one lipophilic pocket while the side chain of **6** may bind with a different lipophilic pocket as a consequence of the stereogenic center.

In attempts to rationalize competitive binding of L-674,636 and MK-886 for the site on FLAP, a structural resemblance between compound **8** and the more planar MK-886 can be postulated (see Figure 8-5). Both compounds have an acidic moiety and they both contain two lipophilic side chains (alkylphenyls for compound **8** and p-chlorobenzyl and t-butylthio groups for MK-886). According to this model, the intrinsic potency of MK-886 should be enhanced by the addition of a quinolinylmethoxy moiety at the 5-position of the indole.

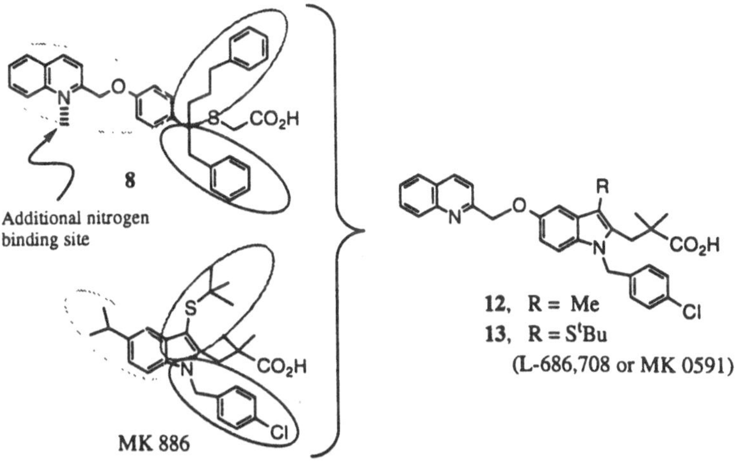

Figure 8-5. The overlay of the indole and the quinoline series.

The phenol **9**, with the MK-886 side chains, was not readily available and thus the phenol **10** (derived from the demethylation of the methoxy compound **11**, which was at hand) was coupled with two equivalents of

2-chloromethylquinoline in the presence of K_2CO_3 followed by aqueous hydrolysis to afford the hybrid compound **12**. The effect of the quinolinylmethoxy moiety in compound **12** was dramatic as compared to the methoxy compound **11**. For instance, the IC_{50} of **11** in the HPMN assay was 870 nM whereas compound **12** was 3 nM. It therefore appeared that the working hypothesis of the overlapping of the acyclic compound **8** and MK-886 is reasonable. The indole ring of MK-886 is acting as a rigid template orienting the acidic and lipophilic side chains into the appropriate region (Structure 8-F).

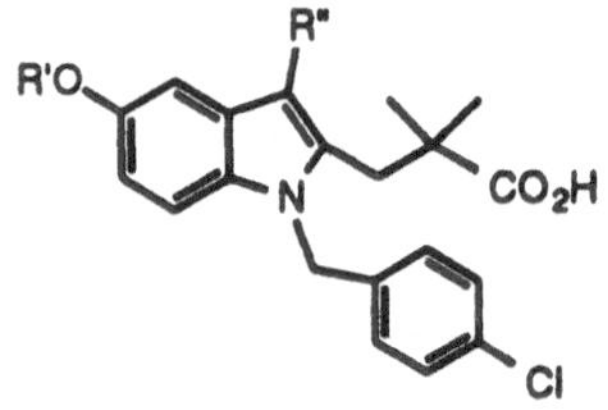

9, R' = H, R" = S^{t}Bu
10, R' = H, R" = Me
11, R' = Me, R" = Me

STRUCTURE 8-F

Stimulated by this observation, the corresponding quinolinylmethoxy analog of MK-886, compound **13** (L-686,708, now designated MK 0591) was prepared. A brief comparison of the intrinsic potency and biological activity of MK 0591 with that of its progenitors MK-886 and L-674,636 are shown in Figure 8-4 and Table 8-1. A radioiodinated photoaffinity analog of this series of inhibitors has also been shown to directly interact with FLAP (Mancini et al., 1992). The amorphous form of the sodium salt of MK 0591 was bioavailable and orally active in a number of animal models. A full biological profile of MK 0591 has been detailed elsewhere (Brideau et al., 1992.), but it is relevant to note the important increase in potency in the human whole blood assay and the radioligand binding assay.

Although a large number of analogs of MK 0591 were prepared, the overall profile of MK 0591 has been such to warrant further development of this compound. It has progressed through safety assessment and is currently being evaluated in human trials for the treatment of asthma and IBD. Results

Table 8-1. Comparison of potency and biological activity of MK 0591 with MK-886 and L-674,636.

Assay (IC_{50}, nM)	MK 0591	MK-886	L-674,636
HPMN	3	3	20
Human whole blood	500	2100	> 22,000
FLAP Binding	2	23	122

from these studies are currently awaited. Preliminary results (Chervinsky, et al., 1993) indicated that when MK 0591 or placebo were administered to mild asthmatics in a rising dose fashion over 4 h, at a cumulative dose of 500 mg, MK 0591 inhibited LTB_4 biosynthesis $\geq 95\%$ for > 36 hours postdose (p < 0.01). Mean forced expiratory volume in one second (FEV_1) after dosing was consistently higher with MK 0591 than placebo at all time points, although not statistically significant. Mean FEV_1 increase was 12.2% for MK 0591 and 7.9% for placebo (p = 0.16).

The ability of MK 0591 to suppress the biosynthesis of both the *ex vivo* whole blood LTB_4 levels and LTE_4 levels in urine essentially completely over a long period of time when administered in a single dose to mild asthmatics is extremely encouraging. However, this inhibition of LT synthesis, both in *ex vivo* whole blood and in the urine does not appear to translate accurately to large increases in the FEV_1. We await the final outcome of these trials to provide some answers on the potential therapeutic utility of FLAP inhibitors, and 5-LO inhibitors in general, in the control of hypersensitivity and inflammatory diseases.

Acknowledgment. The discovery and development of MK 0591 was a very exciting period for Merck Frosst. A lot of blood, sweat and tears went into this program and only highlights and milestones are documented here. Many individuals whose names may not appear in the references provided important contributions that made possible the success of the program. The identification and characterization of FLAP provides an excellent example of a successful interdisciplinary approach to drug discovery and biological research. We sincerely hope that MK 0591 or other LT biosynthesis inhibitors such as zileuton, D 2138 and Bay X1005, or their backups, will prove to be drugs that will greatly improve the quality of life for those who suffer from hypersensitivity and inflammatory diseases.

REFERENCES

Abramovitz M, Wong E, Cox ME, Richardson CD, Li C, Vickers PJ (1993): 5-Lipoxygenase-activating protein stimulates the utilization of arachidonic acid by 5-lipoxygenase. *Eur J Biochem* 215:105–111

Bel EH, Tanaka W, Spector R, Friedman B, Von de Veen JH, Dijkman JH, Sterk PJ (1990): MK-886, an effective oral leukotriene biosynthesis inhibitor on antigen-induced early and late asthmatic reactions in man. *Am Rev Respir Dis* 141:A31

Brideau C, Chan C, Denis D, Evans JF, Ford-Hutchinson AW, Fortin R, Gillard JW, Guay J, Hutchinson J, Jones T, Leger S, Mancini JA, McFarlane CS, Pickett C, Piechuta H, Prasit, P, Riendeau D, Rouzer CA, Tagari P, Vickers P, Young RN (1992): Pharmacology of MK0591 (3-[1-(4-chlorobenzyl)-3-(t-butylthio)-5-(quinolin-2-yl-methoxy)-indol-2yl]-2,2-dimethyl propanoic acid), a potent, orally active leukotriene biosynthesis inhibitor. *Can J Physiol Pharmacol* 70:799–807

Carter GW, Young PR, Albert DH, Bouska J, Dyer R, Bell RL, Summers JB, Brooks DW (1991): 6-Lipoxygenase inhibitory activity of Zileuton. *J Pharmacol Exp Ther* 256:929–937

Chanarin N, Johnston SL (1994): Leukotrienes as a target in asthma therapy. *Drugs* 47(1):12–24

Charleson S, Evans JF, Léger S, Perrier H, Prasit P, Wang Z, Vickers P (1994): Structural requirements for the binding of fatty acids to 5-lipoxygenase-activating protein. *Eur J Pharmacol* 267:275–280

Charleson S, Prasit P, Léger S, Gillard JW, Vickers P, Mancini JA, Charleson P, Guay J, Ford-Hutchinson AW, Evans JF (1992): Characterization of a 6-lipoxygenase-activating protein binding assay: Correlation of affinity for 5-lipoxygenase-activating protein with leukotriene synthesis inhibition. *Mol Pharmacol* 41:873–879

Chervinsky P, Friedman BS, O'Neil M, Rivard S, Tanaka W, Teahan J, Zhang J, (1993): Biochemical effect of MK 0591, a potent oral leukotriene biosynthesis inhibitor in patients with mild asthma. *J Allergy Clin Immunol* 91:225

Dixon RAF, Diehl RE, Opas E, Rands E, Vickers P, Evans JF, Gillard JW, Miller DK (1990): Requirement of a 5-lipoxygenase- activating protein for leukotriene synthesis. *Nature (Lond)* 343:282–284

Evans JF, Léveillé C, Mancini JA, Prasit P, Thérien M, Zamboni R, Gauthier JY, Fortin R, Charleson P, MacIntyre DE, Luell S, Bach T, Meurer R, Guay J, Vickers P, Rouzer CA, Gillard JW, Miller DK (1991): 5-Lipoxygenase activating protein is the target of a quinoline class of leukotriene synthesis inhibitor. *Mol Pharmacol* 40:22–27

Fitzsimmons BJ, Rokach J (1989): Enzyme inhibitors and leukotriene receptor antagonists. In: *Leukotrienes and Lipoxygenases*, Rokach J, ed. Amsterdam: Elsevier

Ford-Hutchinson AW (1989): Evidence for the involvement of leukotrienes and other lipoxygenase products in disease states. In: *Leukotrienes and Lipoxygenases*, Rokach J, ed. Amsterdam: Elsevier

Ford-Hutchinson AW, Bray MA, Doig MV, Shipley ME, Smith MJH (1980): Leukotriene B4, a potent chemotactic and aggregating substance released from polymorphonuclear leukocytes. *Nature (Lond)* 286:264–265

Ford-Hutchinson AW, Girard Y, Lord A, Jones TR, Cirino M, Evans JF, Gillard J, Hamel P, Leveille C, Masson P, Young R (1989): The pharmacology of L-670,596 a potent and selective thromboxane/prostaglandin endoperoxide receptor antagonist. *Can J Physiol Pharmacol* 67:989–993

Friedman BS, Bel EH, Buntinx A, Tanaka W, Han Y-HR, Shingo S, Spector R, Sterk P (1993): Oral leukotriene inhibitor (MK-886) blocks allergen-induced airway responses. *Am Rev Respir Dis* 147:839–844

Gillard JW, Ford-Hutchinson AW, Chan C, Charleson S, Denis D, Foster A, Fortin R, Léger S, McFarlane CS, Morton H, Piechuta H, Riendeau D, Rouzer CA, Rokach J, Young RN, MacIntyre DE, Peterson L, Bach T, Eiermann G, Hopple S, Humes J, Hupe D, Luell S, Metzger J, Meurer R, Miller DK, Opas E, Pacholok S (1989): L-663,536 (MK-886) (3-[1-(4-chlorobenzyl)-3-(t-butylthio)-5 isopropylindol-2yl]-2,2-dimethyl propanoic acid), a novel, orally active leukotriene biosynthesis inhibitor. *Can J Physiol Pharmacol* 67:456–464

Hill E, Maclouf J, Murphy RC, Henson PM (1992): Reversible membrane association of neutrophil 5-lipoxygenase is accompanied by retention of activity and a change in substrate specificity. *J Biol Chem* 267:22048–22053

Isono T, Koshihara Y, Murota S, Fukuda Y, Furukawa S (1985): Measurement of immunoreactive leukotriene C_4 in blood of asthmatic children. *Biochem Biophys Res Commun* 130:486–492

Jones TR, Letts G, Chan C-C, Davies P (1989): Pharmacology and pathophysiology of leukotrienes. In: *Leukotrienes and Lipoxygenases*, Rokach J, ed. Amsterdam: Elsevier

Kargman S, Prasit P, Evans JF (1991): Translocation of HL-60 5-lipoxygenase: Inhibition of A23187- or fMLP-induced translocation by indole and quinoline leukotriene synthesis inhibitors. *J Biol Chem* 266:23745–23752

Knapp HR, Sladek K, Fitzgerald GA (1992): Increased excretion of leukotriene E4 during aspirin-induced asthma. *J Lab Clin Med* 112:48–51

Lam S, Chan H, LeRiche JC, Chan-Yeung M, Salari H (1988): Release of leukotrienes in patients with bronchial asthma. *J Allergy Clin Immunol* 81:711–717

Lewis RA, Austen KF (1984): The biologically active leukotrienes: Biosynthesis, metabolism, receptors, functions, and pharmacology. *J Clin Invest* 73:889–897

Mancini JA, Abramovitz M, Cox ME, Wong E, Charleson S, Perrier H, Wang Z, Prasit P, Vickers P (1993): 5-Lipoxygenase-activating protein is an arachidonate binding protein. *FEBS Lett* 318:277–281

Mancini JA, Coppolino MG, Klassen JH, Charleson S, Vickers PJ (1994): The binding of leukotriene biosynthesis inhibitors to site-directed mutants of human 5-lipoxygenase-activating protein. *Life Sci* 54:137–142

Mancini JA, Prasit P, Coppolino MG, Charleson P, Leger S, Evans JF, Gillard JW, Vickers PJ (1992): 5-Lipoxygenase-activating protein is the target of a novel hybrid of two classes of leukotriene biosynthesis inhibitors. *Mol Pharmacol* 41:267–272

Maycock AL, Pong S-S, Evans JF, Miller DK (1989): Biochemistry of the lipoxyge-
    nase pathways. In: *Leukotrienes and Lipoxygenases*, Rokach J, ed. Amsterdam:
    Elsevier

McMillan RM, Spruce KE, Crawley GC, Walker ERH, Foster SJ (1992): Pre-
    clinical pharmacology of ICI D2138, a potent orally-active non-redox inhibitor
    of 5-lipoxygenase. *Br J Pharmacol* 107:1042–1047

Miller DK, Gillard JW, Vickers P, Sadowski S, Leveille C, Mancini JA, Charleson P,
    Dixon RAF, Ford-Hutchinson AW, Fortin R, Gauthier J-Y, Rodkey J, Rosen R,
    Rouzer C, Sigal IS, Strader CD, Evans JF (1990): Identification and isolation of a
    membrane protein necessary for leukotriene synthesis. *Nature (Lond)* 343:278–
    281

Perrier H, Prasit P, Wang Z (1994): Synthesis of a radioactive photoaffinity arachi-
    donic acid analog. *Tetrahedron Lett* 35(10):1501–1502

Picado C, Ramis I, Rosello T, Prat J, Bulbena O, Plaza V, Montserrat JM, Gelpi E
    (1992): Release of peptide leukotriene into nasal secretions after local instillation
    of aspirin in aspirin-sensitive asthmatic patients. *Am Rev Resp Dis* 145:65–69

Prasit P, Belley M, Evans JF, Gauthier JY, Léveillé C, McFarlane CS, MacIn-
    tyre E, Peterson L, Piechuta H, Therien M, Young RN, Zamboni R (1991):
    A new class of leukotriene biosynthesis inhibitors: the development of ((4-
    (4 - chlorophenyl)-1- (4- (2-quinolinylmethoxy)phenyl) butyl)thio)acetic acid,
    L-674,636. *Bioorg Med Chem Lett* 1:645–648

Rask-Madsen J, Bukhave K, Laursen LS, Lauritsen K (1992): 5-lipoxygenase in-
    hibitors for the treatment of inflammatory bowel disease. *Agents Actions, Special
    Conference Issue*: C37–C46

Rouzer CA, Ford-Hutchinson AW, Morton HE, Gillard JW (1990): MK-886, a
    potent and specific leukotriene biosynthesis inhibitor blocks and reverses the
    membrane association of 5-lipoxygenase in ionophore challenged leukocytes.
    J Biol Chem 265:1436–1442

Rouzer CA, Kargman S (1988): Translocation of 5-lipoxygenase to the membrane in
    human leukocytes challenged with ionophore A23187. *J Biol Chem* 263:10980–
    10988

Rouzer CA, Matsumoto T, Samuelsson B (1986): Single protein from human leuko-
    cytes possesses 5-lipoxygenase and leukotriene A_4 synthase activities. *Proc Natl
    Acad Sci USA* 83:857–861

Rouzer CA, Rands E, Kargman, S, Jones RE, Register RB, Dixon RAF (1988): Char-
    acterization of cloned human 5-lipoxygenase expressed in mammalian cells. *J
    Biol Chem* 263:10135–10140

Samuelsson B (1983): Leukotrienes: Mediators of immediate hypersensitivity re-
    actions and inflammation. *Science* 220:568–575

Shimizu T, Izumi T, Seyama Y, Todokoro K, Rodmark O, Samuelsson B (1986):
    Characterization of leukotriene A_4 synthase from murine mast cells: Evidence
    for its identity to arachidonate 5-lipoxygenase. *Proc Natl Acad Sci USA* 83:4175–
    4179

Tagari P, Rasmussen JB, Delorme D, Girard Y, Eriksson L-O, Charleson S, Ford-
    Hutchinson AW (1990): Comparison of urinary leukotriene E_4 and 16-carboxy-

tetranordihydroleukotriene E_4 excretion in allergic asthmatics after inhaled anti-gen. *Eicosanoids* 3:75–80

Ueda N, Kaneko S, Yoshimoto T, Yamomoto S (1986): Purification of arachidonate 5-lipoxygenase from porcine leukocytes and its reactivity with hydroperoxye-icosatetraenoic acids. *J Biol Chem* 261:7982–7988

Vickers PJ, Adam M, Charleson S, Coppolino MG, Evans JF, Mancini JA (1992a): Identification of amino acid residues of 5-lipoxygenase-activating protein es-sential for the binding of leukotriene biosynthesis inhibitors. *Mol Pharmacol* 42:94–102

Vickers PJ, DeLuca C, Wong E, Abramovitz M (1994): The effect of 5-lipoxygenase-activating protein (FLAP) on substrate utilization by 5-lipoxygenase. In: *Eico-sanoids and Other Bioactive Lipids in Cancer, Inflammation and Radiation Injury: Proceedings of the 3rd International Conference, October 13–16, 1993.* Washington, D.C. Kluwer Publishing, in press

Vickers PJ, O'Neill GP, Mancini JA, Charleson S, Abramovitz M (1992b): Cross-species comparison of 5-lipoxygenase-activating protein. *Mol Pharmacol* 42: 1014–1019

Wong A, Cook MN, Foley JJ, Sarau HM, Marshall P, Hwang SM (1991): Influx of ex-tracellular calcium is required for the membrane translocation of 5-lipoxygenase and leukotriene synthesis. *Biochemistry* 30:9346–9354

Wong A, Cook MN, Hwang SM, Sarau HM, Foley JF, Crooke ST (1992): Stimula-tion of leukotriene production and membrane translocation of 5-lipoxygenase by cross-linking of the IgE receptors in RBL-2H3 cells. *Biochemistry* 31:4046–4053

Woods JW, Evans JF, Ethier D, Scott S, Vickers PJ, Hearn L, Heibein JA, Charleson S, Singer II (1993): 5-Lipoxygenase and 5 lipoxygenase-activating protein are localized in the nuclear envelope of activated human leukocytes. *J Exp Med* 178:1935–1946

Young RN, Zamboni R (1987): US Patent 4,661,499

9

Discovery of BIRM 270: A New Class of Leukotriene Biosynthesis Inhibitors

Peter R. Farina, Carol Ann Homon, Edward S. Lazer, and Thomas P. Parks

Introduction

Several events in the late 1970s and early 1980s helped to shape the research agenda for pharmaceutical companies engaged in the search for a new generation of anti-inflammatory agents. The structural identity of the arachidonic acid (AA) metabolite leukotriene B_4 (LTB_4) (Borgeat and Samuelsson, 1979) and the discovery that it was a potent neutrophil chemoattractant (Ford-Hutchinson et al., 1980) helped to elucidate how leukocytes migrate to a site of inflammation. In addition, the identification of leukotriene C_4 (LTC_4) as a component of slow reacting substance of anaphylaxis (SRS-A) demonstrated that arachidonic acid-derived substances were also potent bronchoconstrictors (Murphy and Hammarström, 1979; Corey et al., 1980; Sirois and Borgeat, 1980). Common to both classes of leukotrienes is the enzyme 5-lipoxygenase (5-LO) which introduces a molecule of oxygen at the C-5 position of AA followed by cyclization to the epoxide intermediate LTA_4. Further reaction with a glutathione S-transferase yields the peptidyl LTs, LTC_4 and its metabolites LTD_4 and LTE_4. Conversion of the epoxide by LTA_4 hydrolase produces LTB_4. Given that AA also serves as a substrate for the cyclooxygenase (CO) pathway to produce pro-inflammatory prostaglandins and thromboxane, interest was certainly reinforced in inhibiting one or both arms of what is often called the arachidonic acid cascade.

Glucocorticoids, aside from their negative side effects, have long been seen as the standard by which other anti-inflammatory drugs should be measured. Therefore it was not surprising that the identification of a

The Search for Anti-Inflammatory Drugs
Vincent J. Merluzzi and Julian Adams, Editors
© Birkhäuser Boston 1995

glucocorticoid-induced protein that was reported to inhibit phospholipase A_2 (PLA_2), an enzyme that releases AA from membrane phospholipids, received so much attention (Hirata et al., 1980). This 40 kD protein called lipomodulin (Hirata, 1981) was also purported to inhibit neutrophil chemotaxis. We now know that much of this work has been either refuted or seen as overly simplistic in explaining the biochemical mechanism of glucocorticoid action. This area of research, however, helped to focus attention on the role of PLA_2 and AA in inflammatory processes.

We became interested in the role of AA in inflammation during a symposium commemorating the opening of our new research facility in Ridgefield, Connecticut during the summer of 1980. Dr. Fusao Hirata, from Dr. Julius Axelrod's laboratory at NIH, reported on his work on lipomodulin. Following the symposium, two research strategies were adopted: (1) corroborate the findings of the NIH group and test the activity of lipomodulin in our own models of inflammation, and (2) establish a program to discover inhibitors of the 5-lipoxygenase pathway. The former objective was abandoned after we and others could not substantiate the role of lipomodulin as a direct PLA_2 inhibitor or anti-inflammatory agent.

Meanwhile, other colleagues in the company were looking at antioxidant compounds as potential antirheumatic drugs. The Riker compound R-830 (Moore and Swingle, 1982) was one of the early leads in this area. Since we had established a 5-LO assay using intact human neutrophils, we eagerly tested newly synthesized compounds in the hope of finding an inhibitor. Two compounds came from this effort: BI-L-93, a nonselective inhibitor of both CO and 5-LO, and BI-L-239, a selective inhibitor of 5-LO (Lazer et al., 1991). Although BI-L-239 was advanced as a preclinical candidate for an asthma indication, it was held back, in part, by company priorities on platelet activating factor (PAF) antagonists.

The Screens

At this time strides were being made by Vadas and Pruzanski in defining the role of an extracellular PLA_2 in inflammatory diseases, such as rheumatoid arthritis and septic shock (Vadas and Pruzanski, 1986). Since so much of the earlier work on PLA_2 was performed using enzymes from porcine pancreas, bee or snake venoms, the prospects of a well-characterized human PLA_2 was welcome news to researchers in the field. We isolated sufficient quantities of this enzyme from human synovial fluid to start a high-throughput screen (HTS) looking for inhibitors. At the same time, some rational design synthesis was being conducted by our colleagues in Germany. Still frustrating to us, however, was the question of which PLA_2 in human neutrophils might be responsible for the release of AA, serving as a substrate for 5-LO. Although PLA_2 activity was measurable in cell supernatants, little was known about the location (membrane or cytosol), activation or number of intracellular PLA_2s found in inflammatory cells. We felt that isolating and characterizing this intracellular enzyme would require more resources than we could afford, particularly since we were already heavily committed with the extracellular PLA_2. Breaking with our philosophy of screening against well-defined molecular targets, a cellular assay was established using human peripheral blood leukocytes isolated daily from in-house donors. LTB_4 produced from calcium ionophore A23187-stimulated human neutrophils was measured by RIA, and used as an indirect screen for intracellular PLA_2 activity. Since we were not interested in discovering direct 5-LO inhibitors, secondary and tertiary assays were put in place to rule them out.

The secreted and intracellular PLA_2 screens were operated in parallel. We were, of course, hoping that one compound would be found that would inhibit the extracellular enzyme and might also show LTB_4 inhibitory activity. In time, the extracellular PLA_2 assay was plagued with uncovering acylating agents and nonspecific inhibitors, probably acting as surfactants. The few uninspiring hydrophobic leads and our inability to convince management of the importance of extracellular PLA_2 in inflammatory processes led to lack of support and the demise of this screen.

The Lead

Screening of our compound library for inhibitors of leukotriene biosynthesis in the neutrophil-based assay soon turned up a number of hits. Many of these were structurally unattractive from a chemist's point of view, consisting of benzene rings branching out from a central chain with little other functionality, and were derisively referred to as propeller compounds. One of the more structurally interesting compounds to emerge in May 1988 was DF-LS 70, a phenylalanine ester fused to a benzothiazole. This compound, originally synthesized at our German affiliate, Karl Thomae Laboratories in Biberach, for a lipid-lowering program, inhibited LTB_4 biosynthesis with an IC_{50} of 230 nM. Structural analogs were tested, with none offering better activity. We later learned that DF-LS 70 was inspired from tazasubrate, a hypocholesterolemic agent. The Thomae compound had weak *in vivo* cholesterol lowering activity in rats and had not been pursued. Some of our scientists thought they could see a resemblance between DF-LS 70 and Merck's five lipoxygenase activating protein (FLAP) inhibitor, MK 886 which had just recently been reported (Gillard et al., 1989). Both consisted of a fused heterocycle with a carboxylic acid (or ester) separated by a two

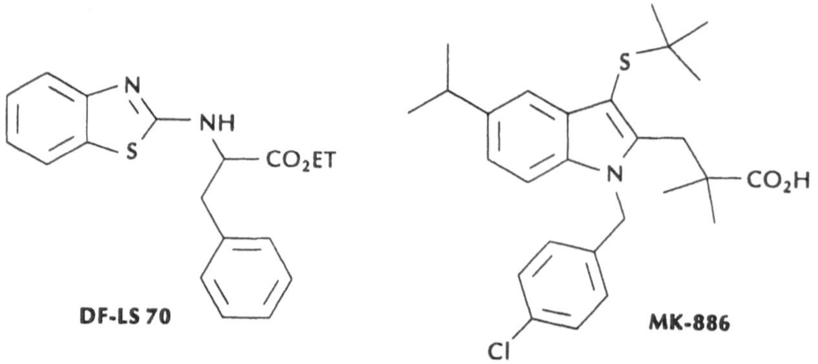

Figure 9-1. The Lead: DF-LS 70 was discovered and initially compared to MK 886.

atom bridge, and had a phenyl ring appended that could conceivably interact at the same location on the protein (Figure 9-1).

The activity of DF-LS 70 compared favorably to our 5-LO inhibitor BI-L-239, an optimized compound, which had an IC_{50} of 340 nM. In both a neutrophil-based 5-HETE assay using exogenously added ^{14}C-AA and a cytosolic 5-LO assay, DF-LS 70 was inactive at concentrations below $10 \mu M$. This was our first signal that the mechanism of action of this hit was unusual. Other than its inhibition of LTB_4 biosynthesis, DF-LS 70 appeared to have no other effect on neutrophil function, was not cytotoxic, and was negative in a general receptor screen. This apparent selectivity was desirable, but to convince ourselves that we had a genuine lead and that the activity was not an artifact, we needed to demonstrate structure-activity relationships.

Our chemistry program began with the assumption that we were acting at the same binding site as MK 886, and that DF-LS 70, an ethyl ester, was actually a prodrug. Therefore we believed that hydrolysis to the carboxylic acid would give the active species and much greater potency. We were surprised to find that the hydrolysis product of DF-LS 70 was completely inactive, as were other carboxylic acid analogs of active esters. Although incubation of DF-LS 70 with human leukocytes resulted in a time-dependent hydrolysis of the ethyl ester to the carboxylic acid, we determined that the t-butyl ester compound BIRL 497 was not hydrolyzed by human leukocytes under the assay conditions. The latter compound retained activity with an IC_{50} of 730 nM. Therefore we concluded that the ester was in fact the active species, and a metabolically stable bioisostere would be needed. To further underscore the difference between our lead and MK 886, the ethyl ester derivative of the latter compound was inactive in our assay. (Later biochemical experiments would confirm that our series did not share the same mechanism as MK 886).

The discovery of enantiomeric selectivity increased our interest in this series. BIRL 455, the racemate of DF-LS 70 (which has the S configuration) was prepared and was found to be less potent (IC$_{50}$ 360 nM). Other examples of enantiomeric selectivity soon followed, ranging from about a five- to 70-fold difference in potency between the more potent S- and the less active R-enantiomers. This suggested to us that there was in fact a molecular target for these compounds, and the activity was not due to some nonspecific interaction.

The presentation of this profile to our research management evoked more questions. Since the mechanism for prevention of the association of 5-LO and FLAP by MK 886 was unclear, it was argued that perhaps we had uncovered a compound with a somewhat related mode of action. In 1990, the world of LT biosynthesis inhibitors was divided into 5-LO and FLAP binding inhibitors, and we fit neither. The lack of a molecular target also became an albatross around our necks, something we recognized could happen when we set up this black box screen.

Medicinal Chemistry—BIRM 270 (Ontazolast) Emerges

Initial structure modifications of the phenylalanine portion of the molecule began in one laboratory and the results are illustrated qualitatively in Figure 9-2. The detailed structure-activity relationships of the entire series has been published elsewhere (Lazer et al., 1994). Replacement of the 2-amino substituent with a carbon or inserting a methylene between the benzothiazole and the amine resulted in a sharp drop in activity. Methylation of the nitrogen also reduced activity. The phenyl glycine analog was inactive, and the homophenylalanine derivative was slightly less active than DF-LS 70. An alpha-methyl group gave a slight increase in potency, but the increased synthetic difficulty did not warrant pursuing this modification. A methyl group on the carbon alpha to the phenyl ring also reduced activity.

At this point our attention shifted to substitution on the benzothiazole and phenyl rings and the search for a stable ester bioisostere. The chemistry capacity was increased to two laboratories. The benzothiazole ring was eventually supplanted by benzoxazole, which generally resulted in more potent compounds. Halogen substituents on the 4-position of the phenyl ring gave a modest improvement in activity (IC$_{50}$s 150–330 nM) as did an alkyl or alkoxy group in the 6-position of the heterocycle.

The two modifications having the most profound effect on potency were soon discovered. These were replacement of the phenyl ring with a

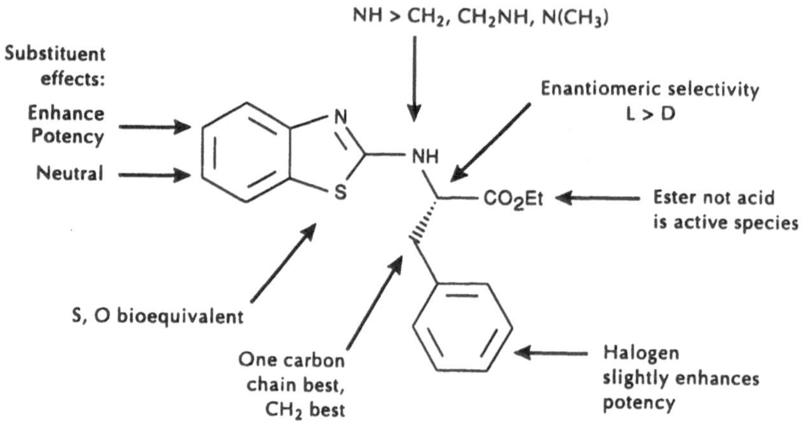

Figure 9-2. Summary of the early structure-activity relationships.

cyclohexyl and the addition of an alkyl substituent in the 5-position. Either substitution alone resulted in a compound with a low-nM IC_{50} (Figure 9-3). The difference between the neutral effect of a 6-substituent and the potency enhancing effect of a 5-substituent was striking. Representative analogs substituted in the 4- or 7-position were also prepared and were less potent than the unsubstituted compound.

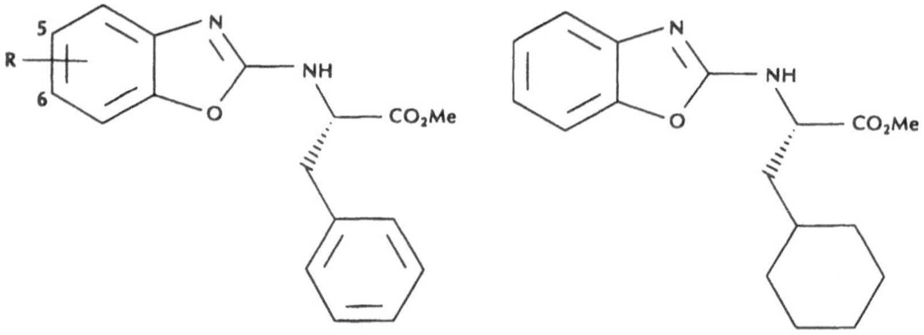

R = 6-iPr: IC_{50} 220 nM
R = 5-iPr: IC_{50} 1 nM

IC_{50} 5.5 nM

Figure 9-3. Cyclohexyl and 5-alkyl substituents enhance potency.

Our search for an ester bioisostere was also proceeding (Table 9-1). Initial attempts with more established ester replacements including a methyloxadiazole (**1**) or amide (**2, 3**) resulted in decreased activity. A methyl ketone (**4**) retained activity, while an alcohol (**5**) or nitrile (**6**) was inactive. We were surprised to find that a phenyl ring, not considered a classical ester bioisostere, resulted in slightly improved activity (**7**) compared to the ester. It was also gratifying that such a hydrophobic molecule (which began to resemble some of the old propeller compounds) retained enantiomeric selectivity (**8–10**), and in fact gave the best separation of potency observed between enantiomers.

Table 9-1. Optimization of the ester bioisostere.

Compound	R_1	R_2	R_3	Stereochemistry	IC_{50} (nM)
1	H	C_6H_5	MO	RS	1,300
2	H	C_6H_5	C(O)NH$_2$	S	3,000
3	H	C_6H_5	C(O)NMe$_2$	RS	3,000
4	6-iPr	C_6H_5	C(O)Me	S	350
5	6-iPr	C_6H_5	CH$_2$OH	RS	>3,000
6	H	C_6H_5	CN	S	>3,000
7	H	C_6H_5	C_6H_5	RS	170
8	H	C_6H_{11}	C_6H_5	RS	8
9	H	C_6H_{11}	C_6H_5	(-)	6
10	H	C_6H_{11}	C_6H_5	(+)	340
11	H	C_6H_{11}	3-C$_5$H$_4$N	RS	53
12	5-Me	C_6H_{11}	2-C$_5$H$_4$N	RS	4
13	5-Me	C_6H_{11}	3-C$_5$H$_4$N	RS	20
14	5-Me	C_6H_{11}	4-C$_5$H$_4$N	RS	190
BIRM 271	5-Me	C_6H_{11}	2-C$_5$H$_4$N	R	40
BIRM 270	5-Me	C_6H_{11}	2-C$_5$H$_4$N	S	1

The extreme lipophilicity of these compounds was a concern to us as we anticipated absorption and bioavailability problems. The 3-pyridyl analog (**11**) was prepared in an attempt to develop more soluble compounds. We believed it would increase solubility particularly in the stomach where we hoped the pyridine would be protonated by the acidic environment. Finding activity with the first analog we prepared the three pyridyl isomers (**12–14**), with the potency enhancing 5-methyl group in place. As seen in the Table 9-1, the 2-isomer was the most potent. The effect on solubility turned out to be marginal and only was observed at pH 2. The more active enantiomer of the 2-pyridyl isomer became our clinical candidate BIRM 270 with 1 nM potency (Figure 9-4).

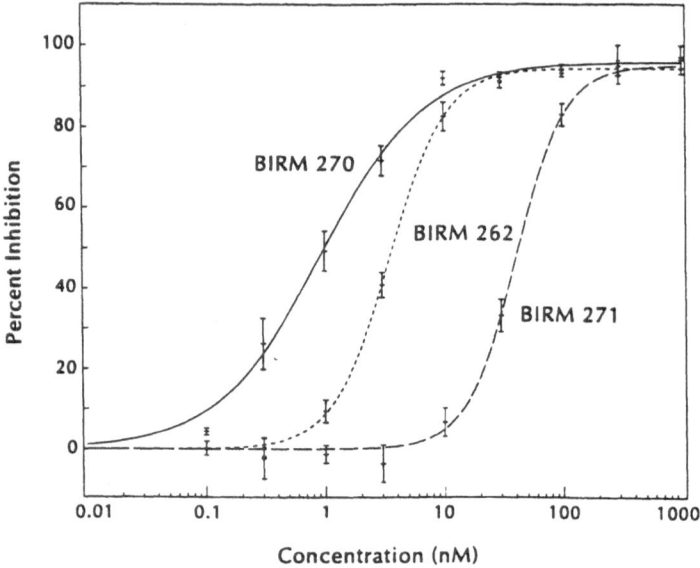

Figure 9-4. Inhibition of LTB$_4$ biosynthesis by BIRM 270 and BIRM 271 compared to racemate in calcium ionophore A23187-stimulated human neutrophils. Reprinted with permission of the American Society for Pharmacology and Experimental Therapeutics, from Farina et al., 1994.

In summary, the time from the synthesis of the first analog, the hydrolysis product of DF-LS 70, to BIRM 270 (Figure 9-5) was 16 months and involved the work of five chemists. About 150 analogs were synthesized and screened in the process. Two additional laboratories were added, and hundreds of additional compounds were made subsequent to BIRM 270, in an effort to further optimize the series.

Bioavailability Problems

Our concerns about the physical properties of these compounds and the effect on absorption and bioavailability proved to be well founded. In pharmacokinetic studies with several of our compounds in primates, doses of 20 mg/kg gave very low and erratic plasma levels, a few hundred ng/mL at best, and often less than 100 ng/mL. Attempts to prepare less lipophilic compounds usually resulted in a significant drop in activity.

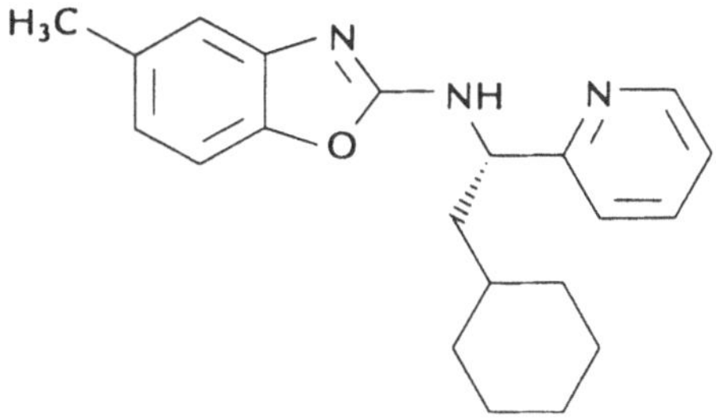

Figure 9-5. The clinical candidate BIRM 270 (ontazolast).

Equally poor results were observed following oral dosing in guinea pigs, in which we were evaluating antigen-induced bronchoconstriction as well as plasma levels. Compound **8** was examined for protein binding in human plasma and was found to be greater than 99% bound. To make matters worse, we discovered that our most potent compounds, those with a cyclohexyl ring at the R_2 position, were metabolically unstable, undergoing hydroxylation on the cyclohexyl ring. Considerable effort went into trying to prepare potent, metabolism-resistant compounds before achieving significant success. These results will be reported in a separate manuscript.

Mechanism of Action

Once active compounds were identified in the black box primary screen, they went through a secondary screening cascade which was designed to discriminate between intracellular PLA_2 inhibitors and 5-LO inhibitors to rapidly determine mechanism of action. Secondary evaluation of the screening hit included testing for inhibition of calcium ionophore A23187-stimulated ^{3}H-arachidonate release and PLA_2 activity. DF-LS 70 inhibited ^{3}H-arachidonate release from human neutrophils at concentrations at least one order of magnitude higher than those which inhibited LTB_4 biosynthesis. The compound did not inhibit the low molecular weight extracellular (synovial fluid) PLA_2 at concentrations up to 10 μM, nor did it inhibit the PLA_2 activity in broken cell preparations of neutrophils. In addition, PAF biosynthesis, as measured by ^{3}H-acetate incorporation into phosphatidylcholine (PC) species, was inhibited relatively weakly by DF-LS 70. These results did not bode well for DF-LS 70 as an intracellular PLA_2 inhibitor.

Several secondary assays were used to identify 5-LO inhibitors found in the screen. High specific activity ^{14}C-arachidonate was added simultaneously with calcium ionophore A23187 to intact human neutrophils, and the formation of ^{14}C-dihydroxyeicosatetraenoic acid (^{14}C-DiHETE) (including LTB_4) was monitored by TLC. This assay was sensitive to both 5-LO and FLAP inhibitors. In addition, a cell-free 5-LO assay was available which was only sensitive to direct 5-LO inhibitors. The DF-LS 70 series of compounds were inactive in both of these assays, suggesting that they did not affect arachidonate metabolism by 5-LO. The results of our secondary evaluation left us in a quandary since we now had no mechanism of action; either our screening cascade had misled us, or we had entered a scientific twilight zone, or both.

The potency of DF-LS 70 analogs and their enantiomeric selectivity made these compounds too interesting to ignore, so the series took on a life of its own, even as the PLA_2 program was being abandoned. The focus of mechanistic studies turned back towards arachidonate metabolism. DF-LS 70 and its derivatives were found by HPLC analysis to inhibit the formation of all 5-LO metabolites, including LTB_4 and its omega oxidation products, 5-HETE, and nonenzymatic 6-trans degradation products of LTA_4 (Figure 9-6). Since direct 5-LO and FLAP inhibitors exhibited the same profile, the burden of proof was still on us to prove or disprove that this series of compounds inhibited 5-LO or FLAP. The secondary assays in the screening cascade had already suggested that the DF-LS 70 series did not behave like the typical 5-LO or FLAP inhibitors, but these results continued to make us uneasy.

A FLAP binding assay was established using ^{3}H-labeled MK 886 and solubilized neutrophil membranes. As expected, unlabeled MK 886 and other indole and isoquinoline FLAP inhibitors competed for binding. The DF-LS 70 series of compounds were uniformly devoid of binding activity. There was still the lingering suspicion that our compounds might bind to an alternative site on FLAP (other than the MK 886 binding site), or bind to an alternative FLAP.

By this time, evidence was accumulating that FLAP was indeed essential for 5-LO activation, particularly for the production of LTA_4 (Dixon et al., 1990). This process appeared to require the calcium-dependent translocation of 5-LO to a membrane compartment (Rouzer and Kargman, 1988; Rouzer et al., 1990). Interestingly, in the human neutrophil the 5-LO/FLAP system could be activated by lower concentrations of calcium ionophore (i.e., lower intracellular Ca^{++} concentrations) than those which

---

Figure 9-6. Reversed phase HPLC chromatograms of 5-LO products generated by A23187-stimulated human neutrophils in the absence (A) or presence of 3 nM (B) or 100 nM (C) BIRM 270. Reprinted with permission of the American Society for Pharmacology and Experimental Therapeutics, from Farina et al., 1994. *See figure on following page.*

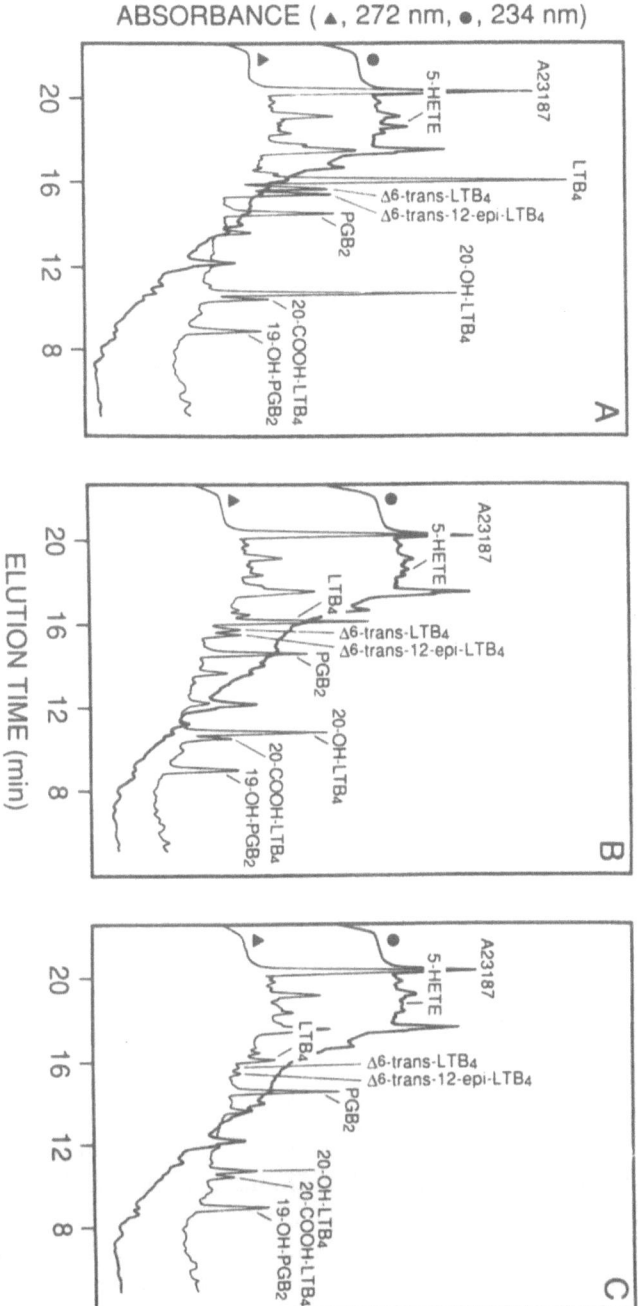

were required for 5-HETE and LT biosynthesis from endogenous substrate. Our collaborator, Dr. Pierre Borgeat of Laval University, had devised an ingenious method for keeping track of these two processes by employing 15-HpETE as an alternative substrate for 5-LO (McDonald et al., 1991). Thus at a low ionophore concentration (50 nM), 15-HpETE could be converted to 5,15-DiHETE by activated 5-LO with very little if any production of 5-HETE or LTB$_4$ from endogenous arachidonate (Figure 9-7). Under these assay conditions, 5-LO or FLAP inhibitors block 5,15-DiHETE formation, whereas BIRM 270 did not (Figure 9-8A). At higher ionophore concentration such as 1 μM, 5-HETE and LTB$_4$ were produced in addition to 5,15-DiHETE. As expected, 5-LO or FLAP inhibitors blocked the formation of products from both exogenous and endogenous substrates. However, BIRM 270 inhibited only the production of 5-HETE or LTB$_4$ from endogenous substrate (Figure 9-8B). This was a strong indication that our compound acted at the level of arachidonate availability, rather than metabolism.

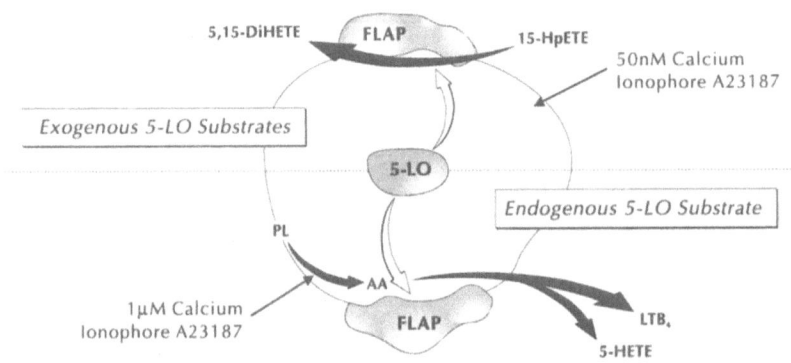

Figure 9-7. Conversion of the exogenous substrate, 15-HpETE to 5,5-DiHETE, compared to the conversion of endogenous AA to 5-HETE or LTB$_4$.

If the compounds prevented AA release, then the addition of exogenous AA should reverse the inhibitory effect of BIRM 270 on LTB$_4$ biosynthesis. We found that exogenous AA did in fact reverse the inhibitory effect of BIRM 270, whereas exogenous oleate, which is not a 5-LO substrate, had no effect (Figure 9-9). The question then was whether BIRM 270 inhibited AA mobilization. We measured the release of endogenous AA from human neutrophils by GC/MS, rather than ^{3}H-AA release from non-equilibrium-labeled cells. We observed potent dose-dependent inhibition which approached 100%, and exhibited enantiomeric selectivity; i.e., BIRM 270 was more potent than BIRM 271 (Figure 9-10). Most gratifyingly, the

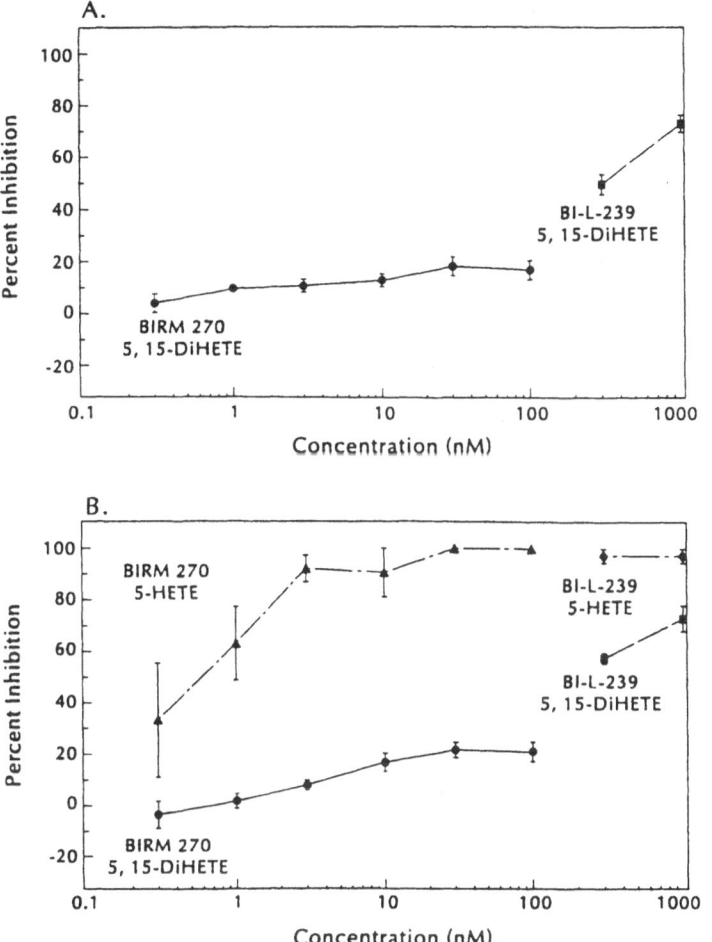

Figure 9-8. Effect of BI-L-239 and BIRM 270 on 5,15-DiHETE biosynthesis using 50 nM A23187-stimulated human neutrophils (A) and on 5,15-DiHETE and 5-HETE biosynthesis using 1 μM A23187-stimulated cells (B). Reprinted with permission of the American Society for Pharmacology and Experimental Therapeutics, from Farina et al., 1994.

inhibition response of AA release was identical and equipotent with the blockade of LTB_4 biosynthesis in this experiment (data not shown; the high cell density used for the AA release assay shifts the dose-response curve of BIRM 270 about tenfold to the right). We concluded that BIRM 270 inhibits LTB_4 biosynthesis by blocking AA mobilization. Ironically, we would also have to conclude that the initial ^{3}H-AA release experiments had seriously misled us on the DF-LS 70 mechanism of action. In retrospect, we now know that ^{3}H-AA is released from different phospholipid pools

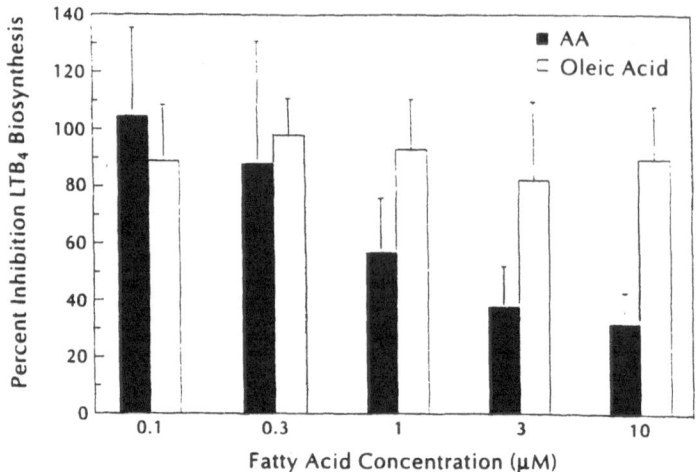

Figure 9-9. Exogenous AA reverses the inhibitory effect of BIRM 270 while oleic acid does not.

than the bulk of the endogenous AA. ^{3}H-AA release followed different kinetics, and was only partially inhibited by BIRM 270, as compared to the release of endogenous AA.

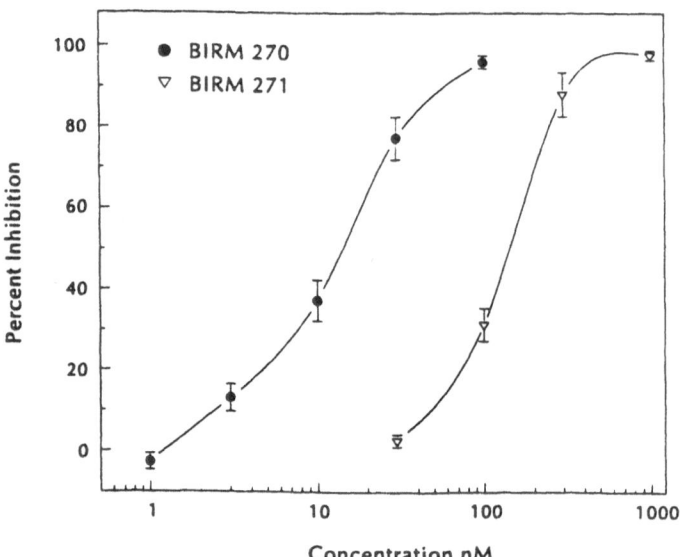

Figure 9-10. Enantiomeric selectivity: BIRM 270 more potently inhibits the release of AA from A23187-stimulated human neutrophils than BIRM 271. Reprinted with permission of the American Society for Pharmacology and Experimental Therapeutics, from Farina et al., 1994.

Since a major portion of AA in neutrophils is esterified in and released from ether-linked phospholipids, including 1-O-alkyl-2-arachidonoyl-glycero-phosphocholine (GPC), a precursor of PAF, we sought to determine whether BIRM 270 inhibited its biosynthesis. We measured the mass of endogenously produced PAF by GC/MS, rather than relying on the indirect and potentially misleading incorporation of ^{3}H-acetate into PC. BIRM 270 potently blocked PAF biosynthesis over the same concentration range that inhibited AA mobilization and LTB_4 production (Figure 9-11). This was the first biochemical evidence linking AA release and PAF biosynthesis, bolstering earlier proposals that these two processes might be coordinately regulated.

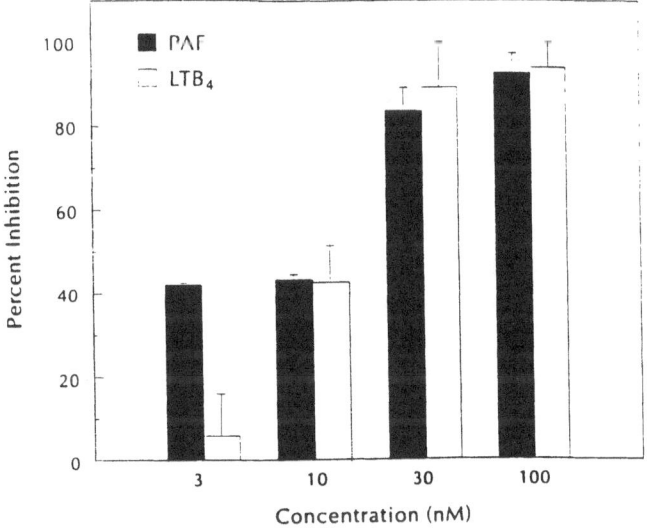

Figure 9-11. BIRM 270 inhibits the biosynthesis of PAF as well as LTB_4. Reprinted with permission of the American Society for Pharmacology and Experimental Therapeutics, from Farina et al., 1994.

The most obvious control point for the coordinate regulation of AA availability and PAF formation would be a PLA_2. However, neither the type I (porcine pancreatic) nor type II (synovial fluid) 14 kDa secretory PLA_2s were inhibited by members of the DF-LS 70 series. The human neutrophil is known to contain both a low molecular weight type II secretory enzyme, and an 85 kDa intracellular PLA_2. BIRM 270 did not inhibit the 85 kDa PLA_2 from human neutrophils, the monocytic cell line U937 or the recombinant enzyme produced in insect cells. Preliminary evidence also indicates that BIRM 270 does not directly inhibit the arachidonyl-selective, CoA-independent transacylase which could also potentially coordinate AA mobilization and PAF biosynthesis. Thus it would appear that BIRM 270 either targets another *sn*-2 acylhydrolase, or indirectly affects the activity of one of the enzymes discussed above (Farina et al., 1994). Since the compound can block AA mobilization almost entirely under certain conditions, identification of the pathway and how it is inhibited by BIRM 270 is a problem of considerable biological interest.

Effect of BIRM 270 on Neutrophil Functions

We clearly showed that BIRM 270 effectively shut down calcium ionophore A23187-stimulated AA release and lipid mediator biosynthesis. However, it was desirable to determine whether the compound could inhibit more physiologically relevant receptor-stimulated LT production, or affect other neutrophil functions. For example, AA has been proposed as a second messenger in the activation of NADPH oxidase (Cockcroft, 1992). We found that PAF could stimulate respectable LTB_4 production if human neutrophils were primed with a brief fMLP pretreatment. BIRM 270 potently and completely suppressed eicosanoid biosynthesis in cells stimulated in this manner (Figure 9-12). However, the neutrophil respiratory burst was not affected by BIRM 270 under the same conditions. In addition, the compound did not inhibit lysozyme or lactoferrin release; it did not affect the upregulation of Mac-1 or neutrophil adhesion to activated endothelium; and it did not inhibit chemotaxis.

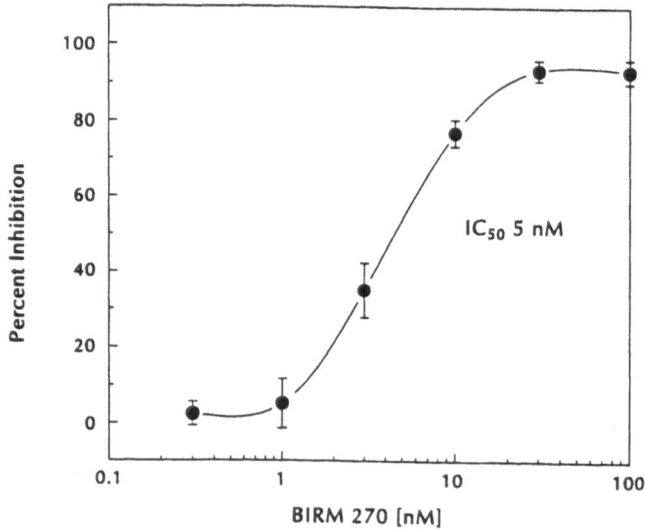

Figure 9-12. LTB_4 biosynthesis in human neutrophils stimulated with fMLP and PAF is inhibited by BIRM 270.

Effect of BIRM 270 on Other Inflammatory Cells

Once it became apparent that the DF-LS 70 series of compounds inhibited LT biosynthesis by neutrophils, we were curious whether these results could be extended to other inflammatory cells. In a collaborative study with Dr. Edward Schulman at Hahnemann University, we determined that BIRM 270 and its less potent enantiomer, BIRM 271, inhibited LTC_4 biosynthesis by mast cells stimulated with calcium ionophore (Figures 9-13 and 9-14). BIRM 270 also inhibited PGD_2 production and, surprisingly, histamine release by these cells. Although the compound completely inhibited LTC_4 biosynthesis, PGD_2 and histamine release were only partially inhibited (Figure 9-14). Our 5-LO inhibitor, BI-L-239, inhibited 5-LO, and CO activity at elevated concentrations, but had no effect on histamine release. Therefore, the effect of BIRM 270 on histamine release did not appear to be due to inhibition of PGD_2 release. In mast cells, it has been suggested that there may be different pools of AA which serve as precursors for LT and PG biosynthesis (Fonteh and Chilton, 1993).

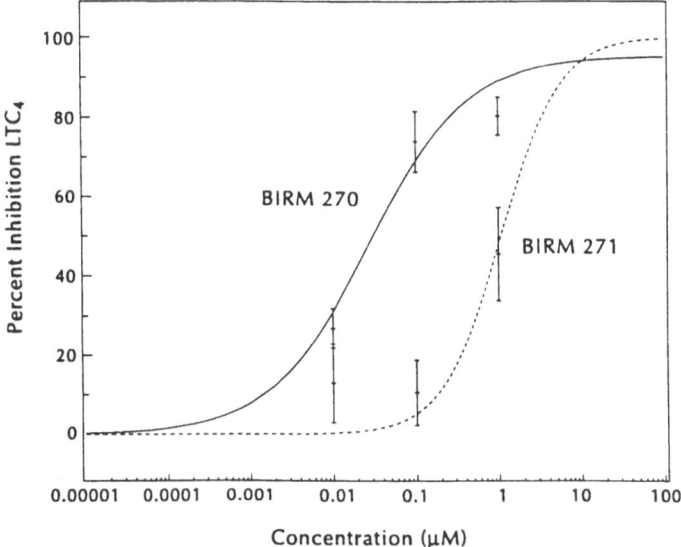

Figure 9-13. LTC_4 production in A23187-stimulated human lung mast cells is inhibited by BIRM 270 and less potently by BIRM 271.

Through our association with Dr. Schulman, we came to know Dr. Donald Raible, who was working on human blood eosinophils from atopic donors. Dr. Raible, who was aware of our mast cell and neutrophil results, conducted experiments on eosinophils with BIRM 270 and BIRM

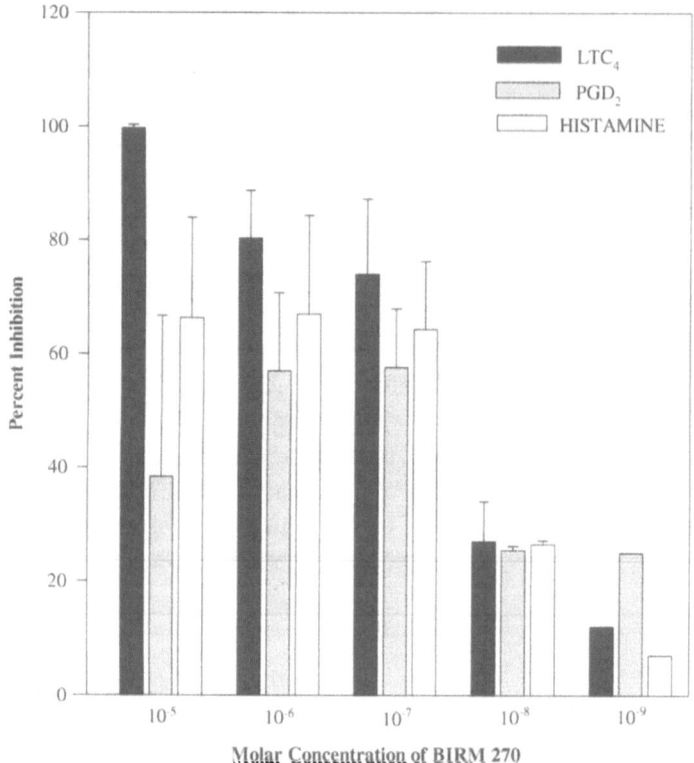

Figure 9-14. BIRM 270 inhibits the release of LTC_4, PGD_2, and histamine from human lung mast cells.

271. BIRM 270 (IC_{50} =13 nM) was approximately 20 times more potent than BIRM 271 in inhibiting the release of LTC_4 from these cells when stimulated with calcium ionophore A23187 (Figure 9-15).

The inhibition of mediator release from these three cell types, which have been shown to play important roles in the inflammatory responses in the lung, supported our decision to pursue BIRM 270 for the treatment of asthma. Interestingly, BIRM 270 did not inhibit leukotriene release from a variety of rodent cells including rat neutrophils and macrophages.

The Final Test

The class of compounds derived from our initial lead DF-LS 70, which ultimately led to BIRM 270, was at times fascinating, frustrating and often perplexing. Findings related to solubility and extensive metabolism limited our options for *in vivo* evaluation. Most difficult was the fact that BIRM 270

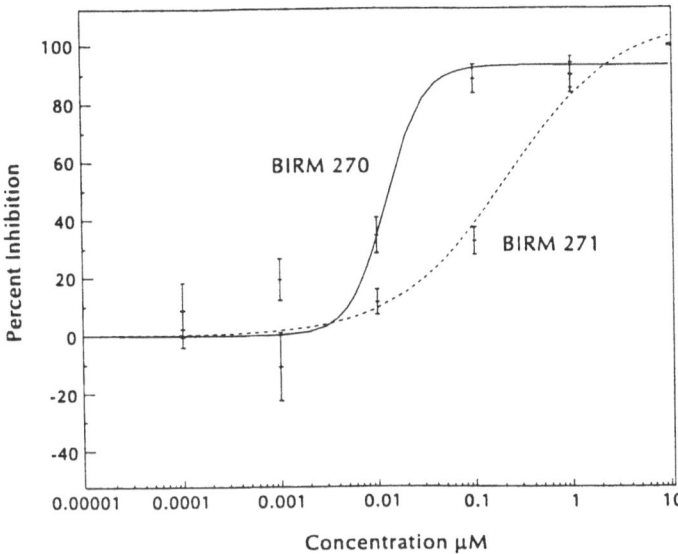

Figure 9-15. LTC_4 release from A23187-stimulated human eosinophils was inhibited by BIRM 270 more potently than by BIRM 271.

inhibited LT biosynthesis in human cells with little or no activity in cells of other species. This clearly hindered pharmacological testing. Rodent models which are the mainstay for evaluating LT biosynthesis inhibitors were not useful with this series of compounds.

Our earlier 5-LO inhibitor, BI-L-239 was an aerosol development candidate that showed activity by topical administration in antigen challenge models in sheep (Lazer et al., 1991). Our obvious choice was to follow along the same path of topical administration. BIRM 270 was therefore evaluated in a sheep model (Abraham, 1990) of antigen-induced late-phase and hyperresponsiveness. It was found to inhibit both responses at total inhaled doses of 1, 3, or 10 mg, thus giving us an *in vivo* link to the *in vitro* efficacy we observed with these compounds (data not shown).

The *in vitro* profile, combined with the results from the sheep studies, plus Boehringer Ingelheim's experience base in aerosol delivery systems for bronchopulmonary drugs shaped the strategy for *in vivo* evaluation of ontazolast. Asthma was chosen as the primary indication. We had good LT inhibitory activity in three cell types: human neutrophils, eosinophils and lung mast cells, all of which are thought to contribute to the inflammatory component of asthma. An added feature is the ability of ontazolast to inhibit PAF biosynthesis in human neutrophils, histamine release and

PGD_2 biosynthesis in mast cells. Bioavailability and presumably metabolism issues could be largely circumvented by topical administration. We concluded that the ultimate test would have to be in humans.

The final chapter on BIRM 270 remains to be written. The compound is in the last phases of safety testing. So far the results are quite favorable. Our Pharmaceutics Department has devised a suitable formulation for aerosol administration and metered dose inhalers can be made for clinical testing. If all goes well, the compound will be tested in humans in 1995.

Acknowledgments. The research and development of a drug requires the efforts of many people and such is the case for BIRM 270. Several people stand out, however, who have made significant contributions to the effort; they are: Anne Graham who uncovered DF-LS 70 and performed most of the secondary testing on the series that led to BIRM 270; Jane Watrous and Regina Mucci who supported the chemistry SAR with the human neutrophil LTB_4 assay; Ann Hoffman who did a great deal of the mechanism of action studies on the series; Gordon Hansen and Roger Dinallo who did the mass spectrometric analyses: Edward Graham for his work on SAS® analysis of data; Dr. Julian Adams, Director of Medicinal Chemistry; medicinal chemists—Dr. Clara Miao, Hin-Chor Wong, and Ronald Sorcek; Dr. Andrew Jayaraj and John Alexander who performed bioavailability and metabolism studies; Dr. Thomas Noonan who performed studies in guinea pigs, looking for efficacy and bioavailability; and Dr. Jim Stevenson, head of the BIRM 270 Development Team. The input and advice of Drs. Alan Rosenthal, Yvan Guindon, Gordon Letts, and Craig Wegner are gratefully appreciated. We very much benefited from our collaborations with Drs. Pierre Borgeat, Edward Schulman and Donald Raible. The effort of Janet Abbott in the preparation of the manuscript is also appreciated.

REFERENCES

Abraham WM (1990): The role of eicosanoids in allergen-induced early and late bronchial responses in allergic sheep. *Adv Prostaglandin Thromboxane Leukotriene Res* 20:201–208

Borgeat P, Samuelsson B (1979): Metabolism of arachidonic acid in polymorphonuclear leukocytes: effect of the ionophore A23187. *J Biol Chem* 254:7865–7869

Cockcroft S (1992): G-protein-regulated phospholipases C, D and A_2-mediated signalling in neutrophils. *Biochim Biophys Acta* 1113:135–160

Corey EJ, Clark DA, Goto G, Marfat A, Mioskowski C, Samuelsson B, Hammarström S (1980): Stereospecific total synthesis of a "slow reacting substance" of anaphylaxis, leukotriene C-l. *J Am Chem Soc* 102:1436–1439

Dixon RAF, Diehl RE, Opas E, Rands E, Vickers PJ, Evans JF, Gillard JW, Miller DK (1990): Requirement of a 5-lipoxgenase-activating protein for leukotriene synthesis. *Nature* 343:282–284

Farina PR, Graham AG, Hoffman AF, Watrous JM, Borgeat P, Nadeau M, Hansen G, Dinallo RM, Adams J, Miao CK, Lazer ES, Parks TP, Homon CA (1994): BIRM 270: A novel inhibitor of arachidonate release that blocks leukotriene B_4 and platelet-activating factor biosynthesis in human neutrophils. *J Pharm Exp Ther* 271:1418–1426

Fonteh AN, Chilton FH (1993): Mobilization of different arachidonate pools and their roles in the generation of leukotrienes and free arachidonic acid during immunologic activation of mast cells. *J Immunol* 150:563–570

Ford-Hutchinson AW, Bray MA, Doig MV, Shipley ME, Smith MJ (1980): Leukotriene B, a potent chemokinetic and aggregating substance released from polymorphonuclear leukocytes. *Nature* 286:264–265

Gillard J, Ford-Hutchinson AW, Chan C, Charleson S, Denis D, Foster A, Fortin R, Leger S, McFarlane CS, Morton H, Piechuta H, Riendeau D, Rouzer CA, Rokach J, Young R, MacIntyre DE, Peterson L, Bach T, Eiermann G, Hopple S, Humes J, Hupe L, Luell S, Metzger J, Meurer R, Miller DK, Opas E, Pacholok S (1989): L-663,536 (MK 886) (3-[1-(4-chlorobenzyl)-3-*t*-butyl-thio-5-isopropylindol-2-yl]-2,2-dimethylpropanoic acid), a novel, orally active leukotriene biosynthesis inhibitor. *Can J Physiol Pharmacol* 67:456–464

Hirata F (1981): The regulation of lipomodulin, a phospholipase inhibitory protein, in rabbit neutrophils by phosphorylation. J Biol Chem 256:7730–7733

Hirata F, Schiffmann E, Venkatasubramanian K, Salomon D, Axelrod J (1980): A phospholipase A_2 inhibitory protein in rabbit neutrophils induced by glucocorticoids. *Proc Natl Acad Sci* 77:2533–2536

Lazer ES, Farina PR, Gundel RH, Wegner CD (1991): 2,6-Disubstituted-4-(2-arylethenyl)phenol 5-lipoxygenase inhibitors: Development and profile of BIL-239. *Drugs Future* 16:641–646

Lazer ES, Miao CK, Wong HC, Sorcek R, Spero DS, Gilman A, Pal K, Behnke M, Graham AG, Watrous JM, Homon CA, Nagel J, Shah A, Guindon Y, Farina PR, Adams J (1994): Benzoxazolamines and benzothiazolamines: Potent, enantioselective inhibitors of leukotriene biosynthesis. *J Med Chem* 37:913–923

McDonald PP, McColl SR, Naccache PH, Borgeat P (1991): Studies on the activation of human neutrophil 5-lipoxygenase induced by natural agonists and Ca^{2+} ionophore A23187. *Biochem J* 280:379–385

Moore GGI, Swingle KF (1982): 2,6-Di-tert-butyl-4-(2′-thenoyl)phenol (R830): A novel nonsteroidal anti-inflammatory agent with antioxidant properties. *Agents Actions* 12:674–683

Murphy RC, Hammarström S (1979): Structure of leukotriene C: Identification of the amino acid part. *Proc Natl Acad Sci USA* 76:4275–4281

Rouzer CA, Ford-Hutchinson AW, Morton HE, Gillard JW (1990): MK886, a potent and specific leukotriene biosynthesis inhibitor blocks and reverses the membrane association of 5-lipoxygenase in ionophore-challenged leukocytes. *J Biol Chem* 265:1436–1442

Rouzer CA, Kargman S (1988): Translocation of 5-lipoxygenase to the membrane in human leukocytes challenged with ionophore A23187. *J Biol Chem* 263:10980–10988

Sirois P, Borgeat P (1980): From slow reacting substance of anaphylaxis (SRS-A) to leukotriene D_4 (LTD_4). *Int J Immunopharmac* 2:281–293

Vadas P, Pruzanski W (1986): Role of secretory phospholipase A_2 in the pathobiology of disease. *Lab Invest* 55:391–404

10

Discovery and Development of the Immunomodulatory Azaspiranes

Alison M. Badger

Introduction

The search for new agents for the therapy of autoimmune diseases and transplantation is a continuous and often frustrating task. For rheumatoid arthritis (RA), a chronic inflammatory immune-mediated disease, that leads to severe damage of the articular cartilage and subchondral bone, there is no one agent or combination of agents fully capable of stopping disease progression or reversing the damage. First line therapy consists of nonsteroidal anti-inflammatory drugs (NSAIDs). This is followed by the use of disease-modifying antirheumatic drugs (DMARDs), also called slow acting antirheumatic drugs (SAARDs), whose mechanisms of action are not well defined. Some DMARDs are currently used because of clinical observations made during treatment of other diseases (Kavanaugh and Lipsky, 1992). For example, gold compounds which were initially used for the treatment of tuberculosis, were introduced for rheumatoid arthritis due to certain similarities between the two diseases and the belief that they might both be caused by a mycobacterium. An improvement in arthritic lesions in malaria patients treated with mepacrine prompted clinical trials with chloroquine and hydroxychloroquine. D-penicillamine, first used in the treatment of Wilson's disease, was found to cause dissociation of macroglobulins, including IgM rheumatoid factor *in vitro* and formed the basis for its use in RA. Although this effect does not appear to be the compound's mechanism of action, it has been established as having therapeutic activity in RA. On the other hand, sulfasalazine, which is a combination of an anti-inflammatory and an antibiotic, was synthesized specifically for use in RA. The cytostatic drugs, azathioprine and cyclophosphamide, are used

The Search for Anti-Inflammatory Drugs
Vincent J. Merluzzi and Julian Adams, Editors
© Birkhäuser Boston 1995

as a last resort in the hope that the nonspecific anti-proliferative effects will halt disease progression through suppression of T-cell proliferation (Rainsford, 1990). Methotrexate, an antimetabolite, is the most recent addition to the list of DMARDs to gain widespread acceptance for the treatment of RA. Again, its mechanism of action in the disease is not well defined. It was previously used for the treatment of malignancy but has proven effective in RA at much lower dosing regimens. Quite often, adverse drug reactions limit the potential benefit of many of these agents, especially as they have to be used for prolonged periods of time.

It is not clear how well the majority of the drugs currently being used to treat RA actually modify the arthritic process in a beneficial manner. Recent efforts have focused on finding disease-modifying agents with more acceptable levels of side effects and that might have protective effects on bone, as well as regenerative effects on bone and cartilage. There was hope that cyclosporin A (CsA) would meet these criteria. CsA has been evaluated in a number of RA clinical trials, and while showing some efficacy, nephrotoxicity has limited its usefulness (Dougados et al., 1988; Boers et al., 1990; Tugwell et al., 1990; Madhok, et al., 1991).

While in the final analysis, the ultimate goal in the treatment and cure of rheumatoid arthritis is an attack on the causative etiological agent, or the specific immune component gone awry, our understanding of the disease is still so limited that, in the short term, the search has to focus on the discovery of new therapeutic agents with disease-modifying activity. There is still a desperate need for drugs that have a novel mechanism of action with a low toxicity profile, and which show bone preservation and/or restorative effects.

Many observations over the last 15 years have indicated that the autoimmune state might be the result of a suppressor cell (SC) defect or dysfunction (for review see Shoenfeld et al, 1993), and we proposed that a compound capable of inducing or activating these cells could have beneficial effects in a number of disease states. There is considerable controversy surrounding the phenotype of these cells, their role or even existence. However, these arguments generally apply to T-cell mediated suppression (Suppressor T cells) (Eichmann, 1988; Mitchison, 1988; Moller, 1988). In fact, a recent article in *Immunology Today* starts with a less than flattering comment on the use of the "S" word.

> There is little doubt that the "S" word (suppression, as in "suppressor T cells") is the nearest thing to a dirty word we have in cellular immunology. Its use is considered by some (not all) to be synonymous with over interpretation of scanty data and phenomenology (the "P" word) bordering on the mystical. (Green and Webb, 1993)

These authors, however, do show some optimism. As progress continues to be made on the cellular and molecular basis of suppression, the topic will once again be the subject of outspoken discussion. Notwithstanding the controversy surrounding the notorious suppressor T cell, one must be aware that T cells are not unique in their ability to down-modulate immune responsiveness. Several other cell types including B cells, macrophages, natural killer cells, and the natural or nonspecific suppressor cell have all been attributed with this activity. The beneficial effects of total lymphoid irradiation (TLI), a treatment that induces natural suppressor cells (Strober, 1984) lends support to the notion that the generation of this cellular activity could have beneficial effects in autoimmune disease (and transplantation). It was therefore deemed desirable to search for compounds, not only active in animal models of autoimmune disease, but that also induced splenic SC activity that could be detected in *ex vivo* cellular assays. It so happened that the screening of metal-containing compounds, for a variety of biological activities, was already underway in SmithKline Beecham (SB) R&D and during our participation in this effort, we made the observation that the metal-containing azaspirane, Spirogermanium (currently in development as an anti-cancer agent by Unimed, NJ, and reported to have some beneficial effect in RA patients) was indeed active in our adjuvant arthritic rat model and also induced the desired SC activity. This was March 1983.

Background

Our interest in metal-containing compounds stemmed from our own efforts on gold compounds (Ridaura®) and the known anti-tumor effects of platinum-containing compounds. A research collaboration was established with Johnson Matthey to investigate the pharmacological activity of a wide range of metal-containing complexes with the primary focus of the evaluation being directed towards determining the potential anti-neoplastic and anti-arthritic activities of the compounds. It was during this random screening effort that we were approached by Unimed with a proposal that SmithKline & French license Spirogermanium (SG) as a potential anti-cancer agent. Spirogermanium (8,8-diethyl-N-dimethyl-2-aza-8-germaspiro[4,5]decane-2-propanamine dihydrochloride) (Figure 10-1) is a member of a class of azaspirane compounds first synthesized and characterized in 1974 by Rice et al., and shown to have *in vivo* anti-tumor activity in experimental rat tumor systems (Rice et al., 1974). However, the drug had been in Phase II clinical trials in human cancer and showed little evidence of efficacy. Later, our own investigations on the anti-tumor activity of SG and related analogs in animal tumor models showed no anti-tumor activity against P388 leukemia or ADJ-PC6 plasmacytoma (Mirabelli et al., 1989). On the other hand, its activity in the adjuvant arthritic rat was of great interest, and, in particular, its ability to generate SC activity in the spleens of treated rats (DiMartino et al., 1986; Badger et al., 1985, 1987).

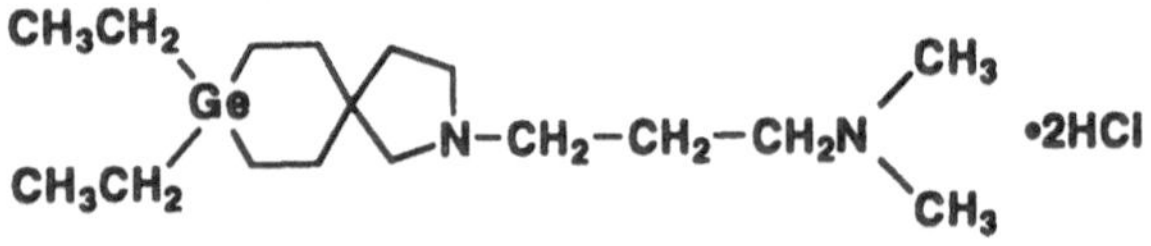

Figure 10-1. Spirogermanium: 8,8-diethyl-N-dimethyl-2-aza-8-germaspiro[4,5] decane-2-propanamine dihydrochloride.

Unfortunately, our enthusiasm surrounding the activity of SG was not shared by everyone. First, there was the general skepticism about SC activity. In addition, the timing coincided with a decision to terminate efforts to discover gold-containing antineoplastic compounds. By implication, this cast a shadow on metals as therapeutic agents; therefore we proposed using SG as a starting point and preparing analogs to demonstrate that the metal was not required for biological activity, and we also proposed that an orally active, patentable compound, with activity equal to or better than SG could be discovered.

This proposal was not received enthusiastically and was turned down. The objections were perfectly reasonable; random screening in animal models of disease was out of the question as the chemical managerial support required to synthesize gram quantities of these specific compounds did not exist. Put another way: without an *in vitro* mechanism based screen, no rational drug design could be undertaken. We did not have any idea how the compound exerted its effects, so it appeared that we were out of business. Surprisingly, our colleagues at Johnson Matthey decided that the idea had some merit and that they would provide the chemistry support for the effort. Despite his concerns regarding direct animal testing but with the realization that there was no other way forward, the Director of Immunology at SK&F reluctantly agreed to evaluating compounds in the AA rat model. Thus, we embarked on the first problem—taking the germanium out of the molecule while maintaining SC activity. It was September 1984.

Chemistry and Selection of the First Lead Compound

Replacing the germanium in the azaspirane nucleus was only the first step in an extensive effort to establish a structure-activity relationship (SAR) with regard to anti-arthritic and immunomodulatory activities in experimental

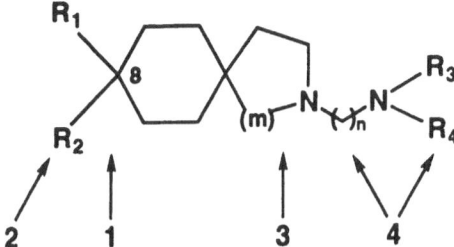

Figure 10-2. Structural modification on the azaspirane nucleus.

animal systems. Four areas of the molecule were selected for synthetic modification (Figure 10-2).

(1) Heteroatom variation at position 8

(2) Alkyl substitutions at position 8

(3) 2-azaspiro[4.5]decane versus 3-azaspiro[5.5]undecane analogs (m = 1 or 2)

(4) Modification of the N-dialkylaminoalkyl group

    A. Cyclic nitrogen-terminal nitrogen alkyl chain analogs (n = 2 to 7)

    B. Terminal alkyl substitutions

        1. Acyclic alkyl substitutions

        2. Cyclic alkyl substitutions

Over 100 structural modifications were made on the azaspirane nucleus, and two models were used to determine the potency of the analogs. First, their ability to inhibit hindpaw inflammation in the AA rat, and second, their suppressor cell-inducing activity in the spleens of treated animals. The assay for measuring SC activity was a straight forward system, whereby irradiated (2000R) spleen cells from treated rats were co-cultured with spleen cells from normal rats in a Con A-induced proliferation assay. For comparison of azaspirane analogs, a program was developed that could be used to compare the area under the curve (AUC) of suppression for different analogs (Badger et al., 1987, 1989). A strong correlation ($r = 0.82$) between AA rat activity and SC generation was observed. As a full description of the SAR of the azaspiranes has been reported (Badger et al., 1990b), only key findings and summaries are described below.

First, our initial SAR studies demonstrated that the substitution of oxygen, N-methyl or sulfur at position 8 led to inactive compounds. The 8,8-diethyl analogs containing carbon or silicon, however, were active in both assays with the 8,8-dimethyl analogs being inactive (Table 10-1).

Table 10-1. Heteroatom variation at position 8.

SK&F No.	X	Dose (mg/kg)	AA[a] activity	SC[a] activity
104036	C(CH$_3$)$_2$	30	0.12	0
104035	Si(CH$_3$)$_2$	30	0	0
104037	C(CH$_2$CH$_3$)$_2$	30	0.56*	88*
105868	Si(CH$_2$CH$_3$)$_2$	15	0.92*	63*

[a] Activity in the adjuvant arthritic rat and in the suppressor cell assay is standardized to the activity of spirogermanium: AA activity = 1.00; SC activity = 100.
* Designates statistically significant activity. Biologically significant activity is considered to be greater than 0.30 in the AA rat and greater than 50 in the SC assay.

Second, the importance of the alkyl chain length at C-8 was observed early in the process of synthesizing analogs of the azaspirane skeleton (Table 10-2). The 8,8-dimethyl analog (SK&F 104036) was inactive but the 8,8-diethyl analog (SK&F 104037) retained approximately 50% of the activity of spirogermanium. Increasing the chain length to three carbons yielded an analog (SK&F 105685) more active than spirogermanium. The 8,8-dibutyl (SK&F 105703) and 8,8-dipentyl (SK&F 105015) analogs were also synthesized but were less active than the dipropyl analog.

The activity seen with SK&F 105685 prompted us to pursue this compound as our primary lead and to establish its immunopharmacological profile and much of the chemical effort from this time focused on structural modifications of the dipropyl analog. This was December 1986.

Third, a series of 8,8-dialkyl-2-azaspiro[4.5]decane and 9,9-dialkyl-3-azaspiro[5.5]undecane derivatives were synthesized, and their pharmacologic activities were compared. It was apparent that increasing the B ring from cyclopentyl to cyclohexyl yielded compounds with decreased activity.

Four, the effects of varying chain lengths between the cyclic and terminal nitrogens in the N-dialkylaminoalkyl moiety was investigated in the 8,8-dipropyl series. Within this series of analogs, activity was optimized in the cases where the nitrogens were separated by three, six, and seven methylene (-CH$_2$-) groups. Reduction to two methylenes between the nitrogens led to an inactive compound and increasing the chain length to four and five reduced activity. The results were consistent with the SAR observed for similar substitutions in the 8,8-diethyl series and suggested a relationship between the spatial distance between the two positively charged nitrogens

Table 10-2. Alkyl substitutions at position 8.

SK&F No.	R1	R2	Dose (mg/kg)	AA[a] activity	SC[a] activity
103811	H-	H-	30	0.27	22
104036	CH_3-	CH_3-	30	0.12	0
104037	CH_3CH_2-	CH_3CH_2-	30	0.57*	88*
105685	$CH_3CH_2CH_2$-	$CH_3CH_2CH_2$-	30	1.32*	172*
105703	$CH_3(CH_2)_2CH_2$-	$CH_3(CH_2)_2CH_2$-	15	0.88*	176*
105015	$CH_3(CH_2)_3CH_2$-	$CH_3(CH_2)_3CH_2$-	7.5	0.41*	118*

[a] Activity in the adjuvant arthritic rat and in the suppressor cell assay is standardized to the activity of spirogermanium: AA activity = 1.00; SC activity = 100.
* Designates statistically significant activity. Biologically significant activity is considered to be greater than 0.30 in the AA rat and greater than 50 in the SC assay.

and optimal pharmacologic activity. While three methylenes may be optimal, six and seven methylene groups may provide sufficient flexibility to fit within a pharmacologic receptor site or perhaps chelate a counter-ion such as sodium, calcium, or potassium.

Variation of the alkyl substitution on the terminal amine significantly influenced activity. Although alkyl substitution at the terminal nitrogen is not required for AA rat or suppressor cell activity (see SK&F 106534; Table 10-3), an SAR was generated to examine the effect of N,N-dialkyl substitution. It was found that the N,N-diethyl analog SK&F 106615 was slightly more active than the N,N-dimethyl derivative (SK&F 105685) but further increasing the chain length beyond three carbons produced inactive analogs.

Terminal cyclic amine substituents such as pyrrolidine, piperidine, and azacycloheptane were incorporated in 8,8-dipropyl analogs. These analogs appeared to slightly increase the pharmacologic activity relative to the N,N-dimethyl analog SK&F 105685 (Table 10-4). Again, it was demonstrated that the 8,8-dipropyl analogs were more active than the 8,8-diethyl analogs. Interestingly, it was observed that incorporation of a morpholino moiety (SK&F 106609) led to a compound that was inactive and toxic at 30 mg/kg.

An acetamino substitution at the terminal nitrogen led to an inactive compound thereby clearly demonstrating the requirement for a dibasic compound for activity.

Table 10-3. Acyclic alkyl substitutions.

SK&F No.	R1	R2	Dose (mg/kg)	AA[a] activity	SC[a] activity
106534	H-	H-	30	0.91*	158*
106535	H-	CH_3-	30	1.55*	217*
105685	CH_3-	CH_3-	30	1.32*	173*
106615	CH_3CH_2-	CH_3CH_2-	30	1.42*	201*
107121	$CH_3CH_2CH_2$-	$CH_3CH_2CH_2$-	30	0.43	89*
107126	$CH_3(CH_2)_2CH_2$-	$CH_3(CH_2)_2CH_2$-	30	0	28

[a] Activity in the adjuvant arthritic rat and in the suppressor cell assay is standardized to the activity of spirogermanium: AA activity = 1.00; SC activity = 100.
* Designates statistically significant activity. Biologically significant activity is considered to be greater than 0.30 in the AA rat and greater than 50 in the SC assay.

Table 10-4. Cyclic alkyl substitutions.

SK&F No.	R	R1	R2	Dose (mg/kg)	AA[a] activity	SC[a] activity
105685	$CH_3CH_2CH_2$	CH_3	CH_3	30	1.32*	172*
107116	$CH_3CH_2CH_2$	-$CH_2(CH_2)_2CH_2$-		30	1.03*	139*
106610	$CH_3CH_2CH_2$	-$CH_3(CH_2)_3CH_2$-		30	1.70*	241*
107111	CH_3CH_2	-$CH_2(CH_2)_2CH_2$-		30	1.53*	77*
105330	CH_3CH_2	-$CH_2(CH_2)_3CH_2$-		30	0	114*
106609	$CH_3CH_2CH_2$	-$CH_2CH_2OCH_2CH_2$-		30	0	71*

[a] Activity in the adjuvant arthritic rat and in the suppressor cell assay is standardized to the activity of spirogermanium: AA activity = 1.00; SC activity = 100.
* Designates statistically significant activity. Biologically significant activity is considered to be greater than 0.30 in the AA rat and greater than 50 in the SC assay.

Summary

In total, over 100 azaspirane analogs were synthesized from 1984–1990 and tested in the AA rat and/or suppressor cell induction models. The results of this screening effort are summarized as follows:

- A statistically significant correlation exists between the degree of activity of a compound in the AA rat and suppressor cell generation assays (Badger et al., 1990).

- Optimal activity is found in analogs with dipropyl substitution at C-8 in the 2- azaspiro[4.5]decane nucleus and 3-carbon separation between the cyclic and terminal nitrogens.

- SK&F 105685 (N,N-dimethyl-8,8-dipropyl-2-azaspiro[4.5]decane-2-propanamine) is among the most active compounds evaluated to date and was chosen as the first lead compound in this program.

- Two other compounds, SK&F 106610 (terminal nitrogen incorporated into a piperidine ring) and SK&F 106615 (diethyl substitution on the terminal amine), which had immunomodulatory activity equal to that of SK&F 105685, were also selected for scale up for pharmacological evaluation (Figure 10-3).

Pharmacological Activity of the Azaspiranes

Animal Models

Over the years we have developed a biological profile of the azaspiranes using SK&F 105685, SK&F 106610, and SK&F 106615 (Figure 10-3). The compounds have been evaluated in a wide variety of *in vitro* assay systems, and *in vivo* in a number of animal models of autoimmune disease. The pharmacological activities of the azaspiranes *in vivo* are summarized below.

- Prophylactic and therapeutic activity on adjuvant-induced arthritis in the Lewis rat (Badger et al., 1989, 1991,1993a) and inhibition of delayed-type hypersensitivity reaction in the AA rat (Badger et al, 1989).

- Experimental autoimmune encephalomyelitis (EAE) in the Lewis rat (Badger et al., 1989).

- Spontaneous diabetes in the BB rat (Rabinovitch et al., 1993).

- Autoimmune uveitis in the Lewis rat (Westeren et al., 1993).

SK&F 105685

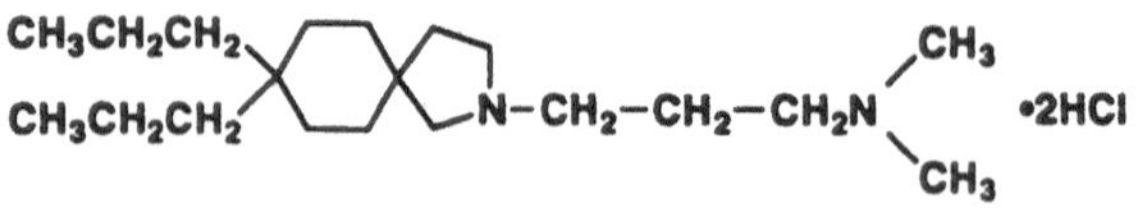

CH₃CH₂CH₂, CH₃CH₂CH₂ spiro N-CH₂-CH₂-CH₂N(CH₃)(CH₃) •2HCl

SK&F 106610

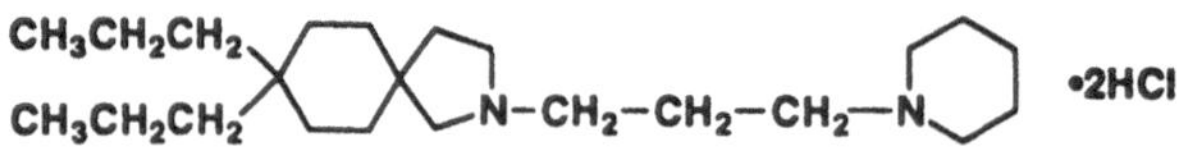

CH₃CH₂CH₂, CH₃CH₂CH₂ spiro N-CH₂-CH₂-CH₂-N(piperidine) •2HCl

SK&F 106615

CH₃CH₂CH₂, CH₃CH₂CH₂ spiro N-CH₂-CH₂-CH₂N(CH₂CH₃)(CH₂CH₃) •2HCl

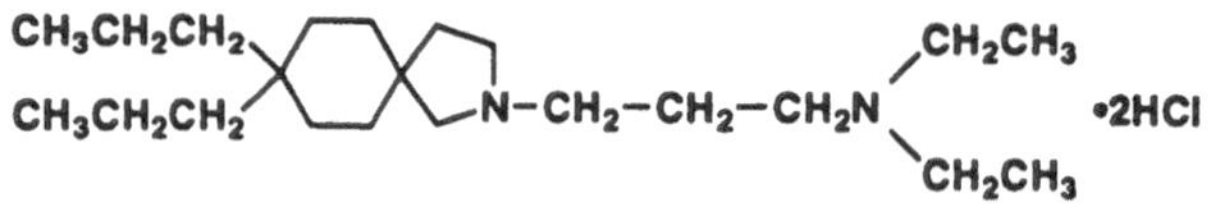

Figure 10-3. Structures of lead azaspiranes.
SK&F 105685: N,N-dimethyl-8,8-dipropyl-2-azaspiro[4.5]decane-2-propanamine dihydrochloride.
SK&F 106610: 2-[3-(1-piperidinyl)propyl]-8,8-dipropyl-2-azaspiro[4.5]decane, dihydrochloride.
SK&F 106615: N,N-diethyl-8,8-dipropyl-2-azaspiro[4.5] decane-2-propanamine dihydrochloride.

- Lupus-like disease in the MRL mouse (Albrightson-Winslow et al., 1990).

Obviously, with a therapeutic target of RA, the focus of our studies with these compounds was to explore their beneficial effects in the AA rat and to define their disease-modifying activity. However, as listed above, the compounds displayed therapeutic activity in a number of other autoimmune disease models including EAE in the Lewis rat (Badger et al., 1989) and spontaneous diabetes in the BB rat (Rabinovitch et al., 1993). Activity in these models suggested additional autoimmune targets besides RA for the future clinical evaluation of an azaspirane.

In addition to autoimmune disease models, the potential activity of SK&F 105685 was examined in experimental organ transplantation. The compound was shown to ameliorate renal allograft rejection in the rat when administered orally (po) at 20 mg/kg/day for five days. There was a substantial improvement in hemodynamic function of allografts and this improvement was associated with reduced intensity of interstitial inflammatory cell infiltrates within the kidney (Fan et al., 1993). In a rat model of cardiac allograft rejection, short-term therapy (20 mg/kg/day) proved effective in both pretreatment (days −7 to −1; allograft day 0) and treatment (days 0 to 6) protocols. Of particular interest was the observation that SK&F 105685 pretreatment exerted at least additive effects with subtherapeutic doses of CsA (1.5 mg/kg/day for seven days) given intramuscularly (im) after transplantation (Badger et al., 1991a; Schmidbauer et al., 1993). The results of these experiments are summarized in Figure 10-4. From these studies we have concluded that, while the prolongation of heart allograft survival was significant, it was not dramatic, and that SK&F 105685 could probably not be used alone for transplantation therapy. Combination therapy with low doses of immunosuppressive agents such as CsA or FK506, however, might be considered later in a development program following the demonstration of safety and efficacy in RA.

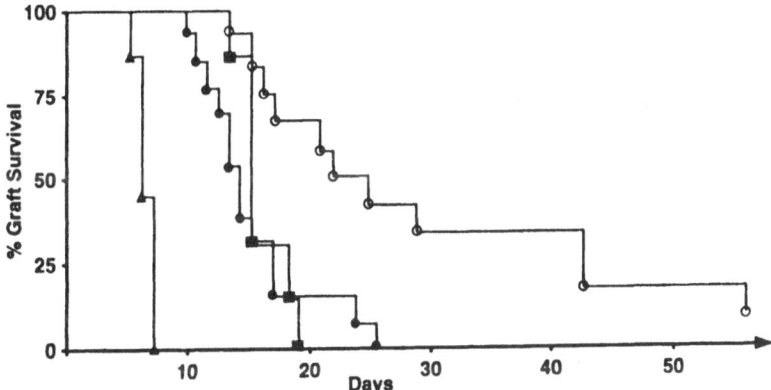

Figure 10-4. The effects of SK&F 105685 therapy alone (20 mg/kg/day, po) and in combination with low-dose CsA (1.5 mg/kg/day im) on the survival of LBNF-1 cardiac allografts in LEW rats. There were 5–13 recipients in each group. Modified with permission of Williams and Wilkins from Schmidbauer et al. (1993).

▲-▲ Control untreated allografted rats
•-• SK&F 105685 pretreatment regimen, days −7 to −1, allograft day 0.
■-■ SK&F 105685 treatment regimen, days 0 to 6, allograft day 0.
o-o SK&F 105685 pretreatment (days −7 to −1) + CsA (days 0 to 6).

Beneficial Effects in the Rat Model of Adjuvant-induced Arthritis

The three compounds selected for pharmacological evaluation (Figure 10-3) all had similar biological activity in the Lewis rat AA model. When administered orally starting on the day of adjuvant injection, maximum inhibition of the inflammatory response (hindpaw swelling) was observed at 30 mg/kg/day (Figure 10-5) and ED_{50}s ranged from 10 to 20 mg/kg (Badger et al, 1989 and unpublished observations). When the compounds were administered for 2–5 weeks prior to adjuvant injection, SK&F 105685 and SK&F 106615 displayed activity at 3 mg/kg and 1 mg/kg, respectively. Therapeutic protocols for the evaluation of anti-arthritic compounds are clearly more relevant for treatment of this disease where intervention will necessarily take place at some point into the disease process. When dosing was delayed until day 10 of the disease process and continued until day 30, SK&F 105685 (20 mg/kg/day po) halted the progress of the inflammatory lesion compared to untreated controls after four days of treatment (Badger et al., 1993a). SK&F 106615 also displayed therapeutic activity in the AA rat at 10 and 20 mg/kg/day (Table 10-5).

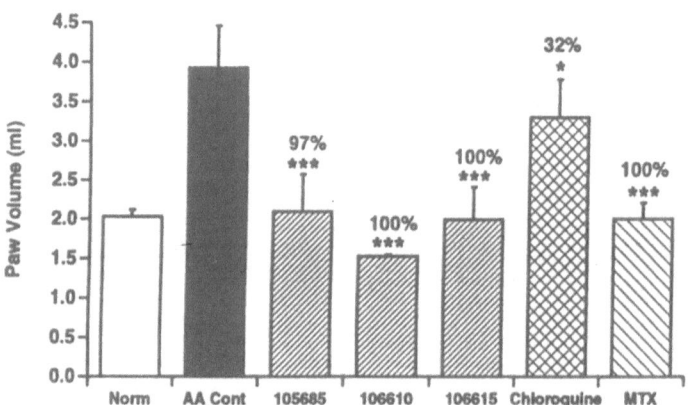

Figure 10-5. Anti-inflammatory activity (hindpaw volumes) of (□) normal rats; (■) AA rats; (▨) rats treated with SK&F 105685, SK&F 106610, and SK&F 106615 at 30 mg/kg per day, po; (▧) chloroquine at 100 mg/kg/day; and (▨) MTX at 0.3 mg/kg per day. Drugs were administered five times a week for a total of 12 doses. Data are mean ± standard deviation of 12 animals/group.

$$\% \text{ inhibition} = 1 - \left[\frac{\text{AA treated} - \text{normal}}{\text{AA control} - \text{normal}} \right] \times 100$$

Studies performed up to this point clearly defined the azaspiranes as having prophylactic and therapeutic activity in the AA rat but did not dissociate them from NSAIDs or characterize them as disease-modifying agents.

Table 10-5. Therapeutic activity of SK&F 106615 in the AA rat.

Treatment		Paw volume	Inhibition[a] (%)	
Experiment 1[b]				
Normal		1.60 ± 0.04^c		
AA control		3.56 ± 0.64		
SK&F 106615	10 mg/kg	2.96 ± 0.51	31	$p < 0.05$
	20 mg/kg	2.60 ± 0.6	49	$p < 0.01$
Experiment 2[a]				
Normal		1.61 ± 0.10		
AA control		3.45 ± 0.49		
SK&F 106615	20 mg/kg	2.13 ± 0.31	72	$p < 0.001$

[a] % inhibition calculated as in Figure 10-5.
[b] In experiment 1 rats were dosed (po) from day 10 to day 23 and from day 10 to day 30 in experiment 2.
[c] Results are mean ± standard deviation for 4 animals/group in the normal rats, and 8 animals/group in the AA control and treated groups.

The Azaspiranes Are Not Classical NSAIDs

Although the azaspiranes have anti-inflammatory activity in the AA rat, as measured by a dose-dependent inhibition of hindpaw edema, this is not the result of cyclooxygenase inhibition as would be the case for a compound such as indomethacin. In the AA rat, treatment with SK&F 105685 (20 mg/kg/day starting on the day of adjuvant injection) did not alter eicosanoid production by peripheral blood cells (measured on day 24). Although there was a sixfold increase in PMNs and an almost twofold increase in platelets in AA rats, there was no change in either LTB_4 production from PMNs or 12(S)-HHT from platelets in A23187-stimulated whole blood (Badger et al., 1993a). Similar results were observed with SK&F 105685 and SK&F 106615 in a therapeutic dosing protocol (20 mg/kg/day on days 10 to 30) (Figure 10-6). NSAIDs show potent anti-inflammatory activity *in vivo* in the carrageenan-induced paw inflammation model. Indomethacin (5 mg/kg po), administered to Lewis rats 30 minutes prior to challenge with carrageenan, inhibited paw inflammation by 50 to 70% whereas the azaspiranes showed minimal to no effect on the inflammatory paw swelling (Table 10-6).

In support of the *in vivo* findings, we have not observed any inhibition of LTB_4 or PGE_2 from A23187-stimulated human monocytes *in vitro* at nontoxic concentrations, nor was there any inhibition of cyclooxygenase (RAM seminal vesicle prostaglandin H synthase type I) at concentrations

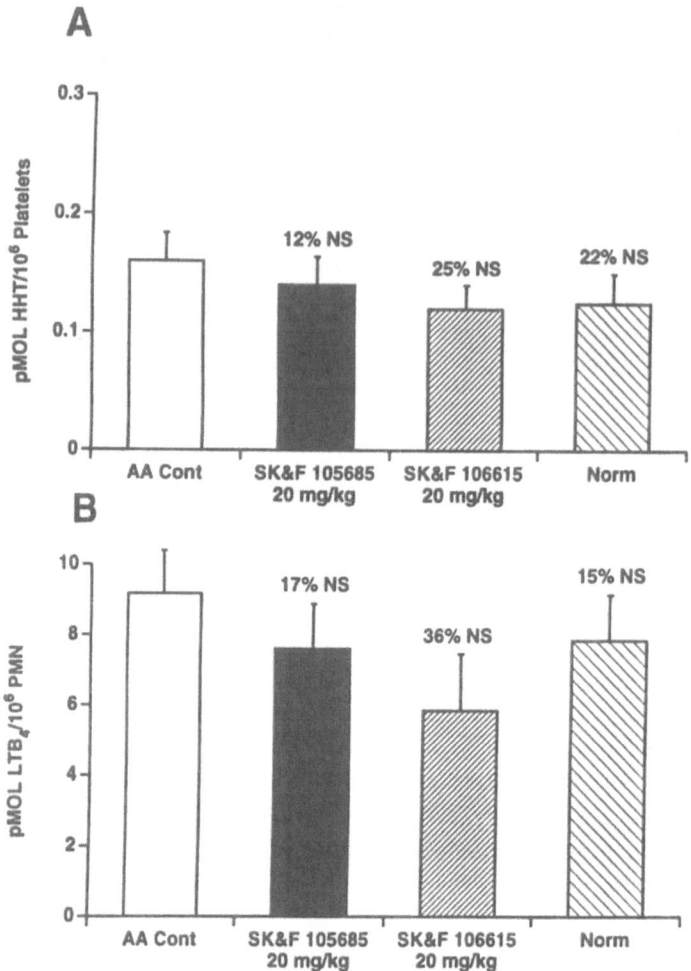

Figure 10-6. Treatment with SK&F 105685 or SK&F 106615 does not affect eicosanoid production in whole blood from adjuvant arthritic rats. Rats were treated from day 10 to day 30 (20 mg/kg/day five times per week po). Data is mean ± standard error of 5 animals/group. A. 12(S)-HHT from platelets. B. LTB$_4$ production from PMNs.

less than 100 μM compared to naproxen (IC$_{50}$ = 13 μM). Very weak activity on 5-lipoxygenase (RBL isolated enzyme system) was observed with SK&F 106615, whereas an IC$_{50}$ of 1 μM is observed with Zileuton (see Chapter 5).

Taken together, these *in vitro* and *in vivo* data indicate that the pharmacological activity of the azaspiranes in the AA rat is not due to eicosanoid inhibition and therefore they cannot be classified as NSAIDs. The pharmacological activity of the azaspiranes is more characteristic of disease-modifying agents.

Table 10-6. SK&F 106615 does not inhibit carrageenan-induced paw inflammation in Lewis rat.

Treatment	Inflammation[a] (ml)	Inflammation[b] (%)	
Control	1.09 ± 0.15^c		
Indomethacin			
5 mg/kg, 30 min	0.38 ± 0.16	65%	$p < 0.001$
SK&F 106615			
20 mg/kg, 30 min	0.93 ± 0.23	14%	N.S.
24 h	0.90 ± 0.16	17%	$p < 0.05$

[a] Inflammation = carageenan-injected paw volume minus non-injected paw volume.

[b] % inhibition $= 1 - \left[\frac{\text{inflammation treated}}{\text{inflammation control}} \right] \times 100$

[c] Data is mean ± standard deviation of 10 animals/group.

The Azaspiranes Have Disease-modifying Activity

While the AA rat is not an ideal model for human RA, the majority of compounds that have shown any efficacy at all in RA have activity in the AA rat. However, for a disease-modifying agent, this is not sufficient. NSAIDs, such as indomethacin, while having a dramatic effect on the inflammatory response have little or no protective effects on joint integrity. In our studies with the azaspiranes, we have made an extensive radiological investigation of the arthritic joint destruction in the model and have demonstrated almost complete protection when SK&F 105685 is administered on the same day as adjuvant. This protection was determined by evaluating radiographic and histological changes, as well as scanning electron microscopy (High et al., 1994). Although less dramatic, protective effects on joint integrity were also observed when the compound was administered therapeutically starting on day 10 of adjuvant disease (Badger et al., 1993a). In our studies with SK&F 106615, we have shown a similar protection of joint integrity and have extended our original observations by showing normalization of bone mineral content (BMC) and bone mineral density (BMD) in treated animals (Figure 10-7). In addition, *ex vivo* magnetic resonance imaging (MRI) was used to compare tibiotarsal joints from normal, AA, and treated rats. The joints from normal rats showed clearly defined architecture, including well-defined physis of the distal tibia and talus, thick cortical bone and epiphyseal and metaphyseal trabeculae were also visible. All the AA control rats showed significant damage of the joint, whereas treatment with SK&F 106615 (20 mg/kg) provided complete protection in 60% of the joints evaluated and partial protection in 40% (Badger, 1996).

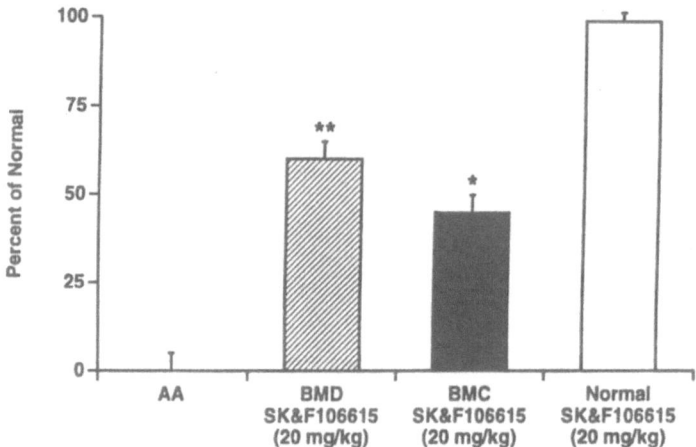

Figure 10-7. Bone densitometry evaluation of distal tiba of AA rats treated with SK&F 106615: normalization of BMD and BMC. Rats were administered 20 mg/kg per day po from day 0 to day 23, five times per week. Data are presented as percent of normal (100%). Mean ± standard error of 7 animals/group. * $p < 0.05$; ** $p < 0.01$.

The profile of activity of the azaspiranes in the AA rat is clearly one of a disease-modifying agent. Not only is a clear beneficial effect observed on the inflammatory response, but there is also a marked decrease in the severity of bone lesions. Although we have not been able to identify the molecular mechanism of action of the azaspiranes in this disease, at the cellular level, the induction of SC activity, which accompanies therapeutic efficacy in the disease models, may well be, in part, responsible for the observed beneficial effects (Badger et al., 1989; Albrightson-Winslow et al., 1990; Rabinovitch et al., 1993; Schmidbauer et al., 1993).

Induction of Suppressor Cell Activity

The discovery of SC activity in the spleens of treated rats was the key early observation that aroused our interest in this class of compounds, and over the years, much effort has been directed to their characterization. Despite our efforts, we have not been able to specifically phenotype the cell responsible for the SC activity observed in the co-culture assay system, but as is apparent in the following summary, there are not only a number of similarities between the azaspirane-induced SC activity and that of the natural or nonspecific suppressor cells described by Strober (1984), as well as cells of the macrophage lineage. The characteristics of azaspirane-induced SC activity are summarized below.

- The azaspiranes induce radiation-resistant (2000R) SC activity in mice, rats, and dogs in the absence of myelotoxicity.
- SC activity is present in the spleen, bone marrow, and lymph nodes but not in the thymus of treated rats.
- Peripheral blood mononuclear cells occasionally have low SC activity.
- Rat alveolar macrophages but not peritoneal macrophages have potent SC activity.
- Induction of SC activity occurs in the absence of myelotoxicity.
- SC activity appears following a single dose and peaks following 6–10 doses.
- SC activity produces enrichment in the low-density (1.07 g/ml) fraction of a Percoll density gradient.
- Analysis of cell subsets has not revealed any isotype switch in the 1.07 g/ml fraction.
- The phenotype of the SC has not been determined, but this activity is not characteristic of mature T or B cells, adherent splenic macrophages, or NK cells.
- SC activity is not mediated by prostaglandins or nitric oxide.

Much of the profile of the suppressor cell activity is described in Badger et al., 1990a; Thiem et al., 1992; and Kaplan et al., 1991.

Splenic SC activity is present in both drug-treated normal rats and in rats with adjuvant arthritis. The activity is measured in a co-culture assay whereby effector cells from treated animals are irradiated and added to normal cells in a Con A-stimulated co-culture assay (Figure 10-8) (Badger et al., 1987, 1989). This data, when converted to an area under the curve (AUC) of suppressive activity, is used to compare the efficacy of different compounds (Badger et al., 1990b) (Figure 10-9). The AUC for suppression has also been utilized to examine SC activity in various lymphoid organs, and it is apparent from the distribution (Table 10-7) that the activity resides in areas where one might expect activated and/or tissue macrophages to reside. The most compelling evidence for this is that alveolar macrophages exert the most potent activity.

However, as neither N^G-monomethyl'-L-arginine (NGMMA), a nitric oxide synthesis inhibitor, nor indomethacin is able to interfere with azaspirane-induced SC-activity, the antiproliferative activity does not appear to be mediated by nitric oxide or prostaglandins (Badger et al., 1990a; Kaplan et al., 1991).

In vitro, SK&F 105685 inhibits the proliferation of rat spleen and lymph node cells to mitogens with an IC_{50} of ≈ 100 nM (Kaplan et al., 1991).

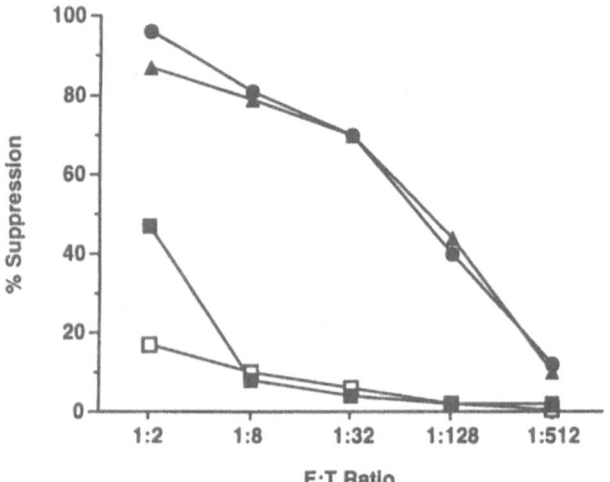

Figure 10-8. Suppressor cell activity in the spleens of AA rats treated with SK&F 105685, SK&F 106615 or methotrexate (MTX). Compounds were administered orally to AA rats from day 10 to day 30 (20 mg/kg/day, five times a week, po). Spleen cells from AA control rats or AA treated rats were added to 96-well microtiter plates at different dilutions and irradiated (2000R). Normal rat spleen cells (2.5×10^5) were then added and incubated for 72 h with an optimum concentration of Con A (5 μg/ml). ^{3}H-Thymidine incorporation was determined over the last 16 h of culture. Data was calculated as the percent inhibition of each E:T ratio (6 replicate wells) from pooled spleen cells. Inhibition of 15% or more is significant. (□) control co-culture; (△) SK&F 105685 co-culture; (o) SK&F 106615 co-culture; (□) MTX co-culture.

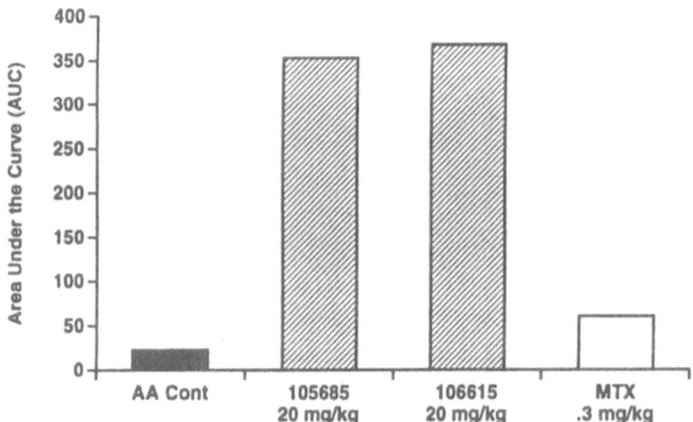

Figure 10-9. Splenic suppressor cell activity (AUC) in AA rats treated with SK&F 105685, SK&F 106615, and methotrexate (MTX).

Table 10-7. Distribution of suppressor cell activity in azaspirane-treated Lewis rats.

Lymphoid organ	SC activity
Spleen	+++
Bone marrow	++
Lymph nodes	
cervical	+/–
mesenteric	+++
Alveolar macrophages	++++
Peritoneal macrophages	+/–
Peripheral blood	+/–
Thymus	–

However, unlike immunosuppressive compounds such as CsA, FK-506, or rapamycin, which show reproducible consistent and dose-related inhibitory effects on lymphocyte proliferation, the inhibitory activity of the azaspiranes is more variable. This would indicate that the compound is not directly immunosuppressive on T-cell proliferation but requires the presence of a suppressor cell, which is stimulated to become inhibitory by the compound. Rat peripheral blood lymphocyte responses to mitogen are poorly inhibited at nontoxic concentrations of SK&F 105685 (Kaplan et al., 1991) and human peripheral blood lymphocyte responses to mitogen and antigen are also poorly inhibited. Again, we surmise the reason for this is the lack of the putative suppressor cell in this cell population. Unlike CsA, the compound does not inhibit IL-2 production by Con A-stimulated spleen cells or the Jurkat cell line, but in a manner similar to rapamycin reduces the IL-2-induced proliferation of Con A blast cells (Kaplan et al., 1991). The distribution of the SC activity and the *in vitro* data support the concept that a poorly adherent macrophage population is responsible for the SC activity induced by SK&F 105685 and other active azaspiranes. This cell population is present in spleen and lymph nodes and in broncho-alveolar lavage cells but is absent, or present at very low levels, in the thymus, peritoneal cavity and in peripheral blood.

In summary, the azaspiranes induce SC activity in a number of lymphoid organs in a variety of species. This SC activity is measured in *ex vivo* assay systems, so its role *in vivo* in the previously described animal models remains to be defined. However, adoptive transfer of spleen cells from azaspirane-treated rats was effective in prolonging graft survival in the heart transplant studies (Schmidbauer et al., 1993) indicating an active role *in vivo*. Current studies are targeted towards defining the mechanism of action of this cellular activity with focus on the alveolar macrophage suppressor function.

The Azaspiranes Are Not Immunosuppressive Compounds

Evidence from *in vitro* studies indicates that azaspiranes do not have potent antiproliferative effects, and despite the ability of the compounds to modulate disease processes in a number of animal models, there is much evidence that these compounds do not cause generalized immunosuppression *in vivo*. In rats, where both inhibition of adjuvant-induced disease and DTH responses is observed with SK&F 105685 (Badger et al., 1989), we have been unable to show inhibition of antibody responses to protein antigens (BSA and KLH), regardless of whether they are administered in complete Freund's adjuvant (CFA) or alum. Neither have we been able to demonstrate inhibition of antibody (IgM) synthesis to sheep red blood cells (SRBC) in BDF1 mice in comparison to an immunosuppressive drug such as rapamycin (Figure 10-10).

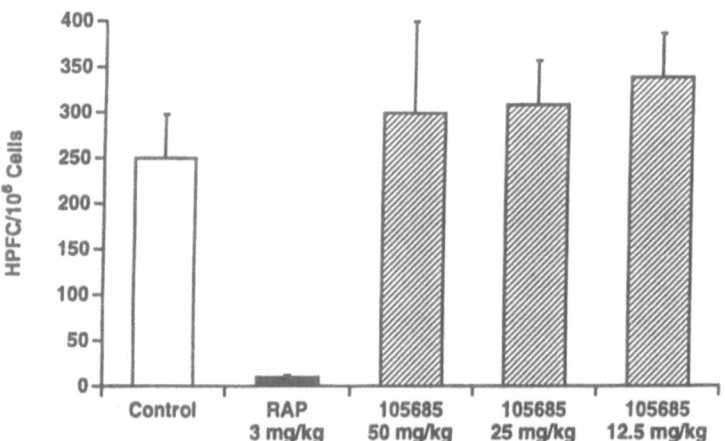

Figure 10-10. Rapamycin, but not SK&F 105685, inhibits the primary antibody response to SRBC. BDF1 mice were treated (ip) with compounds on days −3, −2, and −1 and SRBC (10^9) were injected on day 0. HPFC response was determined on day 5.

Studies to determine the effects of SK&F 105685 on immune function have been extended to the dog, when the animals were treated orally for 29 days with 1 mg/kg, a dose that induced potent SC activity in the spleen (Thiem et al., 1992). The compound did not affect the total number or relative percentages of the various white blood cell types present in peripheral blood. The ability of peripheral blood lymphocytes or spleen cells to produce IL-2 in response to mitogenic stimuli was not impaired. However, when SK&F 105685 was added to spleen cells *in vitro*, it inhibited the generation of specific antibody to SRBC, and we believe this to be a result of the SC activity known to be present in spleen cell cultures (Kaplan et al., 1993).

Many immunosuppressive drugs cause bone marrow depression which can result in extramedullary hematopoiesis and splenic SC activity. However, this is not the case for the azaspiranes. SK&F 105685 and SK&F 106615, in fact, stimulate myelopoiesis in both rats and mice (King et al., 1991a, 1991b, and unpublished observations, 1994), and this may be a function of the upregulation of colony-stimulating factor production by the compound (King et al., 1991a).

Additional support for the concept that the azaspiranes do not cause overt immunosuppression in dogs is provided by the finding that SK&F 105685 treatment failed to affect the nonspecific candidacidal activity of polymorphonuclear leucocytes and spleen cells (Kaplan et al., 1993). In addition, studies carried out with alveolar macrophages isolated from rats treated with SK&F 105685 or SK&F 106610 were not impaired in their ability to kill this intracellular organism. There was, in fact, an enhanced candidacidal effect exerted by these cells (Badger et al., 1993b; Waites et al., 1995).

Taken together, the *in vitro* and *in vivo* observations indicate that in a therapeutic setting, the azaspiranes are unlikely to compromise the immune status of the host, and that they can down-regulate specific immune responses without causing generalized immunosuppression.

SK&F 105685—The First Development Candidate

By July of 1987, we had completed a pharmacological profile of SK&F 105685 adequate for us to present the data/information to the appropriate management committee, and we were successful in that we became an official Feasibility Study, the first step at SB on the road to the clinic. This new status enabled us to further profile the compound and obtain the support of our colleagues in Pathology/Toxicology for initial safety studies.

By January of 1989, we thought we had all the information required to elevate the compound to full Project status. We were not successful in this endeavor, however, and were sent back to the drawing board to determine whether pharmacological activity could be demonstrated in a species other than the rat (after all, our aim was not to cure rat arthritis), and whether we could show activity at doses lower than those currently shown to be active in the AA rat ED_{25} of ≈ 10 mg/kg). This was due to concerns of gastrointestinal (GI) toxicity, now known to be dose-limiting toxicity for the azaspiranes.

In November of 1989, we had our data; suppressor cell induction in dogs at a dose of 3 mg/kg and an ED_{25} of 3 mg/kg in the AA rat following a prophylactic dosing schedule. We also decided (but without supporting data) to include transplantation as a therapeutic target as well as RA. The reason for this was that we thought some minor GI side effects could be more acceptable in the life-threatening position facing a transplant recipient than in the individual with RA. Another criticism was that we did not know the molecular mechanism of action of the compound (true) and therefore would have no basis upon which to select a backup candidate. It was also thought by some (but not all) that we were not far enough advanced to be competitive in the field of immunomodulatory/immunosuppressive agents. This skepticism resulted in the demise of the project, and it was a special irony that this information reached us on December 7, 1989. However, not all members of the management committee(s) felt that an appropriate decision had been made, and in their opinion, the azaspiranes were a novel and unique group of compounds, and SK&F 105685 should be given a chance to be tested in humans. This support translated into dollars which we used to form collaborative efforts at Harvard and Duke University to investigate the effects of our compound in rodent models of transplantation. These collaborations were extremely successful and showed that while we did not have a CsA or an FK506, clear activity could be demonstrated in a rat kidney allograft model (Fan et al., 1993). In addition, as described earlier in this review, several studies in a rat cardiac allograft model showed that the compound extended graft survival when it was administered alone, and also that synergistic activity was observed in a combination therapy with CsA (Badger et al., 1991a; Hancock et al., 1992; Schmidbauer et al., 1993).

Armed with the new data from the experimental transplantation studies, confirmation of biological activity in a second species (SC activity in the dog), and an increasingly expanding profile of the azaspiranes as disease-modifying agents that did not cause generalized immunosuppression, SK&F 105685 was accepted for development by SB's Management Committee on December 7 (just coincidence), 1990.

Development of SK&F 105685 and Discovery of the Backup Compound, SK&F 106615

Elevation of the lead compound to Project Status allowed the Synthetic Program, Pathology/Toxicology Studies, and Drug Metabolism Studies to get underway. Very early on, studies revealed an unexpected finding in that oral administration of SK&F 105685 to either rats or dogs resulted in lowering of serum cholesterol levels. It was not clear to us whether the cholesterol-lowering effects of SK&F 105685 would be an asset or a liability for a drug that would have to be administered chronically to RA patients. Neither did we know whether the activity was a characteristic of the azaspiranes as a chemical class or if it was specific for SK&F 105685. More relevant, was the cholesterol-lowering effect associated with the immunomodulatory activity? Was there a cause and effect relationship? We decided to pursue the mechanism for cholesterol-lowering because it seemed reasonable that should there be a liability associated with this phenomenon, a compound devoid of this activity could be proposed as a backup strategy. A number of studies designed to understand the mechanism of the cholesterol-lowering effect of the compound did not reveal any modulation of citrate lyase, HMG-CoA reductase, ACAT, or cholesterol 7α hydroxylase. While these mechanistic studies were underway, we decided to compare the three lead compounds side by side in beagle dogs for their cholesterol-lowering activity. As expected, following 1 month of treatment at 1 mg/kg/day (a dose that induced high splenic SC activity), SK&F 105685 reduced circulating cholesterol levels by 50%. On the other hand, neither SK&F 106610 nor 106615 had any effect on cholesterol at this dose (Albrightson et al., 1995). Further experimentation revealed that doses of SK&F 106615 as high as 10 mg/kg administered for 4 weeks did not reduce serum cholesterol levels. This exciting finding coincided with the disappointing news that in the Pathology/Toxicology Safety studies, elevation of alanine transaminase (ALT) was observed in dogs treated with high doses of SK&F 105685. However, with the unflagging support of our Pathology/Toxicology colleagues, we were then able to carry out safety studies with the new compound in record time and show that it had an improved safety profile when compared to SK&F 105685. Our pharmacological profile of SK&F 106615 in the AA rat and for SC-induction showed that it was equally potent to SK&F 105685; these results indicated that we had a better therapeutic index with the backup compound. SK&F 106615 was accepted as a development candidate in December 1993. The Investigational New Drug Application (IND) was approved in July 1994 and Phase I clinical trials were initiated in September 1994.

The roller coaster of events described here that took place over the last ten years of azaspirane work at SB, and the dedication of those individuals who believed in the compounds are not unique phenomena in the pharmaceutical drug discovery. The final outcome for these compounds as disease-modifying agents in rheumatoid arthritis, and other autoimmune diseases, will become known over the next few years during their clinical evaluation.

Acknowledgments. The original discovery of the biological activity of the azaspiranes and the subsequent efforts to bring a compound to the clinic have been exciting, stressful, traumatic, depressing, disappointing, and heroic to name but a few of the emotional states through which we have passed. There were many times it seemed that it would be best if these compounds would just go away, but the extraordinary support of certain individuals, although here unnamed, was always enough to keep us going. We overcame, on more than one occasion, what we believed was total defeat, to finally accomplish our goal of gaining approval from the numerous committees required to move an azaspirane into clinical development. We would like to acknowledge everyone who contributed to this success story and thank them for their support.

REFERENCES

Albrightson CR, Bugelski PJ, Berkhout TA, Jackson B, Kerns WD, Organ AJ, Badger AM (1995): Separation of immunomodulatory and cholesterol lowering activities of heterocyclic azaspiranes. *J Pharm Exp Ther* 272:689–698

Albrightson-Winslow CR, Brickson B, King A, Olivera D, Short B, Saunders C, Badger AM (1990): Beneficial effects of long-term treatment with SK&F 105685 in murine lupus nephritis. *J Pharm Exp Ther* 255(1):382–387

Badger AM, Albrightson-Winslow CR, Kupiec-Weglinski JW (1991a): SK&F 105685: A novel immunosuppressive compound with efficacy in animal models of autoimmunity and transplantation. *Trans Proc* 23(1):194–195

Badger AM, DiMartino MJ, Schmitt TC, Swift BA, Mirabelli CK (1987): Suppressor cell induction by the anticancer drug Spirogermanium. *Int J Immunopharmac* 9(5):621–630

Badger AM, DiMartino MJ, Talmadge JE, Picker DH, Schwartz DA, Dorman JW, Mirabelli CK, Hanna N (1989): Inhibition of animal models of autoimmune disease and the induction of non-specific suppressor cells by SK&F 105685 and related azaspiranes. *Int J Immunopharmac* 11(7):839–846

Badger AM, King AG, Talmadge JE, Schwartz DA, Picker DH, Mirabelli CK, Hanna N (1990a): Induction of non-specific suppressor cells in normal Lewis rats by a novel azaspirane SK&F 105685. *J Autoimmun* 3:485–500

Badger AM, Mirabelli CK, DiMartino MJ (1985): Generation of suppressor cells in normal rats by treatment with Spirogermanium, a novel heterocyclic anticancer drug. *Immunopharmacol* 10:201–207

Badger AM, Schwartz DA, Picker DH, Dorman JW, Bradley FC, Cheeseman EN, DiMartino MJ, Hanna N, Mirabelli CK (1990b): Anti-arthritic and suppressor cell-inducing activity of the azaspiranes: Structure-function relationships of a novel class of immunomodulatory agents. *J Med Chem* 33:2963–2970

Badger AM, Swift BA, Bugelski PJ (1993b): Suppressor cell (SC) activity induced by the azaspiranes may reside in a non-adherent or semi-adherent tissue macrophage population. AAI/CIS, Denver, Colorado, April 21–25

Badger AM, Swift BA, Bugelski PJ, High WB, DiMartino MJ, Greig R (1991b): The effect of SK&F 105685, a novel suppressor cell-inducing compound, in the adjuvant arthritic rat. *Br J Rheum* 30:66

Badger AM, Swift BA, Webb EF, Clark RK, Bugelski PJ, Griswold DE (1993a): Beneficial effects of SK&F 105685 in rat adjuvant arthritis: Prophylactic and therapeutic effects on disease parameter progression. *Int J Immunopharmac* 15(3):343–352

Boers M, Dijkmans BAC, van Rijthoven AWAM, Goei Thè HS, Cats A (1990): Reversible nephrotoxicity of cyclosporine in rheumatoid arthritis. *J Rheumatol* 17:38–42

DiMartino MJ, Lee JC, Badger AM, Muirhead KA, Mirabelli CK, Hanna N (1986): Antiarthritic and immunoregulatory activity of Spirogermanium. *J Pharm Exp Ther* 236(1):103–110

Dougados M, Hassane A, Amor A (1988): Cyclosporin in rheumatoid arthritis: a double blind, placebo controlled study in 52 patients. *Ann Rheum Dis* 47: 127–133

Eichmann K (1988): Suppression needs a new hypothesis. *Sc J Imm* 28:273–276

Fan P-Y, Best C, Coffman TM, Howell DN, Albrightson CR, Badger AM (1993): The azaspirane SK&F 105685 ameliorates renal allograaft rejection in rats. *J Am Soc Nephrol* 3:1680–1685

Green DR, Webb DR. (1993): Saying the 'S' word in public. *Immunology Today* 14(11):523–525

Hancock WW, Schmidbauer G, Badger AM, Kupiec-Weglinski JW (1992): SK&F 105685 suppresses allogeneically induced mononuclear and endothelial cell activation and cytokine production and prolongs rat cardiac allograft survival. *Trans Proc* 24(1):231–232

High WB, Bugelski PJ, Nichols ME, Swift BA, Solleveld HA, Badger AM (1994): Effects of a novel azaspirane (SK&F 105685) on the arthritic lesions in the adjuvant arthritic rat: Attenuation of the inflammatory process and preservation of skeletal integrity. *J Rheumatol* 21:476–483

Kaplan JM, Badger AM, Ruggieri EV, Olivera DL, Newman-Tarr T, Bugelski PJ (1991): Inhibition of lymphoproliferative responses by SK&F 105685, a novel anti-arthritic agent. *J Clin Lab Immunol* 36:49–58

Kaplan JM, Badger AM, Ruggieri EV, Swift BA, Bugelski PJ (1993): Effects of SK&F 105685, a novel anti-arthritic agent, on immune function in the dog. *Int J Immunopharmac* 15(2):113–123

Kavanaugh AF, Lipsky PE (1992): Gold, penicillamine, antimalarials, and sulfasalazine. In: *Inflammation: Basic Principles and Clinical Correlates*, Gallin JI, Goldstein IM, Snyderman R, eds. New York: Raven Press

King AG, Badger AM (1991a): Administration of an immunomodulatory azaspirane, SK&F 105685, or human recombinant interleukin 1 stimulates myelopoiesis and enhances survival from lethal irradiation in C57B1/6 mice. *Exp Hematol* 19:624–628

King AG, Olivera D, Talmadge JE, Badger AM (1991b): Induction of non-specific suppressor cells and myeloregulatory effects of an immunomodulatory azaspirane, SK&F 105685. *Int J Immunopharmac* 13(1):91–100

Madhok R, Torley HI, Capell HA (1991): A study of the long-term efficacy and toxicity of cyclosporine A in rheumatoid arthritis. *J Rheumatol* 18:1485–1489

Mirabelli CK, Badger AM, Sung C-P, Hillegass L, Sung C-M, Johnson RK, Picker D, Schwartz D, Dorman J, Martelucci S (1989): Pharmacological activities of spirogermanium and other structurally related azaspiranes: effects on tumor cell and macrophage functions. *Anti-Cancer Drug Design* 3:231–242

Mitchison NA (1988): Suppressor activity as a composite property. *Scand J Immunol* 28:271–276

Moller G (1988): Do suppressor T cells exist? *Scand J Immunol* 27:247–250

Rabinovitch A, Suarez WL, Quin HY, Power RF, Badger AM (1993): Prevention of diabetes and induction of non-specific suppressor cell activity in the BB rat by an immunomodulatory azaspirane, SK&F 105685. *J Autoimmun* 1:39–49

Rainsford KD (1990): Disease-modifying antirheumatic and immunoregulatory agents. *Bailliere's Clinical Rheumatology* 4:405–431

Rice LM, Wheeler JW, Geschickter CF (1974): Synthesis of 4.4-dialkyl-4-germa-cyclo-hexanone and 8.8 dialkyl-8-germaspiro(4.5)decanes. *Journal of Heterocyclic Chemistry* 11:1041

Schmidbauer G, Hancock WW, Badger AM, Kupiec-Weglinski JW (1993): Induction of nonspecific X-irradiation-resistant suppressor cell activity *in vivo* and prolongation of vascularized allograft survival by SK&F 105685, a novel immunomodulatory azaspirane. *Transplantation* 55:1236–1243

Shoenfeld Y, Blank M, Aharoni R, Teitelbaum D, Arnon R (1993): Manipulation of autoimmune diseases with T-suppressor cells: Lessons from experimental SLE and EAE. *Immunology Letters* 36:109–116

Strober S (1984): Natural suppressor (NS) cells, neonatal tolerance, and total lymphoid irradiation: Exploring obscure relationships. *Ann Rev Immunol* 2:219–237

Thiem PA, Kaplan JM, Bugelski PJ, Ruggieri EV, Badger AM (1992): Induction of suppressor cell activity *in vivo* and suppression of lymphoproliferative responses *in vitro* by SK&F 105685 in the dog. *Immunopharmacol* 23:67–74

Tugwell P, Bombardier C, Gent M, Bennett KJ, Bensen WG, Carette S, Chalmers A, Esdaile JM, Klinkhoff AV, Kraag GR, Ludwin D, Roberts RS (1990): Low-dose cyclosporin versus placebo in patients with rheumatoid arthritis. *Lancet* 335:1051–1055

Waites CR, Bugelski PJ, Badger AM (1995): Biochemical and functional analysis of rat bronchoalveolar macrophages containing chemically induced phospholipid inclusions. *Toxicol App Pharmacol* 130:316–321

Westeren AC, Chan C-C, de Smet MD, Nussenblatt RB, Roberge FG (1993): Evaluation of the immunosuppressant SK&F 106610 in the treatment of experimental autoimmune uveoretinitis. *Association for Research in Vision and Ophthalmology*, May 2–7,1993, Sarasota, Florida

Keyword Index

This index was established according to the keywords supplied by the authors.